Technological Innovation Management

Technological Innovation Management

Edited by **Ed Diego**

NY RESEARCH
P R E S S

New York

Published by NY Research Press,
23 West, 55th Street, Suite 816,
New York, NY 10019, USA
www.nyresearchpress.com

Technological Innovation Management
Edited by Ed Diego

International Standard Book Number: 978-1-63238-436-2 (Hardback)

Printed in the United States of America.

Contents

Preface

Technology is advancing at a rapid pace. It has been extensively established that technology is one of the forces motivating economic development. This book is a compilation of various researches. To start with, the chapters are written in developed nations and developing Asian nations. These chapters encompass a broad variety of businesses, inclusive of telecommunication, hygiene and medical, amusement, learning, manufacturing, and economics. The systematic approaches are multi-disciplinary, ranging from arithmetical, financial, logical, experiential and planned. Furthermore, this book also deals with both public and private organizations.

This book unites the global concepts and researches in an organized manner for a comprehensive understanding of the subject. It is a ripe text for all researchers, students, scientists or anyone else who is interested in acquiring a better knowledge of this dynamic field.

I extend my sincere thanks to the contributors for such eloquent research chapters. Finally, I thank my family for being a source of support and help.

Editor

Part 1

Adoption of Technological Innovation

Trends and Directions for Energy Saving in Electric Networks

Gheorghe Grigoraş[1], Gheorghe Cârţină[1] and Elena-Crenguţa Bobric[2]
[1]"Gheorghe Asachi" Technical University of Iasi
[2]"Stefan cel Mare" University of Suceava
Romania

1. Introduction

The existing grids are one-way systems for the delivery of electricity without the self-healing, monitoring and diagnostic capabilities essential to meet demand growth and new security challenges facing us today.

Increasing the efficiency of existing distribution and consumption equates to making additional power available at lower cost. Such efficiencies reduce the need for constructing new generation plants and associated transmission facilities. Smart Grid can provide the communications and monitoring necessary to manage and optimize distributed and renewable energy resources and to maximize the environmental and economic benefits.

The term "smart grid" is hyperbole that seems to imply a future when the grid runs itself absent human intervention. The smart grid concept in many ways suggests that utility companies, executives, regulators and elected officials at all levels of government will indeed face a brutal "pass/fail" future with regard to electric service, a driving force of the U.S. world-leading economy (IEA, 2001).

Intelligent distribution systems are an inevitable reality for utilities as they replace aging infrastructure, deal with capacity constraints and strive to meet the demands of an increasingly sophisticated end-use customer. The benefits of a real-time, single-platform smart distribution network are clear.

The business case must take into account the cost-effectiveness, operational improvements and return on investment of specific initiatives and must consider community-wide benefits. A proactive incremental implementation of smart distribution systems can have a dramatic impact on system improvements and customer satisfaction. A proactive review of smart grid strategy is vital: the utility leadership landscape will reward those who move early.

The essence of the smart grid lies in digital control of the power delivery network and two-way communication with customers and market participants. This intelligent infrastructure will allow for a multitude of energy services, markets, integrated distributed energy resources, and control programs. The smart grid is the essential backbone of the utility of the future (IEA, 2001).

In the nearest future we will have to face two mega-trends. One of them is the demographic change. The population development in the world runs asymmetrically: dramatic growth of population in developing and emerging countries, the population in highly develops countries is stagnating (Breuer et al., 2007).

This increase in population (the number of elderly people in particular) poses great challenges to the worldwide infrastructure: water, power supply, health service, and mobility and so on.

The second mega-trend to be mentioned is the urbanization with its dramatic growth worldwide. In less than two years more people will be living in cities than in the country.

Depending on the degree of development (developing, emerging, industrialized countries) different regions have very different system requirements, Fig. 1.

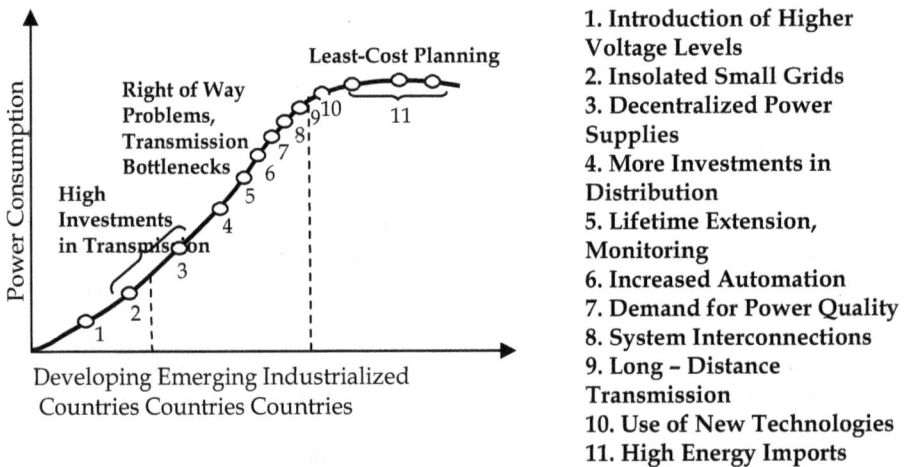

Fig. 1. Development of Power Consumption and System Requirements (Breuer et al., 2007)

Thus, in developing countries, the main task is to provide local power supply. Emerging countries have a dramatic growth of power demand. During the transition, the newly industrialized countries need energy automation, life time extension of the system components, such as transformers and substations. Higher investments in distribution systems are essential as well. At the same time, the demand for a high reliability of power supply, high power quality and, last but not least, clean energy increase in these countries. In spite of all the different requirements one challenge remains the same for all: sustainability of power supply must be provided.

Taking into account these aspects, the energy saving has become a major problem in the worldwide. Numerous studies have indicated that reduction of the power/energy losses in the electric networks is much easier than the increase of generating capacities, and energy efficiency represents the cheapest resource of all. The worldwide experience shows that in utilities with high network loss level, 1 $ expended for loss reduction saves 10 - 15 $ to the utility (Raessar et al., 2007).

But, in evaluation of the energy losses from the electric distribution systems is necessary to know the loads from nodes of the system. Because, in distribution system, except the usual measurements from substations, the feeders and the loads are not monitored, there is few information about the network state. In this situation a modern technique, based on fuzzy set model, it can provide a good operating solution. The core of this technique is the fuzzy correlation model (Cârțină et al., 2003). The combination of the fuzzy approach with the system expert leads to an efficient and robust tool.

2. Strategies for power/energy saving in electric distribution networks

2.1 Minimization of the power/energy losses

Nowadays, power/energy saving has become a major problem in the worldwide. Numerous studies have indicated that reduction of power/energy losses in the electric networks is much easier than the increase of generating capacities, and energy efficiency represents the cheapest resource of all.

Energy losses throughout the world's electric distribution networks vary from country to country between 3.7% and 26.7% of the electricity use, which implies that there is a large potential for improvement. The distribution networks in most countries in the world were significantly expanded during the late 1960s and early 1970s, with different nominal voltages. For example, in distribution networks from Romania there are three levels of voltage: 6, 10, and 20 kV. The 6 kV level is the first who was developed and the availability of this in urban centres and other areas of concentrated demand for power is still quite high. Perspective to maintain the level of 6 kV is full of difficulties because the networks are very old, some distributors are loaded close to maximum capacity and energy losses are very high. The electric equipments installed in these networks now approach the end of their useful life and need to be replaced. But after replacing, the lifetimes of primary components are long and the networks built today will still be in use after several decades. The same problems in electric distribution networks are occurring during past years all over the world. The 20 kV level appeared later and covered the rest of urban and rural distribution areas. The 10 kV level included still very small areas of urban networks (Grigoraș et al., 2010c, 2010d). Thus, in the Figs. 2 and 3, the location by components of energy losses in the electric networks of a Distribution Company from Romania is presented. From Fig. 2 it can observe that a major part of the energy losses of a distribution system are the energy losses in the 6 kV distribution networks. It should be noted that energy losses in the 6 kV networks have about the same percentage as the 20 kV networks (1.25 % vs. ≈ 1 %), Fig 2, even if their total length is much smaller (report lengths, respectively the number of transformers is about 1 to 3). Another issues relates to the energy losses from the 6 kV cables that are very high compared with those on the 20 kV cables, and from the iron of the power transformers.

In the power transformers, the energy losses fall into two components: no-load losses or iron losses (constant, resulting from energizing the iron core; this phenomenon occurs 24 hours per day, 7 days per week, over the lifetime of the transformer, 30 years in average) and load losses (variable, arising when providing power to a user, from the resistance of the coils when the transformer is in use, and for eddy currents due to stray flux) (Eiken, 2007; European Commission, 1999; Grigoraș et al., 2010a).

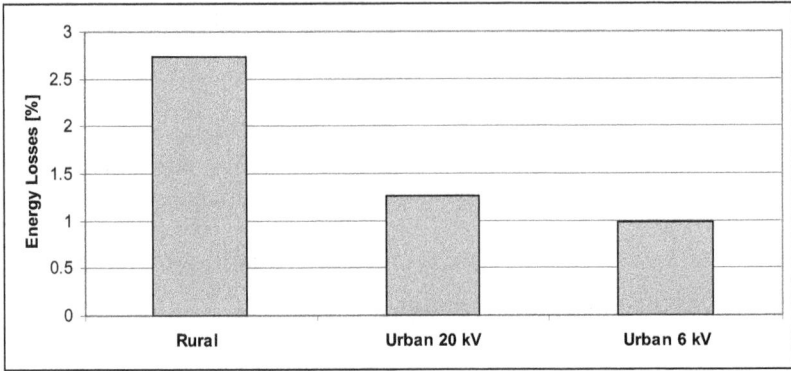

Fig. 2. The total energy losses in electric networks of a distribution company (expressed in percentage of total energy circulating in network)

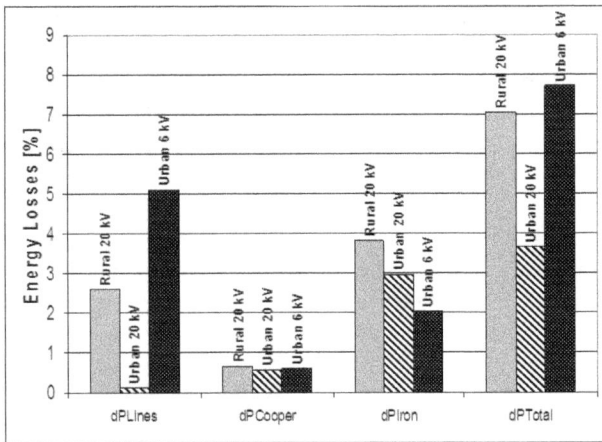

Fig. 3. The total energy losses in a subsidiary of the distribution company (expressed in percentage of energy circulating in the every type of network)

The variable losses depend on the effective operating load to the transformer. The energy consumed in meeting these losses is dissipated in the form of heat, which is not available for the consumers to use.

No-load loss (iron loss) is the power consumed to sustain the magnetic field in the transformer's steel core. Iron loss occurs whenever the transformer is energized; iron loss does not vary with load. These losses are caused by two factors: hysteresis and eddy current losses.

Load loss (copper loss) is the power loss in the primary and secondary windings of a transformer due to the resistance of the windings. Copper loss varies with the square of the load current. The maximum efficiency of the transformer occurs at a condition when constant loss is equal to variable loss. For distribution transformers, the core loss is 15% to 20% of full load copper loss. Hence, the maximum efficiency of the distribution transformers occurs at a loading between 40% – 60%. For power transformers, the core loss is 25% to 30% of full load copper loss. Hence, the maximum efficiency of the power transformers occurs at a loading between 60% – 80%. The efficiency of the transformers not only depends on the design, but also, on the effective operating load.

A policy for the reduction of losses can contain short and long term actions, (Grigoraş et al., 2010a; Raesaar et al., 2007). The some short term measures are following:

- Identification of the weakest areas in distribution network and improve them;
- Reduction the length of the distribution feeders by relocation of distribution substation/installations of additional transformers, and so on.

The long term measures may relate to:

- Mapping of complete distribution feeders clearly depicting the various parameters such as nominal voltage, the length, installed transformation capacity, the number of the transformation points, the circuit type (underground, aerial, mixed), load being served etc.
- Replacement of the 6 kV or 10 kV voltage level with 20 kV voltage level;
- Replacement of the old power transformers with the efficient transformers;
- Compilation of data regarding existing loads, operations conditions, forecast of expected loads etc.

For further development of plans of energy loss reduction and for determination of the implementation priorities of different measures and investment projects, an analysis of the nature and reasons of losses in the system and in its different parts must be done.

From these measures, we will refer only to replacement of the voltage of 6 kV level to 20 kV and the old power transformers with the efficient transformers.

The replacement of the voltage of 6 kV level to 20 kV can be done in order to improve reliability and to minimize power losses in electrical distribution networks. On the other hand, most of the electric distribution infrastructure in urban areas is underground, so if excavation work is done to lay new distribution feeders, it makes much more economic sense to deploy 20 kV distribution lines that have about three times the capacity of 6 kV lines. Other solution that can be applied to minimize the power losses, correlated with the above is the use of efficient transformers. The distribution power transformer is the most important single piece of electrical equipment installed in electrical distribution networks with a large impact on the network's overall cost, efficiency and reliability. Selection and acquisition of distribution transformers which are optimized for a particular distribution network, the utility's investment strategy, the network's maintenance policies and local service and loading conditions will provide definite benefits (improved financial and technical performance) for both utilities and their customers (Amoiralis et al., 2007)

For most electric distribution networks in Europe consist of aged network assets that have reached the end of their original amortized life. Fig. 4 shows a typical asset age profile of such assets and suggests that if original replacement times were to be exercised the majority of gear would have to be replaced in a short interval (Northcote-Green & Speiermann, 2010).

Thus, for an electric utility (Distribution Company) that has numerous distribution transformers in its network, there is an opportunity to install high efficient distribution transformers that have less total energy losses than less efficient transformers, so they pollute the environment less.

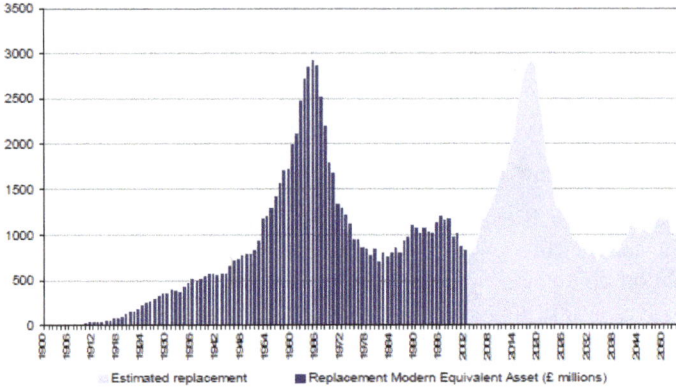

Fig. 4. Typical Electrical Power Distribution Network Asset Age Profile (Northcote-Green& Speiermann, 2010)

2.2 Energy performance standards for power transformers

Worldwide there are programs on Minimum Energy Performance Standard (MEPS) for to reduce energy losses associated with transformer operation in the electricity distribution system. Since the original MEPS levels were specified there has been significant development in transformer efficiency standards and requirements in other countries including the USA, European Union, Canada, Japan, China, Mexico and India. Thus, in Fig. 5 it presents a comparison of the requirements of international standards in terms of performance transformer oil at a loading of 50% (Ellis, 2003).

HD 428 standard imposed by European Union specific levels of energy losses in the transformer core for three different classes: A', B' and C' (C' having the lowest level of energy loss and A' the highest level). Also energy losses in the windings for three categories: A, B and C (C being the lowest level of losses and type A has the highest level of losses) (Ellis, 2003; European Commission, 1999). Some states have used the category of transformers the most efficient C-C' as a necessity while others use transformers less efficient by category B-B'. C-C' category present iron and copper losses of low values compared with other types of categories, presented in Table 1 (Ellis, 2003).

Several European projects have shown the interest in acquiring efficient transformers. A project initiated in collaboration with European Commission from 1999 estimated that energy efficient transformers could save approximately 22 TWh per year by means of C-C' units; amorphous core transformers could save even more. The Prophet project continued this task in 2004 and arrived at similar conclusions; furthermore, it showed a rising trend in the installation of amorphous transformers in Japan and China, and India and USA install them too. In USA, 10% of new transformer sales are amorphous transformers (about 100,000 new amorphous transformers per year); 15% of new pole transformer sales in Japan are amorphous transformers (about 350,000 amorphous transformers were in service in 2003 (Frau&Gutierrez, 2007). Today, another EU project is working to highlight energy efficiency on Distribution Transformers. The SEEDT project represents one of the projects in the Intelligent Energy Europe programme. The aim of this project is to promote the use of energy-efficient distribution transformers, which can be profitable for investors, and, by contributing to European Community energy savings, may help to fulfil EU energy policy targets (Polish Copper Promotion Centre & European Copper Institute, 2008).

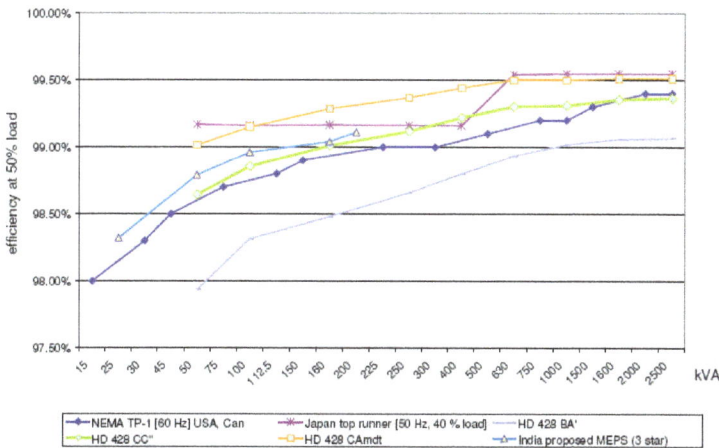

Fig. 5. Requirements of international standards in terms of performance transformer oil at a loading of 50% (Ellis, 2003)

S_n [kVA]	Power losses (Transformers with standard HD428 (<20kV))					
	A [W]	A′ [W]	B [W]	B′ [W]	C [W]	C′ [W]
50	1100	190	1350	145	875	125
100	1750	320	2150	260	1475	210
160	2350	460	3100	375	2000	300
250	3250	650	4200	530	2750	425
400	4600	930	6000	750	3850	610
630(4%)	6500	1300	8400	1030	5400	860
630(6%)	6750	1200	8700	940	5600	800
1000	10500	1700	13000	1400	9500	1100
1600	17000	2600	20000	2200	14000	1700
2500	**26500**	**3800**	**32000**	**3200**	**22000**	**2500**

Table 1. Power losses in transformers according with standard HD 428

There are a number of factors that will enable the achievement of higher efficiencies and support the increase in the current minimum efficiency performance standards levels (Blackburn, 2007):

- Better use of traditional materials to achieve loss reduction and improvement of efficiency;
- Better computer-aided design of transformers to reduce losses and improve efficiency;
- Use of low loss core materials such as amorphous metals;
- New lower loss core configuration designs such as the "Hexaformer";
- Improved operational applications of transformers to optimize energy efficiency in operation;
- Consideration of total life cost of transformers: purchase cost plus operational energy losses;
- The effect of increasing harmonic levels from non-linear loads in increasing losses and reducing efficiency;
- Increased transformer life resulting from lower operating temperature with more efficient transformers.

The savings brought about by loss reduction not just about the monetary value of the energy saved: the released capacity of the system can serve to delay a costly expansion and reduce ageing of the components.

In the past there was little concern for lowering losses in transformers. This was mainly due to the fact that when compared to motors and other electrical devices, transformers were considered to be very efficient.

Thus, low loss transformers can be called"efficient transformers". Operating losses are less causing less heat generation and effecting longer life. One of the prime components of losses is the no-load loss which can be drastically reduced by better design and using superior grades of electrical steels. The other components of losses are the load loss. Load loss can be reduced by using thicker conductors. With use of superior grades of electrical steels and thicker conductors for the windings, the losses of transformers may be brought down to minimum.

The conventional transformer is made up a silicon alloyed iron (Grain oriented) core. The iron loss of any transformer depends on the type of core used in the transformers. However, the latest technology is to use amorphous material for the core. The expected reduction in energy loss over conventional (Si Fe core) transformers is roughly around 70%, which is quite significant. Electrical distribution transformers made with amorphous metal cores (high efficiency transformers) provide an excellent opportunity to conserve energy right from the installation. Though these transformers are costlier than conventional iron core transformers, the overall benefit towards energy savings will compensate for the higher initial investment.

It must be underline if now for us the objective is replacements of old transformers by efficient transformers, (EU, Fig. 6), in Japan the objective the passing to high efficient transformers (Amorphous).

Thus, the technical solutions exist to reduce transformer losses. Energy-efficiency can be improved with better transformer design (selecting better, lower-core-loss steels; reducing flux density in a specific core by increasing the core size; increasing conductor cross-section to reduce current density; good balancing between the relative quantities of iron and copper in the core and coils; and so on.), or by the adoption of amorphous iron transformers world-wide (distribution transformers built with amorphous cores can reduce no-load losses by more than 70% compared to the best conventional designs).

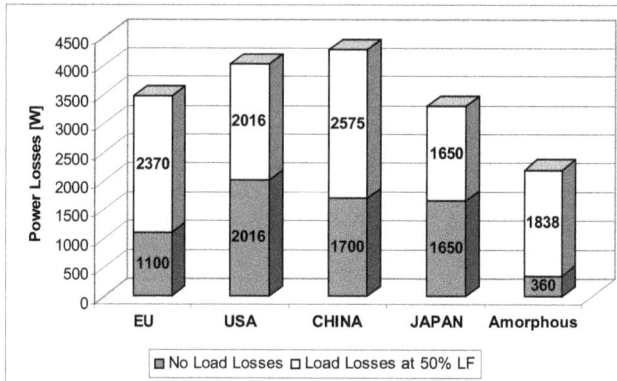

Fig. 6. 1000 kVA Transformers losses from fabrication Norms for several countries (Eiken, 2007; European Commission, 1999)

3. Fuzzy modeling in energy losses determination

3.1 Fuzzy modeling

In the last few years, research in the area of the optimal operation and planning of the electric networks is in expansion. Many papers and reports about new models have been published in the technical literature, due mostly to the improvement of the computer power availability, new optimization algorithms, and greater uncertainty level introduced by the power sector deregulation.

A considerable part of the information is uncertain, i.e. it is vague, fuzzy, and even ambiguous. Uncertainty of the information in distribution planning, as example, is caused by errors in measurements as well as inevitable errors in estimation of future forecasts. Furthermore, since most of the data used for the planning tasks are not based on the direct measurements, the degree of information uncertainty may be quite high. From the descriptive viewpoint, all the initial information may be categorized into the following several classes (Neimane, 2001):

- Deterministic (voltage levels, sites for new substations etc).
- Probabilistic (existing loads, reliability data for the network components, power quality indices etc).
- Fuzzy (information in linguistic form: large, average small, etc). The fuzzy information is often very subjective and is usually based on expert judgment; however it can be a huge aid during the Decision-Making (DM) process.

Since its first presentation in 1965 by L. A. Zadeh, the Fuzzy Techniques (FT) have had an unexpected growth and success. The broad development of mathematical theory especially in areas of Possibility Theory, Fuzzy Control, Artificial Neural Networks, and Pattern Recognition provided the basis for different applications. They finally became the driving force of FT that today is reflected in many different software and hardware products.

The basic idea of FT is to model and to be able to calculate with uncertainty. Mathematical models and algorithms in distribution systems aim to be as close to reality as possible. The required human observations, descriptions, and abstractions during the modeling process are always a source of imprecision, Fig. 7.

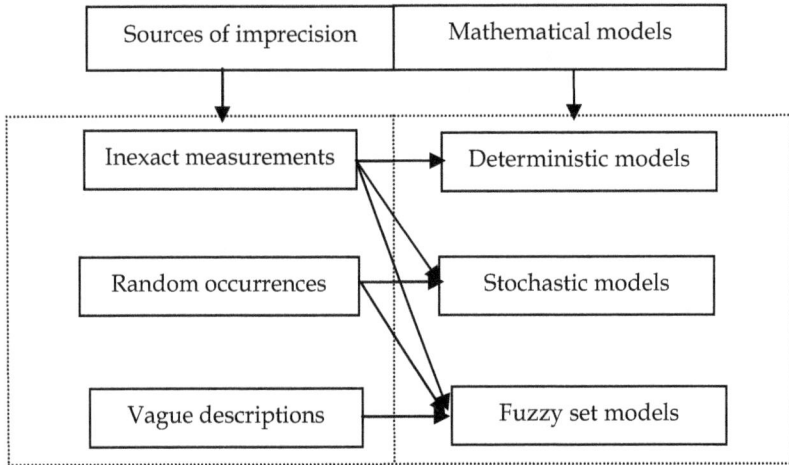

Fig. 7. Mathematical models for imprecision (Steitz et al., 1993)

While the two sources of imprecision have long since led to suitable mathematical models, the last one came in our mind only a few decades ago, although we use it instinctively in our everyday life, e.g.: *The reliability of this component is very high*. Most of linguistic descriptions such as *Small, Medium or High* are in nature fuzzy. These vague descriptions are as well part of modeling process and the algorithm. The system analyzer has to differ between classes, e.g., when classifying system operation states according to certain operational aspects (Steitz et al., 1993; Cârțină et al. 2003).

Uncertainty in fuzzy logic is a measure of nonspecifically that is characterized by possibility distributions. This is, somewhat similar to the use of probability distributions, which characterize uncertainty in probability theory. Linguistic terms, used in our daily conversation, can be easily captured by fuzzy sets, for computer implementations. A fuzzy set is a set containing elements that have varying degrees of membership in the set. Elements of fuzzy set are mapped to a universe of a membership function.

Fuzzy sets and membership functions are often used interchangeably. There are different ways to derive membership functions. Subjective judgment, intuition and expert knowledge are commonly used in constructing membership function. Even though the choices of membership function are subjective, there are some rules for membership function selection that can produce well the results. The membership values of each function are normalized between 0 and 1.

The uncertain of the load level, the length of the feeders or loading of the power transformers and so on will be represented as fuzzy numbers, with membership functions over the real domain $\Re$. A fuzzy number can have different forms but, generally, this is represented as trapezoidal or triangular fuzzy number, Figs. 8 and 9.

In the case of triangular and trapezoidal representations, a fuzzy number $\tilde{A}$ is usually represented by its breaking points (Cârțină et al., 2003).

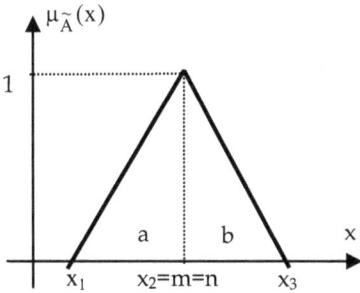

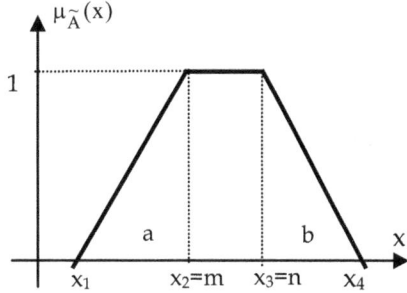

Fig. 8. Triangular fuzzy number Fig. 9. Trapezoidal fuzzy number

$$\tilde{A} \Leftrightarrow (x_1, x_2, x_3) = [m, a, b] \tag{1}$$

$$\tilde{A} \Leftrightarrow (x_1, x_2, x_3, x_4) = [m, n, a, b]$$

The usual algebraic operations with numbers can be extended to fuzzy sets:

$$\tilde{A} = \left\{ \left(x_1, \mu_{\tilde{A}}(x_1) \right), x_1 \in X \right\} \tag{2}$$

$$\tilde{B} = \left\{ \left(y_1, \mu_{\tilde{A}}(y_1) \right), y_1 \in X \right\} \tag{3}$$

$$\tilde{A} + \tilde{B} = \left\{ \left(z_1, \mu_{\tilde{A}+\tilde{B}}(z_1) \right), z_1 \in X \right\} \tag{4}$$

$$\mu_{\tilde{A}+\tilde{B}}(z_1) = \max_{z_1 = x_1 + y_1} \{ \min[\mu_{\tilde{A}}(x_1), \mu_{\tilde{B}}(y_1)] \} \tag{5}$$

$$\tilde{A} \cdot \tilde{B} = \left\{ \left(z_1, \mu_{\tilde{A} \cdot \tilde{B}}(z_1) \right), z_1 \in X \right\} \tag{6}$$

$$\mu_{\tilde{A} \cdot \tilde{B}}(z_1) = \max_{z_1 = x_1 \cdot y_1} \{ \min[\mu_{\tilde{A}}(x_1), \mu_{\tilde{B}}(y_1)] \} \tag{7}$$

$$c \cdot \tilde{A} = \left\{ \left(x_1, \mu_{c \cdot \tilde{A}}(x_1) \right), x_1 \in X \right\} \tag{8}$$

$$\mu_{c \cdot \tilde{A}}(x) = \mu_{\tilde{A}}(c \cdot x) \tag{9}$$

If for three factors $\tilde{A}, \tilde{B}, \tilde{C}$, defined as fuzzy variables, we accept the trapezoidal form, represented by breaking points, Fig. 9:

$$\tilde{A} = [m_1, n_1, a_1, b_1], \; \tilde{B} = [m_2, n_2, a_2, b_2], \; \tilde{C} = [m_3, n_3, a_3, b_3] \tag{10}$$

the resulting expressions for addition and multiplication are (11), respectively, (12):

$$\tilde{A} = \tilde{B} + \tilde{C} = [m_2 + m_3, \; n_2 + n_3, \; a_2 + a_3, \; b_2 + b_3] \tag{11}$$

$$\tilde{A} = [m_1, n_1, a_1, b_1] = \tilde{B} \cdot \tilde{C} = \\ [m_2 m_3, \; n_2 n_3, \; m_2 a_3 + m_3 a_2 - a_2 a_3, \; n_2 b_3 + n_3 b_2 + b_2 b_3] \tag{12}$$

In particular case of the triangular fuzzy number representation, m = n, Fig. 8, and from (11) and (12), we have:

$$\tilde{A} = \tilde{B} + \tilde{C} = [m_2 + m_3, \; a_2 + a_3, \; b_2 + b_3] \tag{13}$$

$$\tilde{A} = [m_1, a_1, b_1] = \tilde{B} \cdot \tilde{C} = [m_2 m_3, \; m_2 a_3 + m_3 a_2 - a_2 a_3, \; m_2 b_3 + m_3 b_2 + b_2 b_3] \tag{14}$$

In certain conditions, it is necessary to define the radical operation for a fuzzy number. Considering the triangular representation:

$$m_3 = m_2, \quad a_3 = a_2, \quad b_3 = b_2 \tag{15}$$

(14) becomes:

$$\tilde{A} = [m_1, a_1, b_1] = [m_2^2, \; 2 m_2 a_2 - a_2^2, \; 2 m_2 b_2 + b_2^2] \tag{16}$$

From (16), we have the calculation expressions for m_2, a_2 and b_2:

$$m_2 = \sqrt{m_1} \tag{17}$$

$$a_2 = \sqrt{m_1} \pm \sqrt{m_1 - a_1} \geq 0, \; a_2 \leq \sqrt{m_1} \tag{18}$$

$$b_2 = -\sqrt{m_1} \pm \sqrt{m_1 + b_1} \geq 0 \tag{19}$$

Then, for radical operation we can write:

$$\sqrt{\tilde{A}} = [\sqrt{m_1}, \; \sqrt{m_1} - \sqrt{m_1 - a_1}, \; -\sqrt{m_1} + \sqrt{m_1 + b_1}] \tag{20}$$

Similarly, for division operation, from (14), we can write the equations system:

$$m_2 m_3 = m_1 \tag{21}$$

$$m_2 a_3 + m_3 a_2 - a_2 a_3 = a_1 \tag{22}$$

$$m_2 b_3 + m_3 b_2 + b_2 b_3 = b_1 \qquad (23)$$

where from the values of m_3, a_3, b_3 are:

$$m_3 = \frac{m_1}{m_2} \qquad (24)$$

$$a_3 = \frac{a_1 - \dfrac{m_1}{m_2} a_2}{m_2 - a_2} \geq 0 \qquad (25)$$

$$b_3 = \frac{b_1 - \dfrac{m_1}{m_2} b_2}{m_2 + b_2} \geq 0 \qquad (26)$$

Considering the significance of the parameters, Fig. 8, the conditions (25) – (26) can be written:

$$\frac{a_1}{a_2} > \frac{m_1}{m_2} ; \quad \frac{b_1}{b_2} > \frac{m_1}{m_2} \qquad (27)$$

For defuzzification process, the most used method is the center of gravity (CG) method. According to this method, the crisp value is calculated with relation:

$$\text{Crisp} = \frac{\displaystyle\sum_{i=1}^{4} x_i \cdot \mu(x_i)}{\displaystyle\sum_{i=1}^{4} \mu(x_i)} \qquad (28)$$

3.2 Fuzzy modeling in determination of the energy losses

In electrical distribution networks, except the usual measurements from stations, there is few information about the state of network. The loads are not usually monitored. As a result, there is at any moment a generalized uncertainty about the power demand conditions and therefore about the network loading, voltage level and power losses. The effects of the load uncertainties will propagate to calculation results, affecting the state estimation and the optimal solutions of the various problems concerning the operation control and development planning.

Therefore, the fuzzy approach may reflect better the real behavior of a distribution network under various loading conditions. For modeling of the loads, two primary fuzzy variables are considered: the loading factor K_L (%) and power factor $\cos\varphi$, so that the representation of the active and reactive powers result from relations:

$$P = \frac{K_L}{100} \cdot S_n \cdot \cos\varphi, \quad Q = P \cdot \tan\varphi \qquad (29)$$

where S_n is the nominal power of the distribution transformer from the distribution substations.

Thus, the hourly loading factor of a particular distribution transformer can be employed to approximate the nodal load. And, because the most utilities have not historical records of feeders, it is proposed to use linguistic terms, usually used by dispatchers, to describe the uncertain hourly loading factor. These linguistic terms are defined in function by the loading of the transformers at the peak load. Each loading level represented by a linguistic variable is described by a fuzzy variable and its associated membership function.

The loading factor K_L and the power factor $\cos\varphi$ were divided into five linguistic categories with4 the trapezoidal membership function, Table 2 (Cârţină et al., 2003; Grigoraş et al., 2010b).

The fuzzy models used in this case for the loading factor and power factor correspond to urban residential loads. Also, active power and power factor must be correlated as it is shown in Fig. 10 (Cârţină et al., 2003).

Linguistic Categories		x		Linguistic Categories		x	
		$K_L[\%]$	$\cos\varphi$			$K_L[\%]$	$\cos\varphi$
VS	x_1	10	0.75	M	x_3	55	0.87
	x_2	10	0.77		x_4	65	0.89
	x_3	15	0.79	H	x_1	55	0.87
	x_4	25	0.81		x_2	65	0.89
S	x_1	15	0.79		x_3	75	0.91
	x_2	25	0.81		x_4	85	0.93
	x_3	35	0.83	VH	x_1	75	0.91
	x_4	45	0.85		x_2	85	0.93
M	x_1	35	0.83		x_3	95	0.95
	x_2	45	0.85		x_4	95	0.97

Table 2. Values of the primary variables for each linguistic loading level

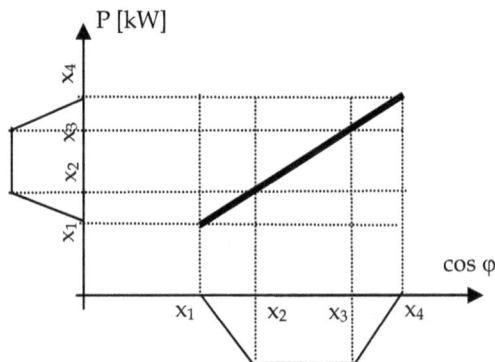

Fig. 10. Fuzzy Correlation between active power (P) and power factor (cos φ)

For estimation of the annual energy losses in the distribution networks, the following empirical formula can be used:

$$\Delta W_T = \left(\Delta P_{Cable} + \Delta P_{TrCo} \right) \cdot LF \cdot 8760 + \Delta P_{TrIr} \cdot 8760 \qquad (30)$$

where:

ΔP_{Cable} – the power losses at the peak load in the cable;

$\Delta P_{Tr\,Co}$ – the cooper losses at the peak load in the transformers;

$\Delta P_{Tr\,Ir}$ – the iron losses in the transformers;

LF – loss factor.

The values of the ΔP_{Cable}, $\Delta P_{Tr\,Co}$, and $\Delta P_{Tr\,Ir}$ are calculated as fuzzy variables using the modeling presented above.

Determination of the loss factor (LF) can be done for each distribution feeder, using the following formulae (Albert&Mihailescu, 1998; Grigoraş et al., 2010d):

$$LF = \left(0.124 + \frac{T_{max}}{10000} \right)^2 \qquad (31)$$

$$T_{max} = \frac{\sqrt{W_P^2 + W_Q^2}}{S_{max}} \qquad (32)$$

where:

W_P - active power measured during a period T (usually a year), (kWh);

W_Q - reactive power measured during a period T (usually a year), (kVAr);

S_{max} - peak load of the distribution feeder, (kVA);

T_{max} - peak load hours.

4. Case study

4.1 Technical analysis

In this paragraph it's presented as example a strategy for energy saving based on the replacement of the 6 kV voltage level with 20 kV voltage level, in correlation with the extent of using efficient transformers. Thus, it considered an urban distribution network with 8 electric stations (110/20/6 kV), which supplies 102 distribution feeders (52 feeders by 6 kV and 50 feeders by 20 kV). The characteristics of this urban distribution network are presented in the Tables 3 and 4.

An analysis of the information from the Table 3 indicates that the length of the distribution networks for two voltage levels is about the same, but the sections between 150 and 185 mm² predominates at the 20 kV. For the 6 kV level the length of the sections less than 150 mm² is close to that of sections between 150 and 185 mm². Regarding the number of transformers, Table 4, it can observed that the average installed power (S_i) of a transformer at the 6 and 20 kV voltage levels is about the same (510 vs. 550 kVA) for a ratio of about 2 to 3. More than eighty percent of the transformers have an installed power above 400 kVA.

Level Voltage	Length, [km]			
	< 150 mm²	≥150 mm² & ≤ 185 mm²	> 185 mm²	Total
6 kV	91.76	126.2	25.58	243.54
20 kV	59.66	227.07	0	286.73

Table 3. The length of cables in function by section for the analyzed distribution network

Level Voltage	< 400 kVA		≥ 400 kVA & ≤ 630 kVA		> 630 kVA		Total	
	Transformers		Transformers		Transformers		Transformers	
	No. [pcs]	S_i [kVA]	No. [pcs]	S_i [kVA]	No. [pcs]	S_i [kVA]	No. [pcs]	S_i [kVA]
6 kV	77	16536	254	124830	50	53000	381	194366
20 kV	78	16569	358	178620	82	89800	518	284989
Total	155	33105	612	303450	132	142800	899	479355

Table 4. Distribution transformer populations for the analyzed distribution network

In order to check the technical profitability of the implementing the strategy, two variants were analyzed:

- Variant I – the 6 kV and 20 kV voltage levels with the old transformers;
- Variant II – the replacement of 6 kV voltage level with 20 kV, in correlation with the use of the efficient transformers.

The technical characteristics for the distribution (old and efficient) transformers (the cooper and iron power losses) are presented in the Table 5.

Nominal power [kVA]	Cooper Losses		Iron Losses	
	Old [W]	Efficient [W]	Old [W]	Efficient [W]
100	2760	1475	600	210
160	3720	2000	890	300
250	5040	2750	1100	425
400	6850	3850	1470	610
630	9720	5400	1920	860
1000	13900	9500	2700	1100
1600	20200	14000	4350	1700

Table 5. Nominal power losses of the distribution transformers (Old vs. Efficient)

For appropriate loading level, Table 2, the power losses of the each feeder can be calculated. Using these power losses (in cables and distribution transformers) and the loss factors, the energy losses can be calculated with the relation (8). For example, in the Table 6 the crisp annual energy losses, as function of the linguistic loading level, for the urban feeders by 6 kV which leave from an electric station (electric station no. I), were presented.

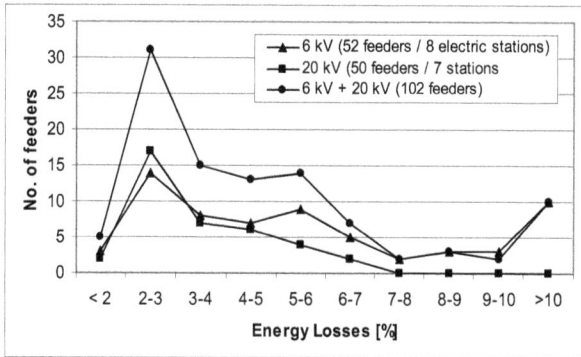

Fig. 11. Annually energy losses' variation in function by number of distribution feeders, variant I

	Loading level	dW_{cable} [MWh]	$dW_{Tr\ Co}$ [MWh]	$dW_{Tr\ Ir}$ [MWh]	dW_{Tr} [MWh]	dW_{Total} [MWh]
1	S	0.77	2.75	88.98	91.73	92.49
2	H	17.24	35.51	133.31	168.82	186.06
3	S	2.40	12.57	101.83	114.40	116.79
4	S	2.00	13.97	110.53	124.50	126.50
5	M	8.39	23.40	149.69	173.09	181.48
6	H	1.34	18.48	63.28	81.76	83.10
7	S	1.75	11.88	105.78	117.66	119.41
Total		**33.89**	**118.56**	**753.39**	**871.95**	**905.84**

Table 6. Crisp values of the energy losses on the feeders which leave from a distribution station, as function of the linguistic loading level, variant I

In the following, the results obtained by making the energy balance of 6 kV feeders/electric stations (crisp values) are presented in the Tables 7 – 9 and Figs. 12 – 16.

ST	dW_{cable} [MWh]	$dW_{Tr\ Co}$ [MWh]	$dW_{Tr\ Ir}$ [MWh]	dW_{Tr} [MWh]	dW_{Total} [MWh]	dW_{cable} [%]	$dW_{Tr\ Co}$ [%]	$dW_{Tr\ Ir}$ [%]	dW_{Tr} [%]	dW_{Total} [%]
I	33.89	118.56	753.39	871.95	905.84	0.31	0.33	3.21	3.54	3.84
II	30.61	53.91	169.26	223.17	253.78	1.13	1.99	6.25	8.24	9.37
III	3639.70	522.65	950.82	1473.46	5113.11	6.73	0.97	1.76	2.73	9.46
IV	1538.50	228.09	1103.89	1331.98	2870.52	3.13	0.46	2.25	2.71	5.85
V	292.79	87.17	443.50	530.67	823.46	1.40	0.42	2.12	2.54	3.94
VI	313.81	58.58	368.70	427.28	741.09	2.31	0.43	2.71	3.14	5.44
VII	21.20	65.83	268.73	334.55	355.75	0.23	0.70	2.87	3.57	3.80
VIII	706.10	119.13	837.38	956.51	1662.60	1.93	0.33	2.29	2.62	4.56
Total	6576.60	1253.92	4895.67	6149.57	12726.15	3.16	0.58	2.33	2.91	6.07

Table 7. The total annually energy losses of 6 kV feeders/electric stations, Variant I

ST	dW_{cable} [MWh]	$dW_{Tr\,Co}$ [MWh]	$dW_{Tr\,Ir}$ [MWh]	dW_{Tr} [MWh]	dW_{Total} [MWh]	dW_{cable} [%]	dW_{TrCo} [%]	$dW_{Tr\,Ir}$ [%]	dW_{Tr} [%]	dW_{Total} [%]
I	3.66	30.92	322.59	353.51	357.17	0.02	0.13	1.40	1.53	1.55
II	15.91	26.30	68.00	94.30	110.21	0.59	0.97	2.51	3.48	4.07
III	240.34	264.84	428.94	693.78	934.12	0.44	0.49	0.79	1.28	1.73
IV	86.46	97.40	499.11	596.52	682.98	0.18	0.20	1.02	1.21	1.39
V	21.80	51.79	190.50	242.29	264.19	0.10	0.25	0.91	1.16	1.27
VI	10.50	19.93	164.72	184.65	195.15	0.08	0.15	1.21	1.36	1.43
VII	1.63	30.52	112.36	142.89	144.51	0.02	0.33	1.20	1.52	1.54
VIII	34.97	60.86	503.85	564.70	599.67	0.10	0.17	1.38	1.55	1.64
Total	415.25	582.56	2290.07	2872.64	3287.99	0.20	0.28	1.09	1.37	1.57

Table 8. The total annually energy losses of new 20 kV feeders/ electric stations, Variant II

ST	ΔdW_{cable} [MWh]	$\Delta dW_{Tr\,Co}$ [MWh]	$\Delta dW_{Tr\,Ir}$ [MWh]	ΔdW_{Tr} [MWh]	ΔdW_{Total} [MWh]	ΔdW_{cable} [%]	$\Delta dW_{Tr\,Co}$ [%]	$\Delta dW_{Tr\,Ir}$ [%]	ΔdW_{Tr} [%]	ΔdW_{Total} [%]
I	66.89	44.38	418.88	463.27	530.15	0.29	0.19	1.81	2.01	2.30
II	14.71	27.61	101.26	128.88	143.58	0.54	1.02	3.74	4.76	5.30
III	3399.36	257.81	521.88	779.68	4178.99	6.29	0.48	0.97	1.44	7.73
IV	1452.04	130.69	604.77	735.45	2187.55	2.96	0.27	1.23	1.50	4.46
V	270.99	35.38	253.00	288.37	559.27	1.30	0.17	1.21	1.38	2.68
VI	303.30	38.65	203.98	242.63	545.94	2.23	0.28	1.50	1.78	4.01
VII	19.57	35.31	156.36	191.67	211.24	0.21	0.38	1.67	2.05	2.25
VIII	671.14	58.26	333.53	391.81	1062.93	1.84	0.16	0.91	1.07	2.91
Total	6198.00	628.08	2593.67	3221.76	9419.65	2.96	0.30	1.24	1.54	4.50

Table 9. The energy saving in case of Variant II

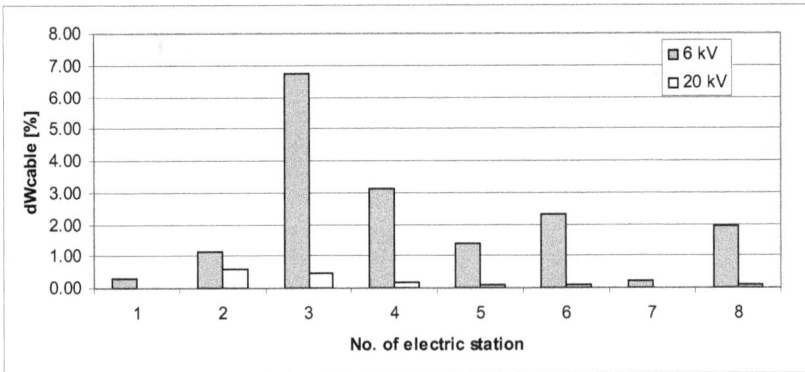

Fig. 12. The annually total energy losses in the cables/electric stations

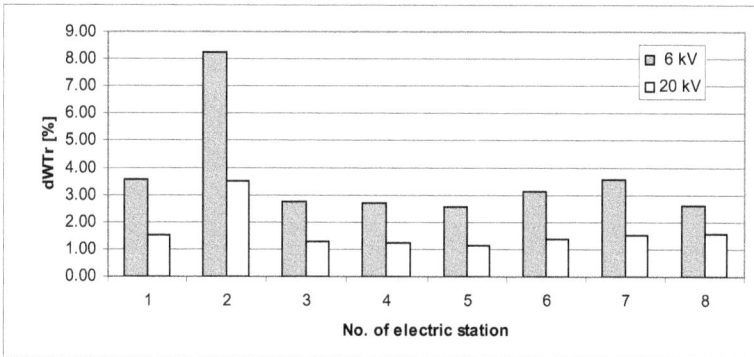

Fig. 13. The annually total energy losses in the transformers/electric stations

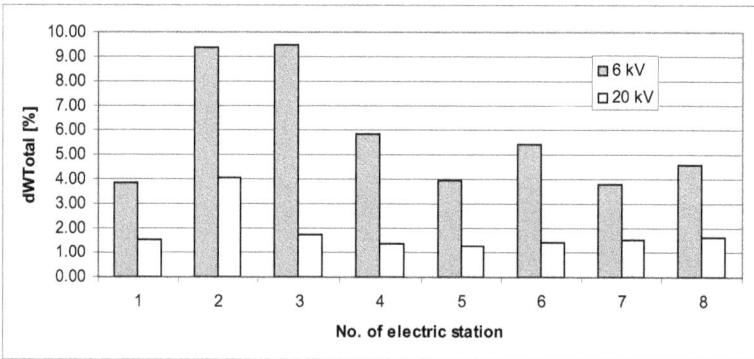

Fig. 14. The annually total energy losses/electric stations

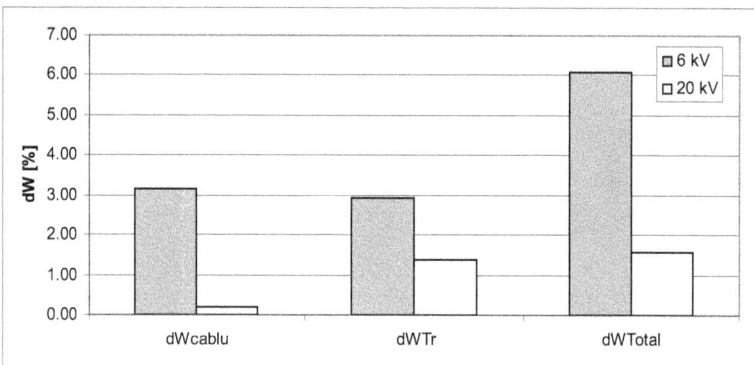

Fig. 15. The annually total energy losses/network elements (cables and transformers)

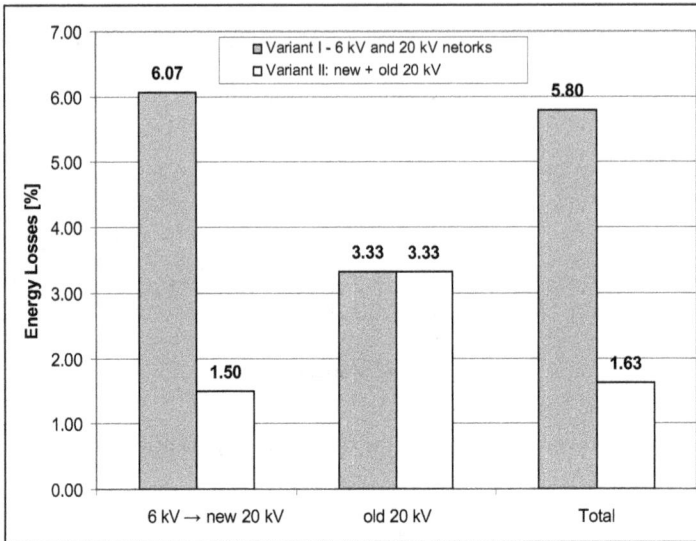

Fig. 16. The annually total energy losses/voltage levels

From the analysis of the results it can be seen that by implementing this strategy, a reduction in losses (which translates into energy savings) of about 9420 MWh /year (4.5% from total energy that entering in the 6 kV network) was obtained. Total energy losses (old and new networks by 20 kV) in the whole analyzed network decrease from 5.8 to 1.63 %, as can be seen in Fig. 16. In this figure, the energy losses for every voltage level and whole distribution network were calculated in percents from the total energy that entering in the every voltage level, respectively from the circulating total energy in network.

4.2 Economic analysis

For economic analysis of the strategy for energy saving, the payback time method can be used. This method is quite simple. The relationship for calculating the payback time of investments is:

$$P_T = \frac{N_{tr} \times C_{tr} + L_{line} \times C_{km}}{W_S \times C_{kWh}} \text{ , (years)} \tag{33}$$

where:

W_S – energy savings realized fom the replacement of the lines and transformers, [kWh];

N_{tr} – number of efficient transformers;

C_{tr} – price of an effcient transformer, (euro);

L_{line} – the length of the cable, (km);

C_{km} – price/km of the cable, (euro);

C_{kWh} – price of a kWh, (euro);

At today's commodity prices (low loss magnetic steel 2 500 - 3 000 euro/tonne, copper 6 000 - 7 000 euro/tonne) the indicative transformer price for AC' class 100 kVA typical distribution transformer is around 3 000 euro, 400 kVA is around 7 000 euro and 1 000 kVA around 12 000 euro. The price/rating characteristics can be roughly described as (Eaton Corporation, 2005):

$$C_1 = C_0 \cdot \left(\frac{S_{in}}{S_{0n}}\right)^x \qquad (34)$$

where:

C_i - is cost of transformer "i"

C_0 - is cost of transformer "0"

S_{in} - is rated power of transformer "i"

S_{0n} - is rated power of transformer with the nominal power by 100 kVA;

x - exponent (cost factor).

The x factor is about 0.4 to 0.5. For more efficient units this factor has a tendency to increase up to 0.6 or even higher.

Also, the price for one km of electric cable with section of 150 mm² was considered 4700 euro/km, and for a section of 185 mm², the price is 5900 euro/km.

In Table 10, the payback times of investment, in the case of the urban distribution network with 8 electric stations (110/20/6 kV) considered in the above paragraph, are presented.

The payback times of investment vary different from one to another distribution feeder in function by the loading level, power installed and the length. In Fig. 17, the variation of the payback time of investment in function of energy savings is shown.

ST	Energy losses/voltage level [MWh]		W_S [%]	P_T [years]
	6kV	20 kV		
I	905.84	357.16	60.57	6.91
II	253.78	110.20	56.57	12.87
III	5113.1	934.11	81.73	2.37
IV	2870.52	682.97	76.21	4.82
V	823.45	264.08	67.93	9.46
VI	741.09	195.15	73.66	4.78
VII	355.75	144.51	59.37	2.56
VIII	1662.6	446.23	73.16	9.26
Total	12726.13	3134.41	75.37	6.63

Table 10. Energy saving and the payback time of the investment/electric stations

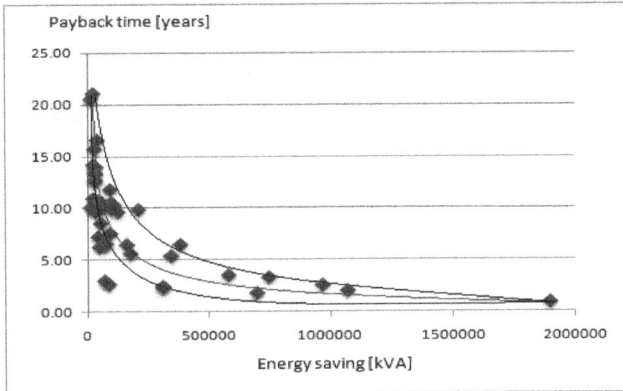

Fig. 17. The payback time of investment in function by saving energy

From the figure it can be seen that the distribution feeders with high energy saving have a payback time more reduced than the feerders with the small values of the energy saving.

5. Conclusions

Power/energy losses have a considerable effect on the process of transport and distribution of electrical energy and thus the strategies for saving energy are a concern to electrical companies in the country and abroad. In this chapter, a strategy for energy saving based on the minimization of the power/energy losses in electric networks, especially by replacement of the 6 kV voltage level with 20 kV voltage level in correlation with using efficient transformers, is presented.

This strategy can lead to increased capacity of electric distribution lines (by switching from 6 kV to 20 kV), to increase network reliability and minimize energy losses (the annually energy saving is about 9400 MWh, 2.67% from the circulating total energy in network). In terms of the environmental impact, the strategy can have a control and management of energy use not entailing the use of supplementary resources.

The economic analysis revealed that the payback time of initial investment in the network elements (lines and transformers) is on average 10 years, depending on the loading level, power installed and the length.

6. References

Albert, H. & Mihăilescu, A. (1998). *Minimization of the power/energy losses in electric networks*, Ed. Tehnica, Bucharest, Romania, ISBN 973-31-1071-X.

Alexandrescu, V.; Cârţină, G. & Grigoraş, Gh. (2010). An Efficient Method to Analyze Distribution Network Operation, *Proceedings of the 6th International Conference on Electrical and Power Engineering*, Vol. 1, pp. I-125 – I-131, ISBN: 978-606-13-0077-8, Iaşi, România, October 28-30, 2010.

Amoiralis, E. I., Tsili, M. A., Georgilakis, P. S. & Kladas, A. G. (2007). Energy Efficient Transformer Selection Implementing Life Cycle Costs and Environmental Externalities, *Proceedings of the 9th International Conference Electric Power Quality and Utilisation*, Barcelona, Spain, October 9-11, 2007, Available from:

http://users.ntua.gr/mtsili/tsili_files/articles/conferences/B15_formal.pdf.

Blackburn, T. R. (2007). *Distribution Transformers: Proposal to Increase MEPS Levels, Technical Report*, Available from: http://www.energyrating.gov.au/library/pubs/200717-meps-transformers.pdf.

Breuer, W.; Povh, D., Retzmann, D., Urbanke, Ch. & Weinhold, M. (2007). Prospects of Smart Grid Technologies for a Sustainable and Secure Power Supply, *The 20th World Energy Congress &Exhibition*, pp. 1-30, Rome, Italy, November 11-15, 2007, Available from: http://www.worldenergy.org/documents/ p001546.pdf.

Cârțină, G.; Song, Y.-H. & Grigoraș, G. (2003), *Optimal operation and planning of power systems*, VENUS Publishing House, Iași, Romania, ISBN 973-7960-09-2.

Cârțină, G.; Grigoraș G. & Bobric, E.C. (2008). Power/Energy Saving Potential Evaluation in Distribution Networks by Fuzzy Techniques, *Proceedings of International World Energy System Conference 2008*, CD, ISSN: 1198-0729, Iași, Romania, July 1-2, 2008.

Cârțină, G.; Grigoraș, G. & Bobric, E.C. (2010) Opportunities to conserve energy right from distribution installations, *Scientific Bulletin of the University Politehnica of Bucharest, Series C*, Vol. 72, No. 1, (March 2010), pp. 81 – 90, ISSN 1454-234x.

Eaton Corporation (2005), Energy Efficient Transformers Reduce Data Center Utility Costs, Available from: http://www.ecomfortohio.com/Webstuff/Eaton %20Whitepapers /Energy%20Efficient%20Transformers%20in%20PDUs.pdf.

Eiken, S. (November 2007). Energy Saving by Amorphous Metal Based Transformers, Carbon Forum Asia, Singapore, November 6-7, Available from: http://bic. go.jp/japanese/base/topics/ 071127_2/pdf.

Ellis, M. (2003), *Minimum Energy Performance Standards for Distribution Transformers*, Available from: http://www.eeca.govt.nz/sites/all/files/consultation-paper-distribution-transformers-03.pdf.

European Commission (December 1999). The scope for energy saving in EU through the use of energy-efficient electricity transformers, Available from: http://www.leonardo-energy.org.

Foote, C., Future direction for Distribution Networks, Available: http://www.ee.qub. ac.uk/blowing/activity/belfast2/foote.pdf.

Frau, J. & Gutierrez, J. (2007). Energy Efficient Distribution Transformers in Spain: New Trends, *Proceedings of the 19th International Conference on Electricity Distribution*, Paper 0141, Vienna, Austria, May 21-24, 2007.

Grigoraș, G.; Cârțină, G. & Bobric, E.C. (2010a). Strategies for Power/Energy Saving in Distribution Networks, *Advances in Electrical and Computer Engineering*, Vol. 10, No. 2, (Febraury 2010), pp. 63-66, ISSN 1582-7445.

Grigoraș, G.; Cârțină, G., Bobric, E.C. & Rotaru, Fl. (2010b). Evaluation of the performances of efficient transformers in distribution networks by fuzzy techniques, *Proceedings of the 12th International Conference on Optimization of Electrical and Electronic Equipments*, pp. 1281 – 1284, ISBN: 9878-973-131-7018-1, Brasov, Romania, May 20-22, 2010.

Grigoraș, G. & Cârțină, G. (2010c). Strategies Regarding Operating Voltage Levels in Distribution Networks, *Journal of Energy and Power Engineering*, Vol. 4, No.6, (June 2010), pp. 60-63, ISSN 1934-8975.

Grigoraș, G.; Bobric, E.C., Cârțină, G. & Rotaru, Fl. (2010d). Strategies for minimization of power losses in electric distribution networks, *Energetica Magazine*, Vol. 58, No. 7, (July 2010), pp. 314 – 318, ISSN 1453-2360.

Grigoraș, G. & Cârțină, G. (2011). Improved Fuzzy Load Models by Clustering Techniques in Distribution Network Control, *International Journal on Electrical Engineering and Informatics*, Vol. 3, No. 2, 2011, (July 2011), pp. 207–216, ISSN 2085-6830.

IEA Secretariat Energy Efficiency Working Party (September 2001), Proposal for an International Energy Association Initiative to Promote Energy-Efficient Distribution Transformers, Available from: http://www.copperinfo .com/energy/transformers. proposal.html.

International Electrotechnical Commission (2007). Efficient Electrical Energy Transmission and Distribution, 2007, Available from: http://www.iec.ch/about/brochures/ pdf/technology/transmission.pdf.

Kikukawa, S.; Tsuchiya, K., Kajiwara, S. & Takahama, A. (2004). Development of 22 kV Distribution Systems and Switchgear, Hitachi Review, Vol. 53. No. 3, (September 2004), pp.158-164, ISSN 0018-277X

Miranda, V; Pereira, J. & Saraiava, J. (2000). Load Allocation in DMS with a Fuzzy State Estimator, *IEEE Transaction on Power Systems*, Vol. 15, No. 2, (May 2000), pp. 529 – 534, ISSN 0885-8950.

Moore, D. & McDonnell, D. (March 2007), Smart Grid Vision Meets Distribution Utility Reality, Available from: http://www.uaelp.peenet.com/display-article/289077/34.

Munasinghe, M. (1984). Engineering Economic Analysis of Electric Power Systems, *Proceedings of the IEEE*, Vol. 72, No. 4, (April 1984), pp. 424.461, ISSN 0018-9219.

Neimane, V. (2001). *On Development Planning of Electricity Distribution Networks*, Doctoral Dissertation, Royal Institute of Technology, Stockholm.

Northcote-Green, J. & Speiermann, M. (2010). Third generation monitoring system provides a fundamental component of the Smart Grid and next generation power distribution networks, Available from: http://www.powersense.dk/ Download/DEMSEE_ DISCOS_Paper_Cyprus.pdf.

Polish Copper Promotion Centre & European Copper Institute (2008). Selecting Energy Efficient Distribution Transformers A Guide for Achieving Least-Cost Solutions, Available from: http://www.copperinfo.co.uk /transformers /downloads/seedt-guide.pdf.

Quittek, J.; Christensen, K. & Nordman, B, Energy-efficient networks, *IEEE Network*, Vol. 25, No. 2. (March-April 2011), pp. 4-5, ISSN 0890-8044.

Raesaar, P.; Tiigimägi, T. & Valtin, J. (2007). Strategy for Analysis of Loss Situation and Identification of Loss Sources in Electricity Distribution Networks, *Oil Shale*, Vol. 24, No. 2 Special, (2007), pp. 297–307, ISSN 0208-189X.

Ramesh, L.; Chowdhury, S.P., Chowdhury, S., Natarajan, A.A. & Gaunt, C.T. (2009). Minimization of Power Loss in Distribution Networks by Different Techniques, *International Journal of Electrical and Electronics Engineering*, Vol. 3, No. 9, (Summer 2009), pp. 521-527, ISSN 2010-3964.

Rotaru, Fl.; Grigoraş, G. & Cârţină, G. (2010), Opportunities Related to Implementing a Development Strategy of Electric Distribution Networks, *Proceedings of the 6th International Conference on Electrical and Power Engineering*, Vol. 1, pag. I-147 - I-150, ISBN: 978-606-13-0077-8, Iaşi, România, October 28-30, 2010.

Rotaru, Fl.; Grigoraş, G., Comănescu, D. & Cârţină, G. (2011). Economic Efficiency of the Solutions for the Renewals/Reinforcements on Distribution Networks, *Proceedings of 8th International Conference on Industrial Power Engineering*, pp. 103 – 108, ISSN 2069-9905 2011, Bacau, Romania, April 11-15, 2011.

Seitz, Ph.; Hubrich, H.J. & Bovy, A. (2003). Reliability Evaluation Of Power Distribution Systems With Local Generation Using Fuzzy Sets , *Proceedings of the 11th Power Systems Computation Conference*, pp. 39-45, Avignon, France, August 30 –September 3, 2003.

Targosz, R.; Belmans, R., Declercq, J.,0 De Keulenaer, H., Furuya, K., Karmarkar, M., Martinez, M., McDermott, M. & Pinkiewicz I. (2005). *The Potential for Global Energy Savings from High Efficiency Distribution Transformers*, The European Copper Institute, 2005, Available from: http://www.leonardo-energy.org.

2

RF Sounding: Generating Sounds from Radio Frequencies

Claudia Rinaldi, Fabio Graziosi, Luigi Pomante and Francesco Tarquini
Center of Excellence DEWS, University of L'Aquila
Italy

1. Introduction

In this chapter we present an innovative way of exploiting technological innovations brought by Wireless Sensor Networks (WSNs) functionalities and cellular communications in two important social fields: art and education. The result is an artistic installation built through the use of innovative technologies that can also be used with educational purposes.

In recent years there has been a growing interest on the theme of multidisciplinarity as well as exploitation of new technologies in the artistic field. Artistic installations and performance have exploited the most recent technological innovations, introducing new needs and requirements to be satisfied from a scientific point of view.

On the other hand the problem of the electromagnetic pollution has recently become an important theme of discussion for both scientific and media communities and it has often been faced with incomplete, where not wrong, knowledge of the phenomena involved. This is especially true while concerning with cellular networks where the most popular belief is that transmitting antenna produce the most of the electromagnetic radiation. RF Sounding will provide the user with the possibility of understanding the real behaviour of cellular communications in terms of RFs transmission power.

A classification of interactive art installations, organized according to which software technology (like for example MAX/MSP) and software engineering processes have been used, is presented in (1). (2) presents the software engineering process that has lead to the development of *Flyndre*, an installation at the coast of Norway, where the composition is changed according to weather parameters. In this respect, (3) presents the technological and artistic processes around *Sonic Onyx*, an installation in Trondheim where the audience can send Bluetooth files, which are reconstructed to new sound compositions and played.

The usefulness of technology into art expressions and education has also been demonstrated in two projects from MIT. *Musicpainter*, (4), is a networked, graphical composing environment that aims to encourage sharing of music creation and collaboration within the composing process. Indeed a composer could start with a small idea (e.g. only rhythmic or melodic), share it in the network and receive suggestions from other composers. *TablaNet*, (5), is an intelligent system that listens to the audio input at one end and synthesizes a predicted audio output at the other, when live music event where musicians are located remotely from each other. Other examples of artistic performances that are possible because of technology are the ones presented by the Opera of the Future Group from MIT media lab, (6). For instance *"Death and the Powers"* is a groundbreaking opera that brings a variety of technological, conceptual, and aesthetic innovations to the theatrical world. Musical performances with educational purposes too are also possible by using *Reactable*, (7), an electronic music instrument realized by the Music Technology Group of Pompeu Fabra University of Barcelona. While concerning with the possibility of enabling a joint reserach in both artistic and technological fields, the *Allosphere* project has also to be cited, (8). Allosphere is the result of 26 years of research and it allows visualizing, hearing and exploring complex multi-dimensional data. Scientifically, the AlloSphere can help provide insight on environments into which the body cannot venture. Artistically, the AlloSphere can serve as an instrument for new creations and performances fusing art, architecture, science, music, media, games, and cinema. Another related multidisciplinary application is the field at the intersection between adaptive music and games. For example, (11) presents an experimental application for individualized adaptive music for games.

Between the different technological innovations suitable for artistic purposes, *Wireless Sensor Networks* (WSNs) are used very often. For instance *Spheres and Splinters* is a new work composed by Tod Machover for hypercello, electronics, and responsive visuals that exploits audio analysis and a multitude of wireless sensors on the cello and the bow that capture how the instrument is being played. WSNs are used for personal purposes in (9) where the basic idea is to generate soundtracks for portable music players basing on the activity the user is carrying on. This way a proper music for exercise can be available to the user without the need for him to choose.

The main advantages deriving from the use of WSNs in such an application field is given by their flexibility and suitability for temporary network setups. Moreover implementation cost is cheaper than wired network even if they are more complex to configure and sometimes may be affected by surrounding events.

Advantages of using WSNs can be exploited for various purposes such as military target tracking and surveillance, natural disaster relief, biomedical health monitoring, and hazardous environment exploration and seismic sensing, (10). In military target tracking and surveillance, a WSN can assist in intrusion detection and identification. With natural disasters, sensor nodes can sense and detect the environment to forecast disasters before they occur. In biomedical applications, surgical implants of sensors can help monitor a patient's health. For seismic sensing, ad hoc deployment of sensors along the volcanic area can detect the development of earthquakes and eruptions. With the main goal to provide secure living environments for people in the world, a class of approaches refers to the exploitation of wireless sensor networks for surveillance purposes, (16).

2. RF sounding overview

This section is intended to introduce the main components of the project as summarized in figure 1. Innovations introduced and novelty will be described in detail in the next sections.

RF Sounding is an artistic installation whose aim is twofold. Indeed from one side we want to increase end users knowledge of the strength of the power emitted by their cellular phones

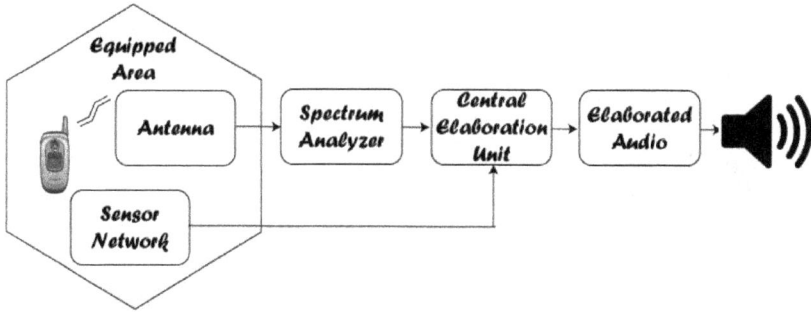

Fig. 1. RF Sounding-dataflow diagram.

with respect to the electromagnetic fields produced in the environment, on the other hand we want to provide for an artistic and interactive installation that can also be remotely joined through a web interface. It is worth noting that everything is based on the conversion of Radio Frequencies (RFs) that can be registered inside the area, into audible sounds that are spread all over the installation space through a sound diffusion system.

Here follows a brief description of the installation's architecture. Axonometric views of the project are provided in figures 2 and 3, while a block diagram is given in figure 1.

RF Sounding is built inside a hexagonal area that is accessible through at least 2 entrances equipped with gating sensors; these are intended to provide user entrance information. A loudspeaker is placed on each vertex of the hexagon in order to exploit sound spatialization. Along with the speakers there are three or more wireless sensor nodes that feed a positioning algorithm allowing to evaluate the user position and movement in the equipped area. These sensors thus allow the installation to interact with user's movements by changing lights and sounds conditions. In the center of the hexagon, at a level of 2.5-3 m from the ground, a receiving antenna is placed in order to gather all signals in the band of interest and to send them to a spectrum analyser. The analyser is linked to an elaboration unit, equipped with an audio processing board, that implements sound's elaboration and spatialization algorithms. This unit also handles the processing of the localization data obtained from the WSN. RF Sounding thus allows a user that enters the installation area bringing his switched off mobile phone, to perceive a low intensity acoustic signal that can be associated to RF signals emitted by far sources such as *Base Stations* (BSs) or other *Mobile Terminals* (MTs). On the other hand if the user inside the equipped area switches on his cellphone, he will sense a much higher acoustic signal.

3. From RF to audible frequencies

In this section we describe the process used to transform radio frequencies coming from cellular networks, into audible frequencies. The first subsection is intended to briefly review frequencies utilization and organizations in GSM and UMTS standards, the second subsection proposes an example of two cellular networks procedures that are transformed into sound events, while the third subsection goes deeper inside the translation procedure for both standards.

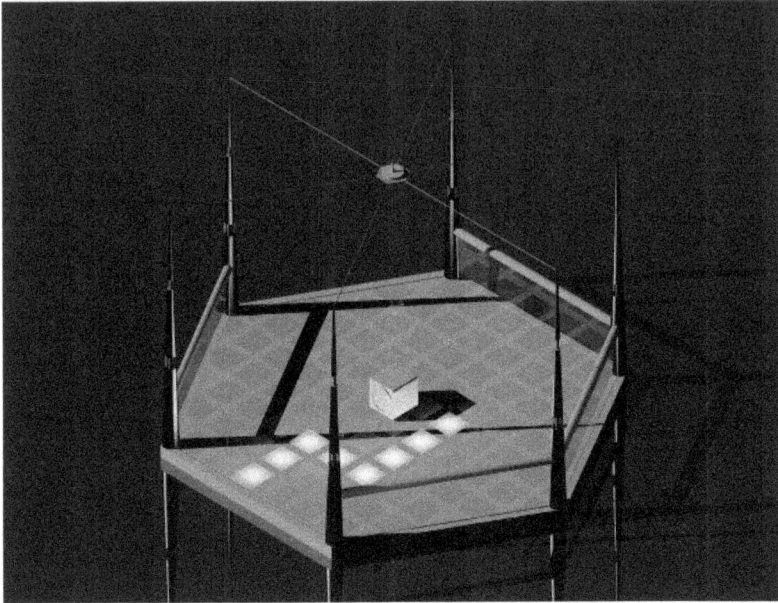

Fig. 2. RF Sounding-axonometric view.

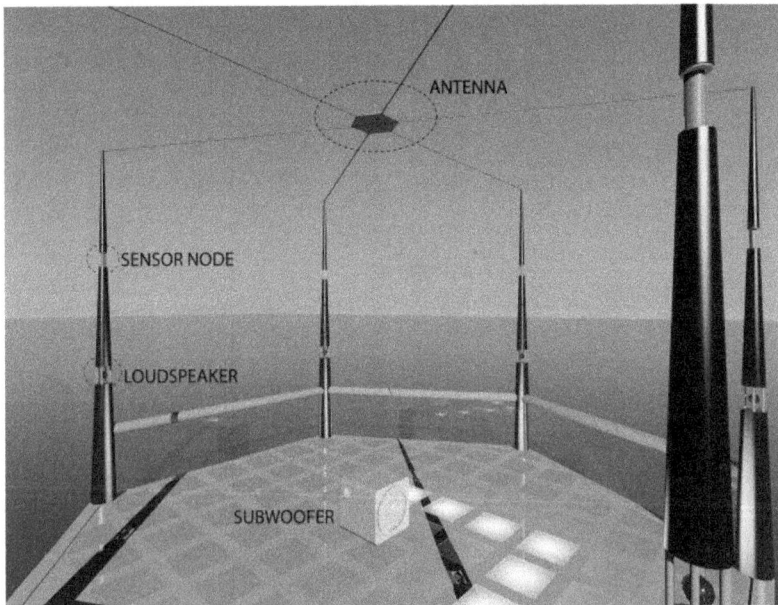

Fig. 3. RF Sounding-axonometric view with details.

System	Band	Uplink (MHz)	Downlink (MHz)	Channel number
T-GSM-380	380	380.2-389.8	390.2-399.8	dynamic
T-GSM-410	410	410.2-419.8	420.2-429.8	dynamic
GSM-450	450	450.6-457.6	460.6-467.6	259-293
GSM-480	480	479.0-486.0	489.0-496.0	306-340
GSM-710	710	698.2-716.2	728.2-746.2	dynamic
GSM-750	750	747.2-762.2	777.2-792.2	438-511
T-GSM-810	810	806.2-821.2	851.2-866.2	dynamic
GSM-850	850	824.2-849.2	869.2-894.2	128-251
P-GSM-900	900	890.0-915.0	935.0-960.0	1-124
E-GSM-900	900	880.0-915.0	925.0-960.0	975-1023, 0-124
R-GSM-900	900	876.0-915.0	921.0-960.0	955-1023, 0-124
T-GSM-900	900	870.4-876.0	915.4-921.0	dynamic
DCS-1800	1800	1710.2-1784.8	1805.2-1879.8	512-885
PCS-1900	1900	1850.2-1909.8	1930.2-1989.8	512-810

Table 1. GSM bands assignement.

It is worth noting that aesthetic considerations about these procedures are taken into account in section 6.

3.1 GSM and UMTS frequencies exploitation

The shared global use of the radio spectrum is established by ITU (International Telecommunication Union) (17), that promotes international cooperation in assigning satellite orbits, works to improve telecommunication infrastructure in the developing world and establishes worldwide standards.

We first focus on GSM, whose radio technology is specified in the 3GPPTM TS 45.-series specifications, (18), (19). We particularly refer to (19), where frequency bands of GSM systems are presented.

A GSM system may operate in 14 frequency bands as presented in table 1.

The carrier spacing, that is the channel bandwidth is of 200 KHz.

GSM-900 and GSM-1800 are the most used in Europe and we thus focus on these 2 bands of interest. It has to be noticed that GSM uses a variety of channels that are distinguished into physical and logical channels.

The method to divide up the bandwidth among as many users as possible, chosen by GSM, is a combination of Time- and Frequency-Division Multiple Access (TDMA/FDMA). FDMA divides the frequency band, which ha a width of (maximum) 25 MHz, into 124 carrier frequencies. Each Base Station (BS) is assigned one or more carrier frequencies. Using a TDMA scheme each carrier frequency is divided in time, which forms logical channels. More in particular a channel number assigned to a pair of frequencies, one uplink and one downlink, is known as an Absolute Radio Frequency Channel Number (ARFCN). GSM divides up each ARFCN into 8 time slots. These 8 timeslots are further broken up into logical channels.

Logical channels can be thought of as just different types of data that is transmitted only on certain frames in a certain timeslot. Different time slots will carry different logical channels. We underline that RF Sounding does not concern with logical channels since it relies on the use of a spectrum analyser.

A first prototype of our project that has been developed on 2010, (20), used a GSM engine (Siemens TC35, (21)) that allows to check only for the downlink channel, returning information concerning the serving Base Station (BS) channel and the power received (Rx power) as well as the same parameters for adjacent BSs. With the introduction of the spectrum analyser we take into account both uplink and downlink physical channels.

GSM standard has been the first to be investigated in the project, subsequently we deepened problems and issues arising while considering 3G networks.

UMTS is an evolution of GSM as well as a complete system architecture, offering substantially higher data rates with the main goal of delivering multimedia services in the mobile domain. The process of reserving and allocating frequency spectrum for 3G networks began on 1992 during the World Administrative Radio Conference and ended on 1997 with Resolution 212 adopted at the World Radiocommunication Conference (Geneva, Switzerland), that endorsed the bands specifically for the International Mobile Telecommunications-2000 (IMT-2000) specification. According to WARC-92 "the bands 1885-2025 MHz and 2110-2200 MHz are intended for use on a worldwide basis, by administrations wishing to implement International Mobile Telecommunications-2000 (IMT-200)". Basing on (22) and (23), UMTS frequencies can be summarized as follows:

- 1920-1980 and 2110-2170 MHz Frequency Division Duplex (FDD, W-CDMA) Paired uplink and downlink, channel spacing is 5 MHz and raster is 200 kHz. An Operator needs 3 - 4 channels (2x15 MHz or 2x20 MHz) to be able to build a high-speed, high-capacity network.
- 1900-1920 and 2010-2025 MHz Time Division Duplex (TDD, TD/CDMA) Unpaired, channel spacing is 5 MHz and raster is 200 kHz. Tx and Rx are not separated in frequency.
- 1980-2010 and 2170-2200 MHz Satellite uplink and downlink.

CDMA (Code Division Multiple Access) is a spread spectrum multiple access technique where Data for transmission is combined via bitwise XOR (exclusive OR) with pseudorandom code whose rate is much higher then the data to be transmitted. Moreover UMTS supports both frequency division and time division duplexing mode.

Carrier frequencies are designated by a UTRA Absolute Radio Frequency Channel Number (UARFCN). The general formula relating frequency to UARFN is: UARFCN = 5 * (frequency in MHz).

3.2 Cellular network procedures examples

In this subsection we present two of the procedures that characterize GSM cellular communications and that are transformed into an audible sound by RF Sounding.

The first procedure is related to the switch on of a MT inside the equipped area. In absence of previous information, the MT starts scanning the radio channels in the service band in order to find the carrier frequencies emitted by each cell with constant power. Once this procedure is completed the MT has a list of all possible BS to which it can connect, and it chooses the BS characterized by the highest receive power (top of the list). Once the carrier frequency is known, the local oscillator (LO) of the mobile terminal is tuned to this frequency through a phase locked loop operating on a periodically emitted Frequency Burst (FB) (40) carried by the Frequency Correction Channel (FCCH) of the Broadcast Control Channel (BCCH) of the GSM signalling frame structure. The information travelling through the FCCH is transmitted at the maximum power since it has to be received by all terminals. This procedure takes place even in absence of mobile terminals and this aspect is exploited in the installation, after a proper elaboration, as a persistent basic element. This basilar aspect represents the starting point of the real performance.

Another interesting procedure is the reception of a call. This event is preceded by a connection sequence that sends a signalling message on a common channel (paging channel), through which the MT is warned about the incoming call. The MT recognizes its code on the paging channel and it makes a random access on the RACH (Random Access CHannel) using the maximum allowed power, since it does not know its distance from the serving BS. If this procedure is successful (i.e. there are not collisions with other transmitting MTs), the network transmits through the AGCH (Access Grant CHannel) mapped on a beacon frequency. The remaining procedure for the prearrangement of call reception is achieved through SDCCH (Stand alone Dedicated Control CHannel). Power control is another procedure that is activated over the SDCCH in order to adapt the transmission power to the distance between the MT and the BS. Other procedures are activated over the SDCCH and they are completed when a TCH (Traffic CHannel) is assigned for the effective phonic communication. Only at this time the MT rings and this is an interesting aspect to be considered in the procedure of translation to audible frequencies. Indeed we have to adapt this fast mechanism to the possibility of our auditory system that can be affected by temporal masking when the ear is stimulated by two successive sounds separated by less than 200 ms, (24), (25).

3.3 Radio frequencies to audible sounds translation procedure

In this subsection we intend to present the process of converting radio frequencies into audible sounds.

First of all we have to keep in mind that the human auditory system is able to sense a range of frequencies between 20 Hz and 20 KHz, even if the upper limit tends to decrease with age and thus most adults are unable to hear above 16 kHz. Moreover, the ear does not respond to frequencies below 20 Hz but signals with lower frequencies can be perceived through body's sense of touch, if they are characterized by a certain level of intensity.

If we just take into account GSM, it is already evident that the bandwidth offered by GSM cellular communication is much larger than the one defined by the human hearing range. Indeed GSM-900 offers a 25 MHz uplink and 25 MHz downlink, GSM-1800 offers a bandwidth of 75 MHz both in uplink and downlink, the theoretical bandwidth of the human ear is just equal to 19980 Hz, that is 0.01998 MHz. If we also take into account that GSM-900 and GSM-1800 provide respectively 124 and 374 channels, the total number of channels offered by the GSM standard is 498, each one of them corresponding to a carrier frequency. In the simplified hypothesis of just GSM-900 and GSM-1800 the best solution would be to

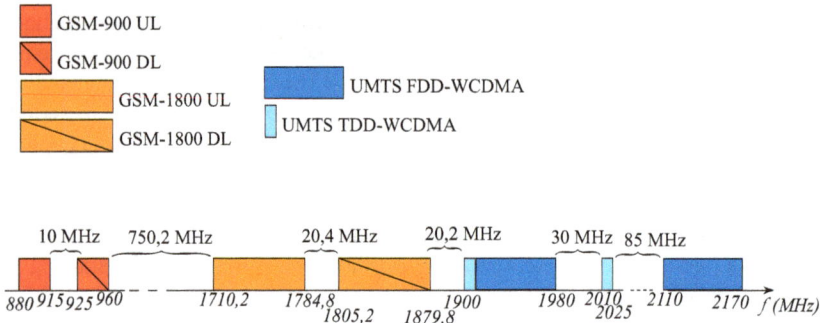

Fig. 4. GSM and UMTS frequencies allocation.

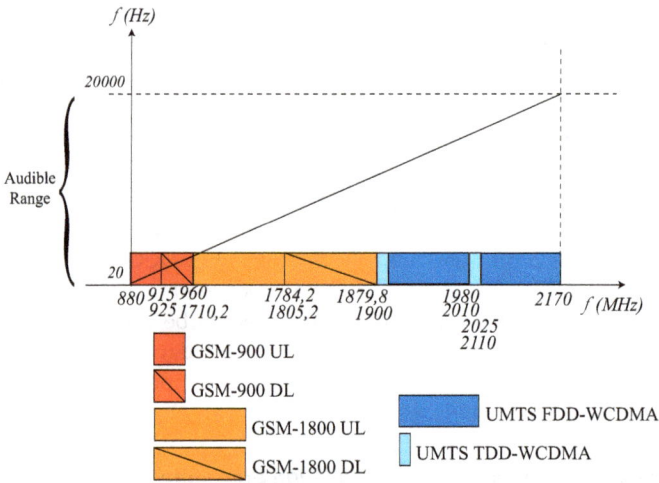

Fig. 5. Assumed relation between audible band and GSM and UMTS frequencies.

convert each carrier frequency into an audible frequency. The overall distribution of cellular frequencies for both GSM and UMTS is shown in figure 4.

By observing this figure it is evident that there are large frequency gaps that imply important gaps in an eventual linear conversion into the audible band. This can represent a problem since the audible range is very small compared to RF range, and this type of conversion would cause a noticeable loss of useful Audio Frequencies (AFs).

For this reason, in order to not lose audio bandwidth, we assumed to not consider the gaps in the RF range of interest while plotting the line corresponding to the relation between AFs and RFs, thus obtaining a conversion as roughly shown in figure 5.

As a consequence of our assumption we obtain 7 different linear relations between RFs and AFs. These relations has been computed as follows:

1. First of all we computed the equation of the line shown in figure 6 where the upper limit of the x-axis is now equal to the highest RF minus the total amount of frequency gap. In

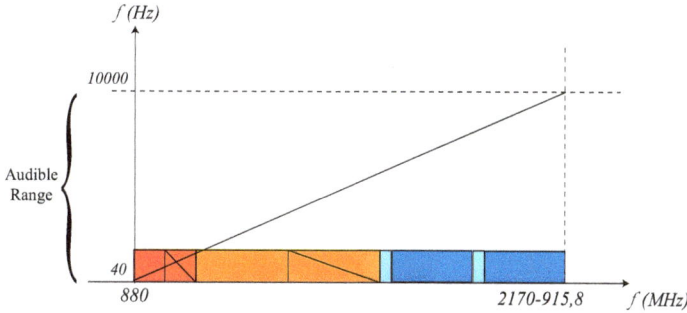

Fig. 6. Conversion procedure.

these procedure we were obliged to make our first aesthetic assumption concerning the range of AFs to be chosen. Indeed, even if the whole audible range was theoretically to be assumed for the conversion, we decided a reduction between 40 Hz and 10000 Hz, because lower frequencies are hard to be reproduced with medium quality devices, while higher frequencies than 10 KHz may result in unpleasant sensations for the human ear.

2. Once the first line equation has been found, we had to take into account that the highest AF computed for the highest RF of the first band (i.e. GSM-900 UL) is also the lowest AF corresponding to the lowest RF of the seconf RF band (i.e. GSM-900 DL). Indeed considering the first band we have:

$$y_{max} = 0.02661 \cdot 10^{-3} \cdot 915 \cdot 10^{-3} - 23.38 = 0.967 \tag{1}$$

While taking into account the second band we had to force this relation:

$$y_{max} = 0.02661 \cdot 10^{-3} \cdot 1710.2 \cdot 10^{-3} + b = 0.967 \tag{2}$$

This way we found the new constant term value.

3. The same procedure described at point 2. was followed for the other bands.

The complete list of relations found following the previously described procedure is presented below.

$$y = 0.02661 \cdot 10^{-3} x - 23.28 \quad \text{GSM-900 UL}$$
$$y = 0.02661 \cdot 10^{-3} x - 23.67 \quad \text{GSM-900 DL}$$
$$y = 0.02661 \cdot 10^{-3} x - 43.61 \quad \text{GSM-1800 UL}$$
$$y = 0.02661 \cdot 10^{-3} x - 44.15 \quad \text{GSM-1800 DL}$$
$$y = 0.02661 \cdot 10^{-3} x - 44.69 \quad \text{UMTS TDD and FDD WCDMA}$$
$$y = 0.02661 \cdot 10^{-3} x - 45.48 \quad \text{UMTS TDD-WCDMA}$$
$$y = 0.02661 \cdot 10^{-3} x - 47.75 \quad \text{UMTS FDD-WCDMA}$$

For a first implementation of RF Sounding we used Agilent HP 8592B Spectrum Analyzer shown in figure 7. The device can be connected to an elaboration unit through a IEEE 488 interface commonly called GPIB (General Purpose Interface Bus) interface, (32). Free

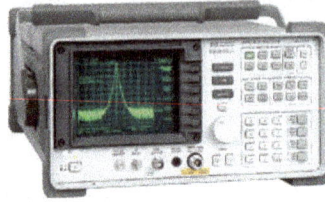

Fig. 7. Agilent HP 8592B Spectrum Analyzer.

Windows utilities that help making and recording research-quality measurements with GPIB-based electronic test equipment are available on the internet. A script written in C-language has been developed to ask the spectrum analyser to take and send its measures once every T seconds, where T can be chosen depending on the purpose of the installation placement and eventually the electronic music composer needs.

4. Localization through a WSN

This section is intended to provide more insight motivations and functioning of WSNs for localization purposes.

WSNs are distributed networked embedded systems where each node combines sensing, computing, communication, and storage capabilities. The nodes constituting the network are inexpensive, consisting of low power processors, a modest amount of memory, and simple wireless transceivers. These properties allowed WSNs to become very popular in recent years for applications such as monitoring, communication, and control. One of the key enabling and indispensable services in WSNs is localization (i.e., positioning), given that the availability of nodes' location may represent the fundamental support for various protocols (e.g., routing) and applications (e.g., habitat monitoring), (26).

The novelty we introduced with this project in the field of WSNs, is the application of one of their main functionalities such as localization, in the field of artistic installation.

In general there exists a variety of measurement techniques in WSN localization such as angle-of-arrival (AOA) measurements, distance related measurements and RSS profiling techniques. Distance related measurements can be further classified into one-way propagation time and round trip propagation time measurements, the lighthouse approach to distance measurements, received signal strength (RSS)-based distance measurements and time difference-of-arrival (TDOA) measurements, (27).

For a first implementation of our specific application we used sensor nodes from Memsic (originally Crossbow) called *Crickets*, (28), (29), (30). Cricket nodes are small hardware platform consisting of a Radio Frequency (RF) transceiver, a microcontroller, and other associated hardware for generating and receiving ultrasonic signals and interfacing with a host device, figure 8. Depending on their configuration, there are two types of cricket nodes: beacons and listeners. Cricket beacons act as fixed reference points of the location system and can be attached to the ceiling or on a vertical wall depending on the application, (28), (29), while cricket listeners are attached to objects that need to obtain their location.

Fig. 8. A Cricket hardware unit.

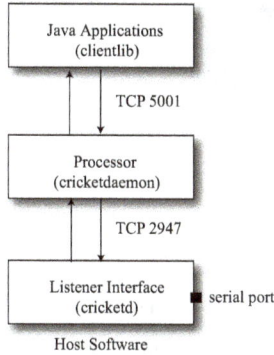

Fig. 9. Cricket software architecture.

Each beacon periodically transmits a radio frequency (RF) message containing beacon-specific information, such as beacon-ID, beacon coordinates, etc. At the beginning of the RF message, a beacon transmits a narrow ultrasonic (US) pulse that enables listeners to measure the distances to the beacons using the time difference of arrival between RF and ultrasonic signals. It is worth noting that the ultrasonic pulse does not carry any data in order to reduce beacon power consumption and ultrasonic hardware complexity.

Cricket listeners passively listen to beacon transmissions and compute distances to nearby beacons. Each listener uses these distances and the information contained in the beacon RF messages to compute their space position.

When beacons are deployed, they do not know their position. To compute beacon coordinates, it is possible to move around a cricket listener that collects distances from the beacons to itself. Using these distances, a host attached to the listener computes inter-beacon distances; the listener has to collects enough distances such that the set of computed inter-beacon distances uniquely define how the beacons are located with respect to each other.

The software for Cricket (embedded software as well as higher-layer software that runs on laptops/handhelds) is under an open source license and can be used for education, research, and commercial purposes as long as the requirements in the copyright notice are followed. The cricket embedded software is written in TinyOS (31), an open source, BSD-licensed operating system designed for low-power wireless devices. The software package includes a library to help developers create Cricket applications in Java. Cricket software architecture is shown in figure 9. At the lowest layer, cricketd allows a Cricket host device to access the Serial Port API to configure low-level Cricket parameters and obtain raw measurements from the Cricket hardware device. CricketDaemon is a server application that connects

Fig. 10. MicaZ mote.

to `cricketd` to filter and process raw Cricket measurements to infer the listener's spatial location and compute its position coordinates. The algorithm lying behind this procedure is based on the localization approach presented in (30). In a nutshell this algorithm combines the benefits of a passive mobile architecture (e.g. scalability, no need of a network infrastructure) with advantages of an active mobile system. This procedure is based on 3 main algorithms:

- *outlier rejection*: it eliminates bad distance samples;
- *extended Kalman filter* (EKF): it maintains the current and predicted device states and corrects the prediction each time a new distance sample is obtained;
- *least-squares solver* (LSQ): it minimizes the mean-squared error of a set of simultaneous non-linear equations.

Java applications may access the processed location information via the Java Cricket client library (`Clientlib`), which interfaces between the application and the `CricketDaemon`.

For our specific application we provided a software with a minimum amount of localization data elaboration based on standard deviation. This was because data coming from the underlying level appeared to be enough stable and rapid for a real time elaboration such as the one required by RF Sounding.

We also used *Memsic MicaZ* motes to provide for a wireless connection between the listener and the central elaboration unit. A MicaZ mote is shown in figure 10. Basically a MicaZ is connected to a listener, while the other one is connected to the central elaboration unit, the motes provide a transparent connection between the listener and the central elaboration unit. This way the user bringing a listener is free to move inside the installation area.

5. Open sound control protocol

In this section we want to provide for a brief overview of methods used to transmit localization data and radio spectrum information to the central elaboration unit shown in figure 1. In particular we focus on the use of *Open Sound Control Protocol* an open, transport-independent, message-based protocol developed for communication among computers, sound synthesizers, and other multimedia devices, (33), (34). OSC was originally developed, and continues to be a subject of ongoing research at UC Berkeley Center for New Music and Audio Technology (CNMAT).

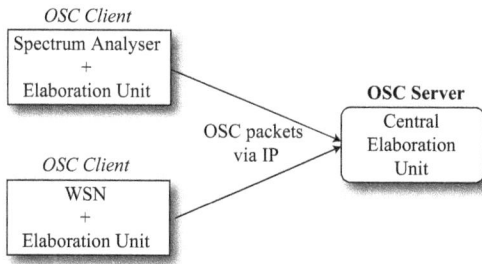

Fig. 11. Communication configuration.

OSC provides some very useful and powerful features that were not previously available in MIDI, including an intuitive addressing scheme, the ability to schedule future events, and variable data types, (35). Moreover the address space could be expanded through a hierarchical namespace similar to URL notation. Using this type of addressing allows different programs to create its own address hierarchy so that the same objects will not need the same addresses from program to program.

The main characteristic that was really appealing from our point of view is that OSC is a transport-independent protocol, meaning that it is a format for data that can be carried across a variety of networking technologies.

OSC data are basically organized into messages, which consists of the following:

- a symbolic address,
- a message name,
- the message payload.

With respect to the well known MIDI protocol, OSC is transmitted on systems with a bandwidth in the 10+ megabit/sec, that is almost 300 times faster than MIDI (31.25 kilobit/sec). Moreover precision is improved and it is much easier to work with symbolic names of objects rather then complicated mapping of channel numbers, program change numbers and controller numbers as in MIDI. On the other hand it has to be noticed that OSC can not replace MIDI, due to missing automatic connect-and play (or plug-and-play) concept, that is connected devices (via Ethernet, WLAN, Bluetooth etc) cannot scan each other and learn about each others capabilities, moreover a file format such as standard MIDI file for exchange of data does not exist.

Any application that sends OSC Packets is an OSC Client; any application that receives OSC Packets is an OSC Server. An OSC server must have access to a representation of the correct current absolute time. OSC does not provide any mechanism for clock synchronization but assumes that the two interacting systems will provide a mechanism for synchronisation. Time tags eliminate jitter introduced during transport by resynchronizing messages in a bundle and setting values for when they should take place.

Given the main advantages coming from the use of the OSC protocol we decided to assume the flexible configuration shown in figure 11.

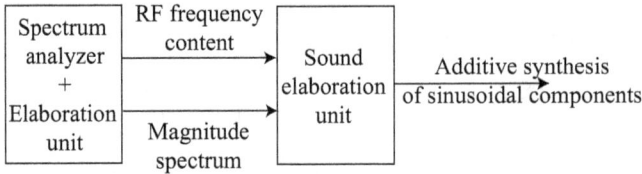

Fig. 12. Sound production for educational purposes.

The Central Elaboration Unit is equipped with Max 5, (36), a visual programming language for music and multimedia that is used in our application in order to implement sound synthesis and elaboration as well as sound spatialization as a function of RFs translation and localization data. The basic language of Max is that of a data-flow system: Max programs (called *patches*) are made by arranging and connecting building-blocks of *objects* within a *patcher*, or visual canvas. These objects act as self-contained programs (in reality, they are dynamically-linked libraries), each of which may receive input (through one or more visual "inlets"), generate output (through visual "outlets"), or both. Objects pass messages from their outlets to the inlets of connected objects.

For our purpose of communication through OSC we exploited the use of two main Max objects: udpsend and udpreceive that allow to realize the configuration shown in figure 11.

6. Sound synthesis and real time processing

In this section we provide for an overview of the main algorithms and techniques used to generate and to spread the sound all over the installation area.

A first distinction has to be done considering the possible applications of the project. Indeed it is clear that if the installation is used with educational purposes, than the sound to be spread through loudspeakers has to represent RFs behaviour as close as possible. On the other hand if the project is considered as an artistic installation rich in technological innovations, than the sound reproduced must be generated taking into account aesthetic considerations.

Let us consider the first applicative field. While using RF Sounding with educational purposes we have to reproduce RFs behaviour such that their effects should be clearly understood by the users. Having this in mind, we decided to assign each translated frequency to a sine wave, whose amplitude is thus defined by the magnitude corresponding to that frequency and computed by the spectrum analyser. Two schemes to better explain this process are presented in figures 12 and 13. The latter is referred to a particular of the procedure of conversion in the Max patch that has been developed. It has also to be noticed that the proportionality between amplitude of the audible signal and emitted power is of fundamental importance to understand how operations on a cellular phone influence the electromagnetic pollution. The effect of this procedure can be easily imagined in the case of mobile switching on and call reception procedures described in section 3.

When the installation is used for artistic purposes it becomes an interesting opportunity for modern music composers to introduce a proper signal processing on revealed RF signals. This way the installation becomes a new instrument for which new music can be composed.

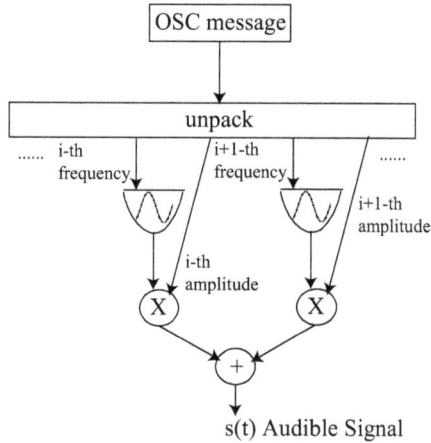

Fig. 13. Max like scheme in sound production for educational purposes.

Basing on our aesthetic point of view we decided to synthesize the sound through 3 main techniques:

1. white noise filtering with resonant filters whose center frequency is defined by revealed RFs and whose Q-factor, (37) is a function of the measured amplitudes;

2. FM parallel modulations with one carrier and one modulating signals that are added to produce the final result, (38);

3. Additive synthesis of sinusoidal components found as in the educational case plus granular synthesis, (39) where envelope and silence interval of duration are defined a a function of revealed frequencies.

It is worth noting that given a set of frequencies and values corresponding to amplitudes, these sets can be used for sound synthesis in an almost infinite number of ways including Amplitude Modulation, Phase Modulation, filtering of complex signals and so on.

We decided to introduce a random variation between the 3 methods of synthesis previously listed, maintaining the randomicity between 3 and 5 minutes. It has to be noticed that the reproduced sound is pretty much variable because of different operations that are exploited by the MT or the BTSs acting on the installation area.

We also exploited the localization data coming from the WSN to produce multichannel spatialiation of sound. While taking into account this procedure, many parameters can be varied, (41), (42), (43):

• power of all the signals emitted in the environment,

• sound motion,

• number of loudspeakers involved in sound emission in a certain instant,

• crossfade envelope between two or more channels,

• residual power of loudspeakers not directly involved in main sound emission.

For what concerns spatialization algorithms it has to be noticed that our installation allows a countless number of possibilities.

A first characterization has to be done as a function of the user's motion inside the equipped area while the installation is used for educational purposes. In this case we made an analysis on the background RF Spectrum with respect to RFs produced by active operations of the user inside the area. Indeed when the spectrum is only defined by periodic downlink operations, the sound produced is not spatialized but instead it is uniformly diffused around the area with a low intensity. On the other hand when user's MT makes its own operation, the sound is spatialized such that to give the user's the impression of being followed.

Obviously when the installation is used with artistic purposes the possibilities allowed by the sound diffusion system can be exploited with a major freedom. In this case we fixed a threshold for the user's speed. Below this threshold we established the sound diffusion system to emit only a low circular spatialized sound. This is achieved by implementing the sound crossfade only between adjacent loudspeaker. The envelope is characterized by sinusoidal panning curves in order to avoid too rapid sound variations between channels. It has to be noticed that this choice requires the introduction of at least one element of randomicity on sound speed in order to avoid an unpleasant "dance" effect on sound motion.

When the user is moving faster we assumed to spatialize the sound with the following methods:

- diagonal motion between single loudspeakers,
- motion between pairs of loudspeakers,
- slow motion between three loudspeakers with respect to a single loudspeaker or a pair of loudspeakers such that to concentrate sound intensity in a certain point in the space,
- offeset variation between loudspeakers not directly involved in sound emission.

It is worth noting that all previously described mechanisms are automated as a function of user position and motion, but also a random element is introduced in order to vary the performance result.

We want also to underline that although the installation is flexible and can be realized both indoor and outdoor, it requires the study of the impulse response of the environment and sound diffusion system characteristics.

7. Web interface

The project provides a Web interface in order to provide for a widespread diffusion of the installation's acoustic results as well as to allow the users enjoying the equipped space to record their performance and eventually use the result for instance as a ring tone. This is possible by leaving the MT just below the antenna connected to the spectrum analyser for a few seconds. This way the installation produces a sort of audio signature that is quite specific for each MT since both transmission power and RFs change as a function of the mobile operator furnishing the service as well as the specific brand and model of the MT itself.

This generated ring tone is associated to the user who can download it by logging in the installation website. This operation can be done in the installation area that is provided with a WiFi access.

8. Conclusions and future works

In this chapter an innovative project integrating technologies and experimental music has been described. RF Sounding is based on the most recent technological innovations that are exploited in an unusual fashion and it focuses on the creation of an interactive installation with a double aim, to increase users' awareness of the spectral occupancy in the cellular networks bands and to provide for a spectral phenomena aesthetic elaboration in order to produce a sounding experience.

A lot of improvements can be achieved on this project. From the WSN point of view it is important to develop a passive localization system through the exploitation of RSSI information as in multistatic RADAR, (44), (45), this way the user is relieved of bringing an external node inside the equipped area. Still concerning the technological aspect we want to make reproducible also other communication standards such as Wifi, WiMax, Zigbee and Bluetooth etc.

On the artistic point of view, we want to improve the installation's adaptability and versatility to different musical contexts, simplicity and rapidity in bringing changes in events, stability of the entire system and overall sound quality.

9. References

[1] Anna Trifonova and Letizia Jaccheri and Kristin Bergaust, "Software Engineering Issues in Interactive Installation Art" *International Journal on Arts and Technology (IJART)*, volume 1, number 1, pages 43-65, 2008.

[2] Trifonova, Anna and Brandtsegg, Øyvind and Jaccheri, Letizia, "Software engineering for and with artists: a case study", in *Proceedings of the 3rd international conference on Digital Interactive Media in Entertainment and Arts*, pages 190–197, 2008.

[3] Salah Uddin Ahmed and Letizia Jaccheri and Samir M'kadmi, "Sonic Onyx: Case Study of an Interactive Artwork. " *ArtsIT 2009: International Conference on Arts & Technology, Springer Lecture Notes of ICST*, 2009

[4] Li, Wu-Hsi, "Musicpainter : a collaborative composing environment", Master's Thesis, Massachusetts Institute of Technology. Dept. of Architecture. Program in Media Arts and Sciences, 2008.

[5] Mihir Sarkar, TablaNet: a Real-Time Online Musical Collaboration System for Indian Percussion, Master's Thesis, MIT Media Lab, Aug. 2007.

[6] MIT Media Lab, *Opera of the Future*, Available: http://www.media.mit.edu/research/groups/opera-future.

[7] Jordà, Sergi, "The reactable: tangible and tabletop music performance", in *Proceedings of the 28th of the international conference extended abstracts on Human factors in computing systems*, pages 2989–2994, Atlanta, Georgia, USA, 2010.

[8] Höllerer, Tobias and Kuchera-Morin, JoAnn and Amatriain, Xavier, "The allosphere: a large-scale immersive surround-view instrument" in *Proceedings of the 2007 workshop on*

Emerging displays technologies: images and beyond: the future of displays and interacton , San Diego, California, 2007.

[9] Robert Jacobs, Mark Feldmeier, Joseph A. Paradiso, "A Mobile Music Environment Using a PD Compiler and Wireless Sensors" *the Proc. of the 8th International Conference on New Interfaces for Musical Expression*, 2008.

[10] I. F. Akyildiz and W. Su and Y. Sankarasubramaniam and E. Cayirci, "Wireless sensor networks: a survey" in *Computer Networks*, Amsterdam, Netherlands: 1999, (38) 4: 393–422, Year 2002.

[11] Eladhari, Mirjam and Nieuwdorp, Rik and Fridenfalk, Mikael, " The soundtrack of your mind: mind music - adaptive audio for game characters", in *Proceedings of the 2006 ACM SIGCHI international conference on Advances in computer entertainment technology*, 2006.

[12] Xavier Amatriain and JoAnn Kuchera-Morin and Tobias Hollerer and Stephen Travis Pope, "The AlloSphere: Immersive Multimedia for Scientific Discovery and Artistic Exploration "*Journal IEEE Multimedia*, IEEE Computer Society, vol. 16, pages 64–75, Los Alamitos, CA, USA, 2009.

[13] A. K. Turza, "Dense, Low-Power Environmental Monitoring for Smart Energy Profiling "*Bachelor Thesis*, 2009.

[14] Paradiso, Joseph and Gips, Jonathan and Laibowitz, Mathew and Sadi, Sajid and Merrill, David and Aylward, Ryan and Maes, Pattie and Pentland, Alex, " Identifying and facilitating social interaction with a wearable wireless sensor network"*Journal of Personal and Ubiquitous Computing*, Springer London, volume 14, pages 137–152, 2010.

[15] Aylward, Ryan and Paradiso, Joseph A., " A compact, high-speed, wearable sensor network for biomotion capture and interactive media" in *Proceedings of the 6th international conference on Information processing in sensor networks*, pages 380–389, Cambridge, Massachusetts, USA, 2007.

[16] R. Alesii, F. Graziosi, G. Gargano, L. Pomante, C. Rinaldi, " WSN-Based Audio Surveillance Systems" in *Proceedings of the European Computing Conference* , Lecture Notes in Electrical Engineering Ed. Springer US, 2009.

[17] ITU official site, http://www.itu.int/en/Pages/default.aspx.

[18] ETSI, 3GPP TR 45.050 version 10.0.0 Release 10 "Digital cellular telecommunications system (Phase 2+); Background for RF Requirements (3GPP TR 45.050 version 10.0.0 Release 10)", *Technical report*, April 2011.

[19] ETSI TS 145 005 V10.1.0, " Digital cellular telecommunications system (Phase 2+); Radio transmission and reception (3GPP TS 45.005 version 10.1.0 Release 10)", *Technical report*, June, 2011 .

[20] C. Rinaldi, L. Pomante, R. Alesii, F. Graziosi , " RF sounding " in Proceedings of the *8th ACM Conference on Embedded Networked Sensor Systems*, pages 363-364, ZÃijrich, Switzerland, 2010.

[21] Siemens TC35, Available at: http://www.alldatasheet.com/view.jsp?sSearchword=TC35.

[22] ETSI TS 125 104 V10.1.0, "Universal Mobile Telecommunications System (UMTS); Base Station (BS) radio transmission and reception (FDD) (3GPP TS 25.104 version 10.1.0 Release 10) "*Technical Specification* , May, 2011.

[23] ETSI TS 125 105 V10.3.0, "Universal Mobile Telecommunications System (UMTS); Base Station (BS) radio transmission and reception (TDD) (3GPP TS 25.105 version 10.3.0 Release 10)", *Technical Specification*, July 2011.

[24] F. Alton Everest, Ken Pohlmann, *Master Handbook of Acoustics*, Fifth Edition, Ed. Mac Graw-Hill, 2009.

[25] Albert S. Bregman, *Auditory Scene Analysis: The Perceptual Organization of Sound*, Cambridge, MA: MIT Press, 1990.

[26] Stefano Tennina, Marco Di Renzo, Fabio Graziosi, Fortunato Santucci, "ESD: A Novel Optimization Algorithm for Positioning Estimation of WSNs in GPSâĂŞdenied Environments âĂŞ From Simulation to Experimentation " in *International Journal of Sensor Networks*, Vol. 6, Pages 131-156, Inderscience Publishers, 2009 .

[27] Guoqiang Mao, BarÄśsÂÿ Fidan and Brian D.O. Anderson, " Wireless sensor network localization techniques" in*Journal of Computing Networks*, Vol. 51, Pages 2529-2553, Elsevier North-Holland, Inc., 2007.

[28] Nissanka Bodhi Priyantha, " The Cricket Indoor Location System"*PhD Thesis*, Massachusetts Institute of Technology, June 2005.

[29] Nissanka B. Priyantha, Anit Chakraborty and Hari Balakrishnan, " The Cricket Location-Support System" in Proceeding of *6th ACM MOBICOM*, Boston, MA, August 2000.

[30] Adam Smith, Hari Balakrishnan, Michel Goraczko, and Nissanka Priyantha, "Tracking Moving Devices with the Cricket Location System " in Proceedings of *2nd USENIX/ACM MOBISYS Conference*, Boston, MA, June 2004 .

[31] TinyOS home page, Available at: http://webs.cs.berkeley.edu/tos/.

[32] International Standard, "IEC/IEEE Standard for Higher Performance Protocol for the Standard Digital Interface for Programmable Instrumentation - Part 1: General (Adoption of IEEE Std 488.1-2003)" in *IEC 60488-1 First edition 2004-07; IEEE 488.1*, pp.1-158, 2004.

[33] Wright, M. and A. Freed, " Open SoundControl: A New Protocol for Communicating with Sound Synthesizers" in *International Computer Music Conference* Thessaloniki, Hellas, 1997.

[34] Open Sound Control Official Website, Available at: http://opensoundcontrol.org/ .

[35] Angelo Fraietta, " Open Sound Control: Constraints and Limitations" in *Proceedings of the oth International Conference of New Interfaces for Musical Expression*,5-7 June 2008, Italy .

[36] Max 5 Help and Documentation, available at:
http://cycling74.com/docs/max5/vignettes/intro/docintro.html

[37] Sanjit K. Mitra, *Digital Signal Processing: A Computer Based Approach*, 4th Edition, McGraw Hill International Edition, 2011.

[38] J. Chowning, "The Synthesis of Complex Audio Spectra by Means of Frequency Modulation " in *Journal of the Audio Engineering Society*, vol 21, 1973 .

[39] Curtis Roads, *The Computer Music Tutorial*, Cambridge: The MIT Press, 1996.

[40] U. S. Jha, "Acquisition of frequency synchronization for GSM and its evolution systems" in *Personal Wireless Communications*, 2000 IEEE International Conference on Volume , Issue , 2000 Page(s):558 - 562.

[41] Vidolin A. "Spazi fisici e spazi virtuali nella musica elettroacustica", in *I Quaderni della Civica Scuola di Musica*, special number dedicated to Music Space and Architecture, y. 13, n. 25, pp.58-63, 1995. Revised and expanded in *Ejecutar el espacio.*, Azzurra, A. VII, n.13-15 Istituto Italiano di Cultura a Cordoba, 2000.

[42] Stockhausen K., "Musik im Raum", 1958, italian translation "Musica nello spazi, in *La Rassegna Musicale*, 32(4), 1961.

[43] Rizzardi V., "L'impiego dello spazio" in *Il Suono e lo Spazio*, Catalogo, RAI Sede regionale per il Piemonte, Turin, June 1987.

[44] Pavel Bezouseki, Vladimír Schejbal,   "Bistatic and Multistatic Radar Systems" in *Radioengineering*, vol 17, n. 3, September, 2008.

[45] Victor S. Chernyak, " Fundamentals of Multisite Radar Systems: Multistatic Radars and Multistatic Radar Systems" *CRC Press*, September 1998.

3

Services Oriented Technologies: A Focus on the Financial Services Sector in South Africa

Mazanai Musara
University of Fort Hare
South Africa

1. Introduction

Over the years, business and our daily lives have been bombarded with various forms on inventions and technology. The public adopted technology at an exceedingly fast rate and the world was transformed. In the 21st century, technology rules our daily lives. We assemble documents and reports on computers, use them for PowerPoint presentations, take laptops on the road and communicate via email and social networks, complete our banking transaction using cell phones and do online banking to mention just a few. The 21st century worker has the freedom to work from any place at any time, with always available access to information. Laptops, notebooks, wireless broadband, Smartphones and social networking have transformed the world into a global market. The shape of the business landscape took a different course because of the proliferation of technology.

The financial services sector in South Africa is one of the sectors, where the impact of the information and connection era is highly felt and ever changing. There is a growing recognition across financial service providers in South Africa that technology and innovation is the cornerstone for surviving in the 21st century economy. This is evident in almost all corporate messages in the banking sector. Therefore, the chapter focuses on technological innovations in the financial services sector. The chapter presents a theoretical review as well as the results of the empirical research that was conducted in South Africa. It is paramount as the a starting point to shed light on the concept of technology and innovation in detail. Understanding the concept of technology in detail is paramount as a basis from which innovation can clearly be articulated. Innovation can be in the form of product innovation (that results in new products or services) or process innovation (that involves the introduction of new ways of performing tasks in an organisation).The following section begins with an explanation of the concept of technology and technological innovations.

2. Literature review and concepts

The section will provide a concise discussion of the key concepts, key events and literature related to technological innovations with a particular focus on financial services sector. This is intended to shed light on the key issues related to technological innovations and their impact in the 21st century economy.

2.1 Technology and technological innovation

The term technology can be defined in many different ways depending on the context in which it is applied. For the purpose of this chapter, technology can be defined as, material objects or tools that are used by human beings such machines, hardware and utensils, in performing different activities (Bain, 1937). Technology furthermore, includes systems, methods of organisation and techniques used by the business. The term technology can be applied generally or to specific area, for example information technology (technologies use in obtaining, storing, retrieval and dissemination of information), medical technology (tools, machinery, utensils, etc, used by medical practitioners), state-of-art technology, etc.

In the financial sector, technologies which are mainly going to be focused on in this chapter includes Automated Tellers Machines (ATMs), computer hardware and software, telephones and mobile phones, the internet, among others. Technological innovations have been attributed to contribute to the distribution channels in the financial services sector. Electronic Banking (E-Banking) is one of the most notable channels. E-Banking merges several different technologies such electronic fund transfer point of sale technologies, internet banking, cell phone banking, etc. Each of these evolved in different ways, but in recent years different groups and industries have recognized the importance of working together. Several technologies were invented with a particular aim to improving service delivery in the business sector. Of particular influence in the financial services sector is the information technology, the following section discuses information technology.

2.1.1 Information Technology

Information technology (IT) refers to hardware and software that are used to store, retrieve and manipulate information. IT comprises computer hardware and software as well as other telecommunication equipment such as telephones, fax machines and mobile communication devices to mention just a few (Jürgen, 2002). The introduction of telecommunications into bank markets dates back to 1846 when the telegraph reduced stock price differentials between New York and regional stock markets (Garbade and Silber, 1978). According to Leslie (2000), the most important IT applications had their origins in US government-sponsored research in the first half of the twentieth century. Interactive IT applications would never have existed without a long and expensive gestation period in which computer power and telecommunication applications were devoted to help the US gain the initiative in science and technology. Indeed, the British experience with computer hardware development would tend to confirm the view of a defence-based technology push. The first stored-program computer in the world was developed in 1948 by academics (Freddie Williams and Tom Kilburn) at Manchester University (Anonymous, 1998).

In brief, early adoptions of telecommunications and computer applications had greatest impact in organised high value wholesale bank markets, that is, those activities that had traditionally been further away from volume transactions through retail bank branches (Anonymous, 1998). Banks absorbed the new technology on the back of a growing market for retail bank services, which expanded as middle income individuals became a growing proportion of the population. Information technology has enormous effects on the functioning of each and every enterprise operating in the 21st century economy. Jürgen (2002) argued that IT facilitates complementary innovation, enabling firms to increase output via the introduction of new processes and altering the competitive environment, thereby creating pressure for firms to adjust. Meeting these challenges require changes in the organisation of firms, for instance in the form of vertical disintegration, streamlining of managerial levels and more decentralised production, and a well educated labour force able and motivated to exploit the opportunities offered by new technology.

Knowledge on the existence and importance of IT to an enterprise alone is not sufficient for the successful running and competitiveness of the enterprise in the market. The ability to develop new and unique ways of doing things is the key to success in the 21st century economy, which is characterised by rapid changes in technology. Hence, to be competitive in the market the enterprise should be innovative. The financial services sector in South Africa is bombarded with information technologies, a differentiation strategy by several banks.

2.1.2 Technological innovation

Innovation involves using new knowledge to transform organizational processes or create commercially viable products and services. The sources of new knowledge may include the latest technology, the results of experiments, creative insights or competitive information.

Innovation can be broadly defined as the process of creating new ideas and putting them into practice. It is the means by which creative ideas find their way into everyday practice in the form of new goods or services that satisfy consumers or as new systems or practices that help organisations to improve the produce of goods or services (Wood, Wallace, Zeffane, Schermerhorn, Hunt and Osborn, 2001:611).

Rogers (1998:2) defined innovation as a process of introducing new ideas to the firm which result in improved performance of the firm. He identified five types of innovation which are:

- Introduction of new products/services or a qualitative change in the existing product,
- Process innovation new to the industry,
- Opening of a new market,
- Development of new sources for raw materials and other inputs, and
- Changes in industrial organisation.

Innovation can be thought of in two contexts: product innovation and process innovation.

a.   Product Innovation

Product innovation results in the creation of new or improved goods or services. Product innovation encompasses development of new products, changes in design of established products, or use of new materials or components in manufacture of established products (Wood et al., 2001). Product innovation can take one of the following ways:

i.   a modified version of an existing product range,
ii. a new model in the existing product range,
iii. a new product outside the existing range but in a similar field of technology,
iv. a totally new product in a new field of technology

b.   Process Innovation

Innovation that results in better ways of doing things, which is new efficient and effective processes and structures, is called process innovation (Wood et al., 2001). Process innovation is aimed at improving the operations of the organisations. This includes increasing productivity, efficiency, safety at work and waste reduction to mention just but a few. Process innovation culminates in the production of high quality products at lower costs, thus process innovation can be viewed as instrumental to product innovation.

Both product and process innovation are important in creating, communicating and delivering superior value to consumers. Technological innovations, which encompass both product innovation and process innovation, should be part of the organisation's culture to survive the competitive pressures of the 21st century economy.

Technological innovation comprise of product or process, continuous or discontinuous, radical or incremental innovations in the financial services sector leading to improved or new products. 'Radical' innovations refer to new products that result from advances in knowledge/technology. 'Incremental' innovations include improvement of process or product designs, with or without up gradation of machinery/acquisition of new machinery.

Two main aspects form the core of innovation namely invention and application. Invention is the process of discovery, while application is the act of use (Wood et al, 2001). Thus;

$$\text{Innovation} = \text{Invention} + \text{Application}$$

Innovation should be viewed as a continuous process in an organisation if the organisation is to continue operating successfully.

Having laid out the major aspect of technology and innovation, the meaning of technological innovations becomes apparent. In this case technological innovations can be viewed in terms of the creation of new tools, machines and processes that are used by human beings in the quest to develop better goods or services, and better ways of doing things. According to Subrahmanya (2005) technological innovation is the transformation of an idea into a new or improved saleable product or operational process in industry or commerce. It is therefore important to examined technological innovations in relation to their deferential impact on the business landscape. The following section provides an overview of a few selected services oriented technologies that impact the financial services sector and business in South Africa.

2.2 An overview of services oriented technologies in the financial services sector

Technological innovations in the financial services sector appeared in the form of the introductions of automated tellers machines, the rise of internet banking, electronic cards, cell phone banking and various other customers tracking and accounts management software to mention just a few. In the financial services industry at large, the banking sector was one of the first to embrace rapid globalization and benefit significantly from Information Technology (IT) developments. Technological developments in the banking sector started in the 1950s with the installation of the first automated bookkeeping machine at banks. Automation in the banking sector became widespread over the next few decades as bankers quickly realized that much of the labour intensive information handling processes could be automated with the use of computers. The first Automated Tellers Machine (ATM) is purported to have been introduced in the USA in 1968 and it was only a cash dispenser (Jayamaha, 2008). The emergence of the ATM marked the beginning of self-service banking as services provided by the bank teller could be performed on a 24-hour schedule and at the customers' convenience rather than during banking hours.

The following case study illustrates the way technology is shaping the financial services sector, in particular banking.

Partnering around technology to tackle poverty

The mobile money transfer service M-Pesa (M stands for "mobile", Pesa means "money" in Swahili) is one of the best examples of transformative power of partnerships. With the initial support of the UK Department for International Development (DfID) through matching a fund, the Kenyan mobile phone service provider Safaricom has created a service which allows users to make and receive payments, transfer money to other users and non-users, and deposit and withdraw money without needing to visit a bank.

By relying solely on the ubiquitous mobile phone, M-Pesa has significantly expanded financial access among Kenya's poor. By bringing the unbanked into the market, it has also created new markets for goods and services tailored to mobile banking. Already serving more than 14 million users in Kenya, this service is being replicated in other African countries including Rwanda, South Africa, Tanzania and Uganda.

Source: Adapted from: African Business, 45th year, No. 377. July 2011

These new technologies in the financial services sector introduced new ways of doing financial transactions and also, had made possible the introduction of other financial products or services. For example, in South Africa, technological innovations made it possible for Standard Bank to introduce the E-Plan (Electronic Plan – a card based transaction product that was introduced in 1994). Also a result of technological innovation is FNB's Smartsave suite of products – card-only account access, often with some savings or other features (such as funeral insurance) built in (Porteous and Hazelhurst 2004). ABSA also offers a Business Essentials package that includes QuickBooks Pro Bookkeeping which also assists in financial management.

Various transformations took place in the banking sector which resulted in new banking products, services and processes. Of notable importance is the introduction of online banking, card based banking and ATMs which played a significant role in increasing convenience to banking services by customers. A number of ATMs can be found in most of rural towns and townships as well as in major towns in South Africa.

Also notable in the South African banking sector is the increasing use of electronic commerce (e-commerce). E-commerce is the sharing of business information, maintaining business relationships and conducting business transactions by means of telecommunication networks (Chaffey 2002). The increasing use of the internet and mobile phones has made possible the implementation of e-commerce.

The arrival and wide use of credit cards in the South African economy is also a notable consequence of technological innovations in the South African Banking Sector. Credit card alliances with non financial institutions have been common practices in South Africa. For example credit card alliances were noted in 2006 between Edgars and Standard Bank, Metropolitan and Mercantile Bank, Kulula and FNB and Virgin and ABSA and Voyager and Nedbank among others (Hawkins 2006). These alliances allow customers to do their trading transactions with these companies anywhere without having to go to the bank or the bank's ATM to make withdrawals, thus increasing easy access to funds for customer whenever on demand.

Apart from consumer intended technologies, technologies such as smart phones, notebooks, broadband access and social networking sites and applications among others, found a place in the day to day operations of businesses. For example, smart phones have increased productivity by enabling employees to work outside the confines of their offices by creating a new phenomenon in business that I would like to term, *'the mobile workforce"*. Besides being an integral part of our daily lives, smart phones play a critical role in businesses and smart phones are fast becoming a replacement for laptops and notebooks.

Furthermore, the use of social networking sites as marketing tools as well as connecting with customers is fast becoming common practice among several businesses in South Africa and elsewhere around the world. According to Intergr8 co-founder, Rob Sussman, an estimate of about 10% of the world population is on Facebook, with 50% active every second day. From a business perspective, Facebook allows you to build a substantially larger social network than one can build in an isolated non-networked geographically restricted environment. *"I use Facebook to connect to the Intergr8 employees and customers for more direct personal communication"* said Rob. The majority of companies in South Africa have Facebook pages where they can connect with customers anytime. Almost all banks in South Africa have Facebook pages where customers can "like" and express their views, raise their concerns and ask questions. Furthermore, banks also use this platform to communicate their corporate messages to their customers and potential customers.

The introduction and use of different technologies in the banking sector is mainly to create, communicate and deliver superior value to consumers with the major objective of being competitive in the market. A question will arise as to how consumers are viewing such a technology wave and whether it is beneficial to them. The next section will account for consumers' use of technology and their reactions to technological changes.

2.3 The use of technology in the services sector

Technological innovations play an important role in increasing the competitiveness of the firm, particularly in the 21st century economy where advances in technology are a common phenomenon. Simple technologies in the financial services sector came in terms of the widespread issuing of magnetic stripe ATM cards and the phasing out of high cost manual methods of banking such as transacting at counters with savings books (Porteous and Hazelhurst 2004). A study by the research company, Gartner (2011), estimated that by 2014 more than three billion of the world's adult population will be able to transact electronically via mobile and internet technology. Gartner (2011) revealed that two trends are emerging that will drastically change the future of the world's trading economy, viz; (i) the rapid rise of the adoption of mobile and internet technology in emerging economies; and (ii) advance in mobile payment, commerce and banking. These trends exhibit the shape of our technological era. Also contributory to the banking sector is the increasing use of cell phones. World Bank analysts have suggested that mobile commerce, (the use cell phone in doing trading transactions) and electronic finance are apparent in developing countries as connectivity levels and reliability increases. They predicted that this trend will allow developing countries to leapfrog in the development of their financial systems. For South Africa, 2006 was projected as the take off year for online banking (Porteous and Hazelhurst, 2004).

Herferman (2001) asserted that the retail banking sector has witnessed rapid process technology, where new technology has altered the way key tasks are performed. In his study in the UK, the number of ATMs in services has risen from 568 in 1975 to 15208 in 1995, a trend he said was also observed in all industrialized countries.

In another study, Berger (2003:143) identified the changes in the use of selected banking technologies, indicating a significant growth in the use of new information technologies (IT) and financial technology. To this extent, the role of technology in the banking sector and financial services sector at large need to be examined critically, as the information and communication technologies (ICT) is changing the banking sector operations and efficiency

The benefits accruing from the introduction of new technologies in the banking sector determines greatly the reactions of consumers to these technologies. In that light this chapter highlight some of the perceptions about new technologies by both consumers and the corporate.

3. Research methodology and design

The empirical study was approached from the perspective of a formal research design through the definition of the study population, the incorporation of suitable measuring instrument and reliable techniques for data analysis as stipulated in Cooper and Schindler (2008). The empirical research for the study was conducted in two ways; a pilot study and the main survey. The measuring instrument was designed to measure the influence to technological innovations in the financial services sector. Following the design of the initial questionnaire, a pilot study on 20 respondents was conducted. The result of the pilot study and a discussion with a panel of experts led to the initial questionnaire being revised accordingly taking into consideration all the flaws identified in the process. The questionnaire was later administered to 613 respondents which consisted of financial services sector customers as well as members of the general public in the Eastern Cape province of South Africa. Following rigorous follow-ups on respondent a response rate of 65.7% was achieved.

Figure 3.1 below illustrated the percentage respondents by age who participated in the survey.

The majority of the respondents (27%) were between the ages 26-35 and the least number of respondents was from the ages of more 65.

Furthermore, to supplement the results of the questionnaire survey, interviews were conducted with banking sector officials. Fifteen employees from banks were interviewed and they were from Standard Bank, First National Bank, ABSA, NedBank and Capitec Bank.

4. Results and discussions

This section presents the results of the study that was conducted in the Eastern Cape Province of South Africa. A discussion of the results, linking them to previous studies was also incorporated in this section.

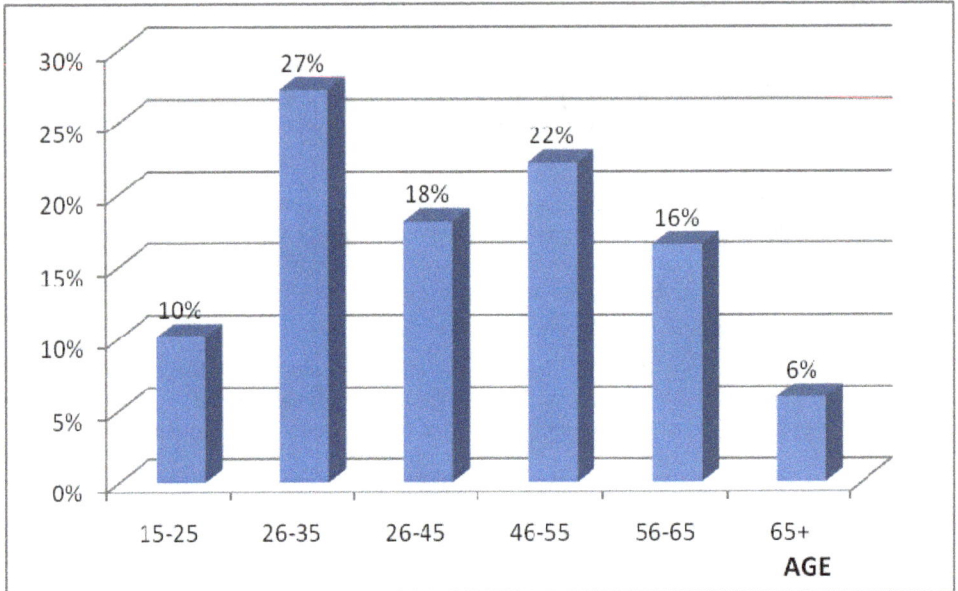

Fig. 3.1. Percentage respondents by Age

4.1 Perceptions about services oriented technology

4.1.1 Customer perceptions

There is a mixed reaction by consumers about the ever changing technological advances. In a study conducted in South Africa, the results reviewed that adult population is not so pleased by rapid changing technological advances. 86% of the adult population (65 or more years old) interviewed argued that technological advances are becoming a burden on their part and in several cases they need a technological savvy or well up individual to assist them in making certain transactions using technology. The reaction was markedly different from that of the younger population with more than 73% viewing technology as a must have necessity to function effectively. Arguments raised include the fact that the use of technology, particularly for business purposes is funny and more convenient.

While the introduction of new technology into the banking sector come with several benefits to the organisation and consumers, the reactions of different groups of people to such changes are different. Some consumers see technological changes as a threat to their old ways of doing things (they have inertia), while other groups of consumers see it as opening doors to learning new things and ways of doing things.

The results of the study further suggested that many people perceive new technologies as a "push" by banks. In this study, 63% of SA adults agreed that banks force them to use new technology due to circumstantial reasons. However a closer analysis of the results suggested that it is not technology per se that is a problem but rather how it is applied. Two thirds of people, rich as well as poor, prefer face to face service to an ATM, a third of the elderly claim to avoid banking machines as much as possible. However 75% of poorer and 83% of the richer people say they are prepared to learn new technology. The rapid acceptance of cell phones across the population is an indication of an ability to learn new technology if the benefit are substantially enough. These results corroborates with findings presented by Porteous and Hazelhurst (2004).

The fact that businesses exist to serve consumers any action should have a customer focus in mind. Business must ensure that their new technologies are received positively by consumers or ensure a thorough market assessment before investing in some sophisticated technologies which may make the consumer worse off.

4.1.2 Corporate perceptions

The impression that we get from the business sector is that technology is a *'necessary evil'*, something that every enterprise operating in the 21st century economy must embrace in order to improve and maintain its competitiveness. An analysis of the banking sector in South Africa revealed that every bank is striving to stand out with a unique technological innovation that distinguishes it from competitors. In every corporate message, every bank is trying to convey the message that, *'We are leaders in technology and we give you the best value for money'*. This message is evident in the adverts by FNB, ABSA, NedBank, Capitec and standard Bank among others.

Furthermore, too often businesses use technology as motivating tools for their employees. Social networking sites like Facebook, Twitter, Myspace and Skype are viewed as tools to boast the employees' moral during their spare times. However, the overuse of such will reduce productivity if employees use company time and resources to pursue their social needs. Using social media can be overwhelming but employees need not be carried away. However, employees can use sites that best fit their goals, market and personality for personal branding purposes. Spending time on social networking promoting yourself online can yield powerful positive results, however you need to be consistent in both promoting your image and your efforts.

If employees approach the social media with a plan and a specific focus of providing and sharing useful information while interacting with fans, friends, family and followers, it can be an effective and even fun part of your marketing mix. Overall, businesses view technological innovations as a tool that must be embraced and be kept improving to remain competitive in the ever-changing market.

4.2 Costs and benefits of technological innovations in the financial services sector

Technological innovations have had major contributions in improving the efficiency of the financial services sector. Of particular note is the fact that technological innovations have reduced significantly the cost of transacting and this will benefit both the bank and the consumer. The use of ATMs, Credit cards, internet banking, among other recent advances in technology in the South African Banking Sector, has seen substantially decreases in the cost of doing banking transactions. The new channels for delivery of banking products and services, such as through the internet, ATMs and cell phone banking, have the advantage for customers of longer hours of service and more efficient and cheaper means of delivering the products (Reixach 2001).

Changes in ways of transacting from the recent advance of information and communications technology are not limited to improving the efficiency of traditional businesses but are also enabling the development of new instruments in specific fields. A good example is the development of a new supply channel for financial services, namely the internet, which has made it possible to establish extensive and low-cost financial networking (Hayami, 2000). To this end, information and communications technology has enabled the more diversified and convenient provision of financial services, including via the internet and ATM networks. These will significantly reduce costs compared to old ways of doing banking transactions, which require face to face-over the counter transactions associated with huge travelling costs for customers to bank branches.

Furthermore, IT innovation affects the competitive position of firms both through production efficiency and changes in the goods markets. Firms that are competitive in the modern market are those that are able to create, develop and implement new and better ways of carrying out tasks. This calls for all the firms to be innovative in order for them to be competitive in the marketplace. Technological innovations enable firms to increase output via the introduction of new processes and altering the competitive environment, thereby creating pressure for firms to adjust (Jürgen et al 2002).

Technological innovations have undoubtedly introduced enormous benefits to banks, particularly in terms of productivity increases, cost reduction through labour saving and increased profitability. The use of new technology has increased outputs and reduced costs as both technological capital investments and technology human resources have a positive relationship to productivity (Jayamaha 2008). Consumers should be awarded the chance to enjoy the full benefits derived from the use of latest technological advances. This would increase banks' competitiveness through differentiation and customer service improvement, reduced transaction costs, better risk avoidance, and maintaining a stable customer base and market share (Jayamaha 2008).

Despite the wide spectrum of benefits identified from the introduction and use of technological innovations in the banking sector, there are some social and economic problems associated with technological innovations some of which have enormous consequences to the society and the nation at large if left unattended.

One of the problems brought in by technological innovations is their strain on the budgets of the banks. Most of the banks' resources are being channelled towards advancing and coping with rapid changes in technology. Apart from personal costs, technology is the biggest item in the budget of the bank, and is the fastest growing one (Reixach 2001). If not managed with caution, emphasis may be put on pursuing technological advancement while compromising other activities which are important to the success of the enterprise.

Technological innovations, particularly in information and communication technology (ICT) have resulted in globalisation of businesses and business operations. Under such conditions, competitive pressures are likely to intensify in many parts of the economy and increasingly affect previously sheltered sectors such as energy, transport, communication and distribution. The result is an increased pressure to adjust and could lead to transition problems in economies which are insufficiently able to change, a common problem of most African economies (Reixach 2001).

To the society, technological innovations, if not managed properly may result in such problems as pollution and unethical behaviours. For example the use of short message services (SMS) had been an instrument in conveying malicious information and images among teenagers which lead to moral decay.

Furthermore, the advances in technology also mean that consumers must learn how to use such technologies. Such learning should furthermore be fast to be up to date with the fast changing technologies. Slow learners and the elders who have inertia may be frustrated with such demands leading to psychological problems on consumers. More importantly, technological innovations have significant effects on service quality, efficiency and profitability.

4.3 Technological innovations vs services quality

Studies have shown that among consumers, service quality and experience is the most spoken about topic, followed by a great product or service which is closely related to something "astonishing". Customer service should focus on understanding and meeting the needs of customers effectively. According to Accenture (2007) the customer service levels in South Africa leaves a lot to be desired. South Africa is rated as one of the worst countries globally in terms of customer service, with a service maturity rating of 6%. Compared to the leader, Singapore with a rating of 89%, there is a need for improvement in customer service in South Africa. This outstandingly poor customer service in the country, however provide a fantastic opportunity for all business owners who want to gain a competitive advantage by tapping into exceptional customer service offering. Providing the best customer service is the best, if not the only solution to gain customer satisfaction, customer retention and creating customer value. Technological innovations plays a significant role is delivering quality customer service. Effective and prompt modes of communication, instant feedback and customer tracking solutions are all ingredients of an excellent customer service package. This has been enabled by the advent of the latest information and communication technologies.

A study into on the banking sector in South Africa revealed that consumers (88.5%) are satisfied with the level of communication they get from banks about their accounts information and in terms of how quick their complaints are addressed due to advanced technologies. Consumer however indicated their displeasure with some bad systems in a few banks. The study revealed that an estimated 80% of the aggrieved customers' problems with technology are caused by bad systems. It became apparent that most of the service problems customers' encounter in trying to do business occurs as a result bad delivery systems, which are out of date, too complex or just customer unfriendly.

Taking for instance, most dysfunctional systems include telephone technology that are characterised by features like call director or voice-mail systems with frustratingly lengthy menu options, or caller holder features that attempt to entertain customers with sometimes annoying music. In several cases some businesses even interrupt the entertainment with an occasional commercial message. Imagine an aggrieved customer on the other end of the line. To aggravate the problem, imagine when a customer finally reaches a real person, too often to find that the person is a company's newly appointed secretary with little knowledge of the company's product and no authority to solve the customer's problem. Such technologies cost companies millions in lost customers due to poor management of technology. It is therefore paramount to ensure that a smooth and efficient management system of technology is embraced in every aspects involving direct contact with customers and elsewhere in the services chain.

4.4 Technological innovations vs efficiency

Efficiency is a general term in economics that describes how well a system is performing in generating the maximum output for given inputs with the available technology. Efficiency is improved if more output is generated without changing the level of inputs, or in other words, the amount of "friction" or "waste" is reduced. The idea of efficiency broadly refers to the fact of using limited resources in the best possible way (Caruana, 2003). Qayyum and Khan (2007) defined efficiency as the ratio output per unit input in which case they suggested that efficiency is decomposed into two components namely, allocative (economic) efficiency and technical efficiency.

Efficiency relates the cost incurred compared to the product obtained. In general terms, the idea of efficiency broadly refers to the fact of using limited resources in the best possible way. In other words, an economic system is efficient if it does not waste its resources, in such a way that it minimises individuals' well-being (Caruana, 2003).

A system can be called economically efficient if no one can be made better off without making someone else worse off. The most output is obtained from a given amount of inputs and when production proceeds at the lowest possible per unit cost.

Efficiency introduces an additional factor to an enterprise, namely risk. Thus in every effort to improve efficiency, the risk factor should be taken into account. This idea factors in significant constraints and makes it necessary to focus on improved efficiency as a balanced process in which attention must be paid to a series of management elements that should not be taken for granted in cost-cutting drives. For example, no clear efficiency gain is achieved if it is done at the expense of increasing operational risk (Caruana, 2003). Operational risk refers to the risk of loss resulting from inadequate or failed internal processes, people and systems, or from external events. The definition includes legal risk, which is the risk of loss resulting from failure to comply with laws as well as prudent ethical standards and contractual obligations (Anon 2003).

These definitions of efficiency are not exactly equivalent. However, they all encompass the idea that nothing more can be achieved given the resources available. Thus efficiency is achieved through making the best use of the available resources.

Measures of technical output efficiency include estimates of banks' scale efficiency. Scale efficiency refers to banks or branches achieving an optimum size for producing financial services and thereby, ensuring operation at the minimum point of the average cost curve.

There is a negative relationship between bank branch size and branch efficiency (Oster and Antioch 1995).

Some of the strategies adopted by banks includes re-engineering its processes (through identifying key business activities that can either be streamlined or eliminated), improving the skills of its labour force through training and increasing the use of technology. Over time, the net effect of these initiatives will result in significant improvements in branch efficiency and elimination of the negative relationship between branch size and efficiency (Oster and Antioch 1995).

Furthermore, the increased competition resulting from financial deregulation may provide impetus for the achievement of further technical output efficiencies through scale economies. Economies of scale are achieved when a bank recognises that the cost of producing a range of outputs is less than the cost of producing them independently (Oster and Antioch 1995). To measure the efficiencies of the banks the major interest is in X-efficiency, which shows whether banks use their inputs efficiently or not (Schure and Wagenvoort, 1999). To customers efficiency is perceived in the sense of the bank completing transactions in the shortest possible time thus minimising the amount spent by customers to complete transactions.

In the financial services sector at large, the concept of efficiency can be thought of in terms of how well a bank employs its resources relative to the existing production possibilities frontier (or, in other words, relative to the current 'best practice') – how an institution simultaneously minimises costs and maximises revenue, based on an existing level of production technology (Oster and Antioch 1995). Investment in technology is an important mechanism for attaining greater efficiency. This is because technological development allows processes to be undertaken more easily, simultaneously eliminating certain time and labour-intensive tasks so that operating cost cuts are achieved.

In addition the capacity to process massive amounts of data efficiently in real time allows for better risk management and also for services better tailored to client needs. However, efforts to improve efficiency must be compatible with the challenges posed by new technologies (Caruana, 2003). In that light, the costs of new technologies should not outweigh the benefits associated with these technologies. Thus a thorough cost-benefit analysis is important in making decisions regarding new technologies. In other words it is imperative that when employing a new technology that the anticipated benefits should outweigh the anticipated costs. Technical efficiency correlates directly to overall company efficiency, this means if a business spends wisely on information technology it will reap the rewards in sales and customer service improvements and ultimately improved profitability.

A detailed account on the determined measurement of efficiency in the South African banking sector is the one provided by Mboweni (2004). According to him, the efficiency of the banking sector can be determined by expressing operating expenses as a percentage of total income.

"Currently, the international benchmark for efficiency is 60 per cent. In the past, South African banks were able to keep this ratio below or close to the international benchmark. The ratio, however, has increased from 60.2 per cent in 1999 to 67 per cent in 2002. The high volatility in efficiency in

2002 indicates that the South African banking sector was indeed experiencing problems with profitability in the first six months of 2002. This deterioration was confirmed by the return on equity of 5.4 per cent (smoothed over 12-months) and the return on assets of 0.4 per cent (smoothed over 12-months) as at the end of June 2002. Since 2002, however, the efficiency ratio has improved and, as at the end of October 2004, the efficiency of the banking sector was 65.2 per cent. By the end of October 2004, the return on equity and the return on assets of the banking sector had similarly improved" (Mboweni 2004).

The historical account of the efficiency of the South African Banking sector above shows a growing trend. This is an indication of various new approaches to service delivery which improve efficiency, most notably advance in new technologies. Looking at the efficiency trend in the South African banking sector, a possibility for further improvements is there if new approaches to improving efficiency are sought.

In a study of the banking sector in South Africa on the view of consumers regarding the improvement of efficiency of the banking sector in the past ten years, 92% of the respondents agreed that the efficiency of the banking sector has improved in the past ten years due to improvement in technology. Only 2% disagree of the fact that the efficiency of the banking sector has improved while 6% neither agree nor disagree.

The perceived improvement was attributed to continued improvements in technologies used in banks. The perceptions of respondents on ATMs, online banking and cell phone banking on their contributions to efficiency and cost reductions were also sought. These also receive a positive response with 53% of the respondents indicating that ATMs are cost effective, 77% of the respondents agreeing that online banking and cell phone banking are cost effective.

4.5 Technological innovations vs profitability

Approaches to increasing efficiency in every venture are aimed at ultimately increasing profitability. This objective is tantamount to that of the firm namely wealth creation for the owners. Thus it is critical to address the link between technological innovations, efficiency improvement and ultimately profitability.

Several studies revealed a positive relationship between efficiency and profitability, that is, the higher the efficiency, the greater the profitability. However in a study carried out by Abbaso˘glu, Aysan and Gunes, (2007) it appears that there is no clear relationship between efficiency and profitability. In their study of the Turkish banking sector, the results indicated that there is no significant evidence from the data that efficiency affects profitability. In the study which took the return on assets into account, foreign banks were found to be significantly more profitable than domestic banks. While the least efficient banks turned out to be foreign with the exception of a few, being foreign increases banks' profitability. This result shows us that foreign banks are less efficient but more profitable compared to the domestic banks. Hence, there is no clear evidence that there is a positive relationship between efficiency and profitability (Abbaso˘glu, Aysan and Gunes, 2007).

Increasing profitability is one of the fundamental objectives of any business operating as a going concern. To achieve increased profitability, the business should aim at reducing costs in the best way possible. The use of advanced and up to date technological innovations is instrumental in this regard. The major focus on improving efficiency is to reduce the costs of producing a desired level of output. Strategies and actions to minimise costs are paramount in the bid to reduce the firm's expenses and this leads to increased profitability of the venture. In the banking sector it is important that the bank be able to employ the most cost effective approaches to complete every transaction. A study on the banking sector in South Africa revealed that 75% of the respondents agree that technological innovations resulted in cost savings. Sixty five percent (65%) of the respondents agreed that cell phone banking is helping in reducing costs of banking. Among the reasons cited included cheapness of cell phone banking, no costs of travelling to the bank and the convenience of cell phone banking (customers can do their transaction anywhere, anytime). 35% of the respondents were of the opinion that cell phone banking does not help in reducing banking costs for customers. The reason for their argument ranged from the need for airtime, the need for cell phones with access to internet (that is, a phone with GPRS) which are expensive and the fact that some transaction are simply not feasible using cell phone banking.

4.6 Linking efficiency to profitability

Approaches to increasing efficiency in every venture are aimed at ultimately increasing profitability. This objective is tantamount to that of the firm namely wealth creation for the owners. Thus it is critical to address the link between efficiency improvement and profitability. Soteriou and Zenios, (1997) suggested a joint analysis of operational efficiency and profitability as shown in Fig 1. In the analysis bank branches were categorised into four categories similar to the (Boston Consultants Group (BCG) matrix, namely stars, dogs, sleepers and cows.

Sleepers are those branches that are highly profitable, while they are inefficient. Hence, their profitability can be further increased if they are awakened and improve their operational efficiency. Stars are the branches that match their superior operational efficiency with profitability, while cows are lagging in profits and a major reason for this is their operational inefficiency. Finally, for the dogs it was concluded from the analysis that enhancement of their profitability can not come from improvements in operations, since they are already efficient on the operational side (Soteriou and Zenios, 1999). The analysis revealed a positive relationship between efficiency and profitability, that is, the higher the efficiency of a bank branch the greater the profitability. However in a study carried out by Abbaso˘glu, Aysan and Gunes, (2007) it appears that there is no clear relationship between efficiency and profitability. In their study of the Turkish banking sector, the results indicated that there is no significant evidence from the data that efficiency affects profitability.

In the study which took the return on assets into account, foreign banks were found to be significantly more profitable than domestic banks. While the least efficient banks turned out to be foreign with the exception of a few, being foreign increases banks' profitability. This result shows us that foreign banks are less efficient but more profitable compared to the domestic banks. Hence, there is no clear evidence that there is a positive relationship between efficiency and profitability (Abbaso˘glu, Aysan and Gunes, 2007).

Increasing profitability is one of the fundamental objectives of any business operating as a going concern. To achieve increased profitability, the business should aim at reducing costs in the best way possible. Thus the use of advanced technologies to improve efficiency and cut costs should the focal strategy to remain competitive in the 21st century economy.

5. Conclusion

Overall, the analysis leads to the conclusion that technological innovations have changed the business landscape of the 21st century. The majority of the respondents interviewed showed positive responses with respect to the improvement of efficiency and cost reductions as a result of the use of advanced technological innovations in the financial services sector.

From the above discussion, it can be noted that the importance of technological innovations in the financial services sector reign supreme for the success of banks operating in the 21st century economy. It is important to lay a foundation to understanding the concept of technology and technological innovations. Understanding the concept of technology in detail is paramount as a basis for which innovation (the creation, introduction and use of new ideas and tools) can take root. Innovation can be in the form of product innovation (that results in new products or services) or process innovation (that involves the introduction of new ways of performing tasks in an organisation).

More contributory to the new developments in the banking sector is the innovation in information technology (hardware and software that are used to store, retrieve, and process and communicate information). This has been discussed in this chapter. Furthermore the acceptance, use and reactions of consumers to new technologies in the financial services sector had been discussed. Finally, the contributions of technological innovations to the banking sector as well as to consumer had been alluded to in this chapter. It can be deduced from the contributions of new technology that, technological innovations has resulted in reduced transaction costs for both the customers and banks, as well as improved efficiency in the banking sector.

6. References

Abbaso˘glu, O. F., Aysan, A. F and Gunes, A., (2007). Concentration, Competition, Efficiency and Profitability of the Turkish Banking Sector in the Post-Crises Period [Online: available http://mpra.ub.uni-muenchen.de/5494/ MPRA Paper No. 5494, posted 07. November 2007 / 04:45, Accessed April 17, 2008.

Abhiman, D., Subhash, C. R. and Ashok, N., (2007). Labour-use efficiency in Indian banking: A branch-level analysis, Omega 37 (2009) pp 411 – 425 [Online: available at www.sciencedirect.com accessed April 17, 2008)]

Anon (2003). Supervisory Guidance on Operational Risk Advanced Measurement Approaches for Regulatory Capital.

Bain, R. (1937). Technology and State Government. American Sociological review Vol. 2 No. 6 p. 860-865.

Caruana, J. (2003). Savings banks - Efficiency and an ongoing commitment to society. Efficiency of financial institutions [Online: available at www.bis.org, Accessed April 17, 2008].

Chaffey, D., (2004). E-Business and E-commerce Management 2nd Edition, Essex: Pearson Education Limited.

Halkos, G. E. and Salamouris, D. S., (1999). Efficiency measures of the Greek Banking Sector: A non-parametric approach for the period 1997-1999. University of Thessaly, Department of Economics, Discussion paper series 01/04

Hawkins, P., (2006). South African banking landscape: Introduction & Background, Greenside: FEAsibility Pty Ltd

Hayami, M., (2000). The impact of innovation in information and communications technology on financial systems [On-line]. Available: www.bis.org [accessed April 3, 2008].

Jayamaha, R, (2008). Impact of IT in the Banking Sector [On-line]. Available: www.bis.org [accessed February 7, 2008].

Jürgen, S., (2002). IT innovations and financing patterns: implications for the financial system [On-line]. Available: www.bis.org [accessed April 3, 2008].

Mboweni, T. T, (2004). The South African banking sector - an overview of the past 10 years [Online: available at www.bis.org, Accessed April 17, 2008].

Oster, A. and Antioch, L., (1995). Measuring Productivity in the Australian Banking Sector [online: available at http://www.rba.gov.au/PublicationsandResearch/Conferences/1995/OsterAntioch.pdf Accessed (May 26, 2008).

Porteous, D. and Hazelhurst, E. (2004). Banking on Change: Democratizing Finance in South Africa, 1994-2004 and beyond, Cape Town: Double Story Books

Qayyum, A. And Khan, S., (2007). X-efficiency, Scale Economies, Technological Progress and Competition: A Case of Banking Sector in Pakistan. Islamabad Pakistan Institute of Development

Reixach, A., (2001). The effect of information and Communication technologies on the Banking Sector and Payment System, University de Girona

Rogers, M., (1998). The Definition and Measurement of Innovation, Melbourne Institute Working Paper No. 10/98

Schatzberg, E. (2006). Technik Comes to America: Changing meanings of Technology before 1930. Technology and Culture, Vol. 47, No, 3 pp. 486-512.

Schure, P. and Wagenvoort, R.(1999). Economies of Scale and Efficiency in European Banking: New Evidence, European Investment Bank Economics and Financial Report, 1999.

Soteriou, A. and Zenios, S. A. (1999). Operations, Quality, and Profitability in the Provision of Banking Services. Management Science, Vol. 45, No. 9, Performance of Financial Institutions, (Sep., 1999), pp. 1221-1238

Subrahmanya, M. H., (2005). Pattern of technological innovations in small enterprises: a comparative perspective of Bangalore (India) and Northeast England (UK). Technovation 25, pp 269-280 [online]

Wallace J., Zeffane R. M., Schermerhorn, J. R., Hunt J. G., and Osborn R. N., (2001). Organisational Behaviour: A global Perspective 2nd Edition, Milton: John Wiley & Sons Inc.

4

Sanitation in Developing Countries: Innovative Solutions in a Value Chain Framework

Meine Pieter van Dijk
UNESCO-IHE Institute for Water Education
DA Delft
The Netherlands

1. Introduction

There are a number of new ways to look to sanitation issues in developing countries, which will be discussed in this paper:

1. *Look as shit as an asset, the beginning of a whole sanitation value chain*
2. *Emphasize the role of the private sector in sanitation, in particular small scale private enterprises*
3. *Pay attention to the economics of investing in sanitation*
4. *Increase the efficiency of the sanitation value chain*
5. *Look at advantages of small scale decentralized versus large scale centralized waste water treatment (WWT) plants*
6. *Considering sanitation as a multi-governance challenge*
7. *Consider the economics of different technological options for sanitation*
8. *Incorporating informality in the sanitation sector*
9. *Tap alternative sources of finance for sanitation*
10. *Be aware of the politics of sanitation*

2. Faeces as an asset, the beginning of a whole value chain

We suggest looking at sanitation as a possibility to make money and will study a number of cases where this is actually happening. In scientific terms this may be called: "resource oriented decentralized sanitation".

In order to achieve the Millennium Development Goals with respect to drinking water and the Johannesburg Plan of Implementation with respect to sanitation, a different approach to these problems is required, including a role for innovations and the private sector, in particular for financing a different approach to sanitation and to provide an alternative for inefficient public schemes. The drinking water and sanitation situation in African slums depends to a large extent on the socio-economic characteristics of the population, such as their income level.

Drinking water issues receive generally much more attention than sanitation issues. The fact that the number of people with no access to toilet facilities is twice the number of people having no access to safe water is the proof that sanitation is very much neglected.[1] Several reasons can be mentioned why not enough attention is paid to sanitation. In the first place most people consider drinking water a priority but they don't always see the need for proper sanitation. Similarly people are willing to pay for drinking water, but are much more reluctant to invest in proper sanitation and pay for using toilets. Connection fees for sanitation also tend to be higher than for drinking water, if only because it is more difficult to recover the investments later, which is certainly easier in the case of drinking water.[2] In the third place drinking water supply is more often characterized as a natural monopoly and hence considered the responsibility of the government. However, if we look at shit as a resource that can be exploited, more investments will come forward.

There are different definitions of sanitation (table 1) and there are many different types of toilets. In table 2 we list criteria for the classification of these toilets, without being exhaustive. To keep it simple we don't mention the management structure, the cost recovery approach or the scale of the facility. Sanitation is defined as safe collection, storage, treatment and disposing in a hygienic way of waste, including human excreta (faeces and urine), household waste water and rubbish at an affordable rate in a sustainable manner. We will deal mainly with the disposal of human excreta (improved sanitation) and leave out what is sometimes included in the wider definition of sanitation (see table 1).

Elements of sanitation	Proposed solutions	Covered by Johannesburg plan of implementation	Covered by Joint monitoring program of UNICEF-WHO
1. Human excreta	Provide access to toilets	Yes	Called **Improved sanitation**
2. Household waste water	Remove used water from within households	Yes, together with human excreta called **Basic sanitation**	No
3. Storm water	Collect and transport	No	No
4. Other sewage effluents to be treated	Reclaim used and dirty water by removing pollution	No	No

Table 1. Different definitions of sanitation

Later on we will distinguish seven types of toilets, but there are of course combinations of the different types listed different varieties exist, like simple pit latrines, or dry urine diversion toilets. The bottom line is that there is enough scope for unbundling in the sanitation value chain and competition and that we have to find the optimal solution for a specific situation.

---

[1] Water and sanitation are linked because contaminated water may result in water borne diseases, such as viral hepatitis, typhoid, cholera, dysentery and other diseases that cause diarrhea. Without adequate quantities of water for personal hygiene, also skin and eye infections, particularly trachoma, spread easily. Finally, drinking water can contain high amounts of harmful chemicals, such as arsenic and nitrates, which can lead to diseases.

[2] In the Buenos Aires concession a water connection would cost the equivalent of 500 US$, while a sanitation connection would cost twice as much.

Three different stages in the case of sanitation can be distinguished, before the product (the raw material) can be 'harvested and manufactured'. They should be separated in the sanitation value chain (figure 1):

1. Building toilets, going for individual or collective solutions (see table 2)
2. Operation and maintenance can be outsourced to small enterprises and also emptying and transport can be done by small private operators
3. Recycling can be done by separate actors, preferably also local enterprises

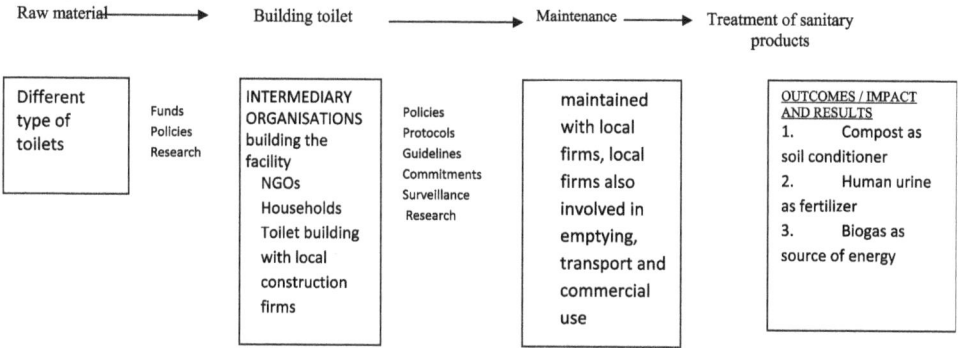

Raw material ⟶   Building toilet ⟶   Maintenance ⟶   Treatment of sanitary products

| Different type of toilets | Funds Policies Research | INTERMEDIARY ORGANISATIONS building the facility NGOs Households Toilet building with local construction firms | Policies Protocols Guidelines Commitments Surveillance Research | maintained with local firms, local firms also involved in emptying, transport and commercial use | OUTCOMES / IMPACT AND RESULTS 1. Compost as soil conditioner 2. Human urine as fertilizer 3. Biogas as source of energy |

Fig. 1. Three levels for unbundling in the sanitation value chain

Figure 1 shows that at each level different operators can be effective and encouraged. Pit latrines need to be built, maintained and emptied. The product can be used for composting, biogas or as fuel, but rarely the activity is considered as a value chain, where each stage built on the previous one and the advantages need to be distributed over the chain in case the chain is upgraded. Upgrading means stimulating the local construction of certain types of toilets, facilitating emptying services and promoting the processing of sanitary products. There are places in the world where there is a whole economy around sanitation, creating employment and income opportunities.

Criteria	Drinking water	Sanitation
Connected to piped system	Individual drinking water connection	Individual sewer connection
Individual or collective	On site or no on site solution	On site or no on site solution
Dry or using water	No pressure, no ground water	Flushing, pour or dry toilet, using chemicals, charcoal or nothing. Also water less urinals to separate phosphor and nitrate
Urine diversion or not	Does not apply	Sophisticated solution
Simple or improved	Well or borehole	Bucket, pit latrine (often inadequate) versus basic VIP latrines
Storage	Yes or no tanks	Septic tank

Table 2. Different criteria to classify technologies in drinking water and sanitation

What is interesting it that each option can be considered a different value chain, with its own operators, technology and distribution channel. Facilitating the supply of finance would be important for the users as well as the providers of these technological options. For that reason we will present alternative ways of financing for water and sanitation before drawing some conclusions from this study at the end of this contribution.

The importance of alternative technological options should be underlined. They provide an alternative to a full fledge sewerage system, which if installed in every African city would contribute to an even higher debt in foreign currency in many African country, given that steel and cement often need to be imported. Finally capacity building is extremely important, to allow local organizations and local small firms to carry out most of the work and to assure the necessary investments will have a maximum effect on the local economy and that they will also be maintained locally.

In sanitation we can distinguish the following technological options:

1. Ordinary or unimproved pit latrines, which is basically a pit with a seat in a shelter. They can be constructed by the people themselves, but may be poorly built and have problems with flies and stench. Small enterprises may do a better job
2. Bucket sanitation systems have the same problems and the buckets may fill rapidly and need to be emptied somewhere by someone
3. VIP or Ventilated pit latrines, where the pit is reinforced with concrete cover and a seat, where the air can circulate, while anti-mosquito screens keep out the flies. A high groundwater table would cause problems, just like a rocky soil.
4. Aqua-privy with on site disposal or simplified network to evacuate the liquid effluent which otherwise needs to soak away. The digester requires periodic emptying and some water is needed for flushing
5. Septic tank is similar to the aqua-privy with on-site disposal, but uses a full flush system. The system is expensive and requires emptying and sludge disposal
6. Intermediate (using less water) and full flush toilets where all waste goes to a sewer. These are expensive systems to construct, using a lot of water.
7. Eco-sanitation, for example composting and composting/urine diversion toilets

3. The role of the private sector in sanitation

Different arguments can be used to explain why private sector involvement (PSI) is better possible and more frequent in sanitation than in drinking water (Van Dijk, 2003). On theoretical and practical grounds it will be argued that there is even less of a natural monopoly in the sanitation sector than in drinking water. Besides the natural monopoly argument it is often said that water and sanitation are public goods, if only because of the negative external effects. Certainly a number of positive and negative external effects of drinking water and sanitation can be mentioned (see table 4). They differ for a chemical toilet or a pit latrine and vary from an open soak away pit to a septic tank. These external effects need to be taken into consideration when considering the choice of a technology.

Hence in many developing countries more can be left to private innovative solutions, often involving the small-scale local private sector. One reason why sanitation problems are often left to individual households, instead of expecting the solution to be supplied by the government, is because many countries have adopted a strategy of decentralization and devolution for water supply and sanitation. The national government puts the responsibility for sanitation at the municipal level, but without providing the necessary means to lower levels of government for this purpose. This means not much is happening since we know that the investments in the construction, or rehabilitation of new water and sanitation systems is many times higher than what is required to extend or upgrade an existing system.

There are many examples of Private Sector Involvement (PSI) in sanitation. In Indonesia Public Private Partnerships (PPPs) in this sector started for example in the 1990s. The Indonesian president even promulgated a presidential decree to promote them (No. 7/1998) and the Asian Development Bank (ADB, 2004) provided technical assistance and loans for this purpose. Even in China in large cities like Chengdu (10.6 million) only about 80 percent of the inhabitants in the centre of the city are served by a piped centralized sewer system. On top of that, only 60 percent of the sewer produced finds its way through the system, because not all buildings are connected. Typically the situation in the periphery of Chengdu is the opposite. There only 20 percent of the area benefits from a networked sewerage system, while the other inhabitants and businesses have to find private solutions. China has a number of Build Operate and Transfer (BOT) contracts in the water and sanitation sector, many with local companies or investors, because this is a municipal responsibility (Tu Shan, 2006). In the case of wastewater treatment the Ministry of Construction has the lead and usually also manages to mobilize the partners and the necessary finance, also form the private sector.

Sanitation was defined and the challenges are listed in table 3. The problems have been classified as institutional, technical, social and financial (ADB 2007).[3] They explain to a large extent the poor performance of many public authorities (table 3). A different approach,

Institutional challenges
1. No regulation to encourage proper sanitation practices
2. Weak institutional framework
3. Lack of clarity of institutional roles and responsibilities
4. Lack of focus on sanitation and waste water
Technical challenges
1. Water resource pollution
2. Deteriorating infrastructure
3. Low sanitation coverage
Social challenges
1. Unsustainable project outcomes
2. Community resistance
3. Low hygiene awareness
Financial challenges
1. Inadequate resources
2. Low or non-existent tariffs
3. Lack of financial sustainability

Source: ADB (2007).

Table 3. Main challenges in the sanitation sector in developing countries

---

[3] One could add operational, commercial, human, and environmental problems.

stressing the ecological aspects of water and sanitation and trying to integrate the urban water cycle is desirable. It is important to consider how Small-scale independent providers (SSIPs) could help in fixing some of the problems mentioned in the table. As will be shown this may require a change in the institutional arrangements, a different look at the technical and social challenges and assistance to deal with the financial challenges in a very different way.

Now the role for the private sector in sanitation can be assessed. Subsequently data on the importance of small-scale independent providers (SSIPs) in the drinking water and sanitation sector will be reviewed. Then the issue how to improve efficiency in sanitation will be raised by looking at the possibility to unbundle this activity, to use technological innovations and to bring in more competition. Finally some dilemmas will be discussed related to the Millennium Development Goals (MDGs) and the role of the private sector in sanitation. If the MDG will not be achieved easily in the sanitation sector it is time for smart sanitation provision for slums and informal settlements.

Sometimes drinking water and sanitation activities are taken up simultaneously in private sector involvement projects. However, the drinking water component usually receives much more publicity (as was the case for the Buenos Aires concession for example). In other cases there is a management contract for sanitation, which doesn't draw as much attention as a concession contract for drinking water, because people do not really know who takes care of the wastewater and the contract period tends to be shorter. For example it is hardly known that the French water company Suez runs a number of wastewater systems in the United States (Mathews, 2003). The relative good performance of these systems is rarely mentioned in the critical discussions about the role of the private sector in water and sanitation.

The Global Water Initiative (GWI; www.globalwaterintel.com) concludes in its March 2005 issue that to date limited progress has been made towards the achievement of financing these MDGs. Only the Eastern Asian countries are ahead of the targets set in 2000, while Sub-Saharan Africa is falling far behind.[4] Lack of clean water and sanitation is the second most important risk factor for people in developing countries, after malnutrition. Problems with public sector supply of water and sanitation services have led to the increasing awareness that more participation of the non state sector is needed in the provision of these services.

4. The economics of sanitation

Who bears the cost of sanitation? What investments are required for different options, ranging from piped systems to collective facilities? Is cost recovery taking place and are the funds used to improve the current system? Small amounts can support small systems.

Sanitation is not really a public good, since people can be (and are) excluded and the system is rivalled, meaning if some households use it the capacity may not be enough for

---

[4] Several programs are active to help African countries to achieve these MDGs. For example the Water and Sanitation Program (WSP, based in Nairobi) with support from the Netherlands and other donors has studied in a number of African countries where they are and what still needs to be done (WSP, 2004).

everybody. One reason for the government to get involved would be the negative and positive external effects (see table 4). However, as such positive or negative external effects are not a strong reason to supply the services by the government. In fact the activity can be outsourced to the private sector and regulated by the government in such a way that these external factors are taken into account. In case of important externalities, there is the need to assure investments in the sector, over and above what private operators are doing because the socioeconomic benefits are larger than the financial benefits resulting from a private cost benefit analysis. Externalities may lead to formulating clear aims for sanitary systems, such as being attractive and hygienic. The challenge is then to make them also affordable to the population and easy to maintain.

Externality	Water	Sanitation
If piped system producing good quality is in place: positive effects	Better health Higher labour productivity	Improved health More dignity and security
If piped system is in place: negative effects	Chemicals in the environment	Such services require space and may smell
If no piped system in place: negative effects	May spread diseases Much time lost, often by women	Diseases can spread easily No dignity and security problems for women

Table 4. Positive and negative external effects of drinking water and sanitation

5. Increasing efficiency through unbundling, technological developments & competition

Increasing access to safe water and sanitation in peri-urban areas of large cities requires (a) increasing the efficiency of urban water supply systems and water demand management, and (b) developing and implementing new sustainable forms of sanitation, including eco-sanitation technologies. Although facilities for collection, treatment, and disposal of waterborne sewage also exhibit significant scale economies it is worth considering when decentralized systems and small-scale private sector providers can be used more. They usually involve other private actors and private capital, although sometimes the large-scale water treatment plants also attract private funding and management.[5]

Increased efficiency in sanitation can be achieved through involving the private sector, which through a combination of unbundling, technological developments and more competition can bring about lower tariffs. In other utility or network sectors these three factors have contributed to lower cost of service provision. For example all three factors have contributed to lower prices in the telecommunication and electricity supply sector and technological progress may currently change the drinking water sector, when desalination is really becoming competitive. The importance of these options in the sanitation sector will now be discussed.

---

[5] Like for example a new wastewater plant in Harnas polder in the Netherlands, which is totally financed by private partners through a BOT formula

5.1 Unbundling

Increased efficiency in the utility sector is often a combination of unbundling, technological development and more competition (Van Dijk, 2003). All this is possible in the sanitation sector and makes the involvement of private sector operators more likely and more effective. They can be involved in small scale construction, or the maintenance and emptying of the sanitary facilities. Unbundling in the sanitation value chain is depicted in figure 1.

5.2 Technological progress

For sanitation, just like in the drinking water sector one can have at least ten technological options and continuing technological progress adds options to this list all the time. In the drinking water sector house connections and yard taps refer to different delivery modes (water vendors or public tanker trucks), wells (communal open or tube wells), tanks (in the yard or on the roof) and solutions outside the plot (water kiosks or communal standpipes). Each option has certain advantages and disadvantages and commands a price.

5.3 Competition

Economists argue that competition will improve the quality of a product or service and drive down the price. Competition in the sanitation sector is possible since in fact a dual system exists in the sanitation sector and different technological options are available. One finds at the same time on site sanitation and large scale centralized water treatment plants and hence there is competition. On site sanitation happens in the periphery of the big cities. The technologies range from pit latrines to the obligation to recycle grey water in each important urban project in the case of Beijing. This is done since piped sewerage system linked to waste water treatment plants are very costly.[6] Because of the unbundling and the technological progress more competition is possible leading normally to lower prices and better services for customers.

6. Look at advantages of small scale decentralized versus large scale centralized WWT

Liang and Van Dijk (2010) have compared the economic cost and benefits of small scale decentralized versus large scale centralized WWT in Beijing and concludes that currently the decentralized systems cannot compete with the centralized because of subsidies and the low price of drinking water.

Big private international water companies are mainly interested in running large scale centralized water treatment systems, if they are combined with drinking water (making charging consumers easier), if they do not have to invest themselves (they do not consider themselves to be the bankers of the water sector any more), or if there is a possibility to recover the investments in another way (for example because a municipality pays for every

---

[6] The duality in Beijing is emphasized by the obligation to reuse the grey water at the level of major construction projects (for example hotels or universities). It allows a comparison of centralized and decentralized water treatment options (Liang and Van Dijk, 2010).

litre of treated water). Attention has shifted from the big centralized systems to the potential of decentralized systems combined with small-scale sanitary improvements and in particular eco-san solutions. The latter allow people, enterprises, or neighbourhoods to take the initiative or participate in it.

To what extent do small-scale private individual providers or operators (SSIP) provide basic services like sanitation in developing countries (Collignon and Vezina, 2000)? Although there are usually economies of scale in networked systems, small-scale operators are tremendously important. The data presented in table 5 concern the role of SSIP in water services. The origin of the data is described in box 1.

The World Bank undertook a literature review of small-scale private operators of water supply and sanitation (Kariuki and Schwartz, 2005), defining small as less than 50,000 customers. The database (over 400 documents) is available under 222.rru.worldbank.org. The 400 documents reviewed provided evidence for about 50 countries and 100 different locations in these countries. In total some 10,000 water SSIP were identified, which maybe still only part of the total, given there are more countries and the SSIP sometimes are informal or illegal. Table 5 summarizes the findings.

Box 1. Data on small-scale private operators in water and sanitation

Both formal and informal small-scale independent private operators are considered, given that they are difficult to distinguish. In table 5 an estimate is provided of the number of people receiving services from SSIP and it is indicated in which regions this is most common. In both the drinking water and sanitation sector there are in fact dual systems and there are reasons to build on that reality, providing more space to the private providers and individual households.

Only 10 to 15 percent of the urban population in developing countries benefits from access to a sewer network according to WUP (2003). The rest depends on on-site or collective facilities. In case small-scale sanitation solutions are adopted, there doesn't have to be a private operator except may be in the construction phase, given the role of the community. When the alternative for a sewerage system is a septic tank or a closed pit latrine, the question of emptying those arises. This may be the responsibility of the household, of a private service provider, or a public operator. Unfortunately no detailed figures for SSIP in sanitation are available. Table 5 provides data for 33 cities concerning water SSIP. Per city the percentage of households being served by the SSIP is given. The huge variation between and even within countries strikes immediately, just like the importance of the SSIPs.

In Senegal, only 21 percent of the households in the capital Dakar are served by SSIP, while in Diourbel, a city more in the interior, it goes up to 90 percent! It usually varies between 0 and 30 percent. The role of SSIP in the water sector is most wide spread in Africa, while for South Asia SSIP are most prevalent in areas with low coverage levels and ineffective public utilities, in particular in India and Pakistan. Also they are important in remote areas. Kariuki and Schwartz (2005) have analyzed the features of these SSIP and classify them according to organizational form (cooperatives to private ventures), technology, staffing (usually less than ten employees), customer service and marketing, financing and pricing, sales and earnings. These activities tend to be outside the legal framework and production is usually at a very small scale. However, a high proportion of local and often unskilled labour is involved and there are very low levels of initial investments. The conclusion is that SSIP are very diverse and often threatened by an extension if the coverage of the formal supply network is extended. The challenge is to consider SSIP as complementary and incorporate informality when formal supply of urban services in not adequate (Van Dijk, 2010). Part of the solution of the MDGs may come from these 'other sources' of supply.

Region and countries	Water SSIPs in	Households served by SSIP	Region and countries	Water SSIPs in	Households served by SSIP
Africa	**City**	**Percentage**	**Latin**	**America &**	**Caribbean**
Benin	Cotonou	69	Argentina	Cordoba	15 – 20
Burkina Fa	Ouagadougou	49	Bolivia	Santa Cruz	100
	Niangolo	68	Colombia	Barranquilla	20 – 25
	Bobo-Dioulasso	33	Guatemala	Guatemala	32
Ivory coast	Abidjan	35	Haiti	Portau-Prince	70
	Boundiali	50	Honduras	Tegucigalpa	30
Ghana	Kumasi	32	Paraguay	Asuncion	30
Guinea	Conakry	66	Peru	Lima	26 – 30
Kenya	Nairobi	60	**East Asia**	**and Pacific**	
Mali	Bamako	63	Cambodia	Ky Cham	50
Mauritania	Nouakchott	51	Indonesia	Jakarta	44
Niger	Guidan	40	Philippines	Manilla	30
Nigeria	Onitsha	95	Thailand	Sawee	10
	Ibi	40	Vietnam	Ho Chi Minh	19
Senegal	Dakar	21	**South Asia**		
	Diourbel	90	Mongolia	Ulaanbaatar	5
Sudan	Khartoum	80	Nepal	Kathmandu	5-7
Somolia	Ali Matan	10	Pakistan	Karachi	40 - 50
Tanzania	Dares Salaam	56	India	Delhi	6 –47
Uganda	Kampala	30	Bangladesh	Dhaka	14

Source: Kariuki and Schwartz (2005).

Table 5. Data for 33 cities concerning water SSIP

Poor people often pay a high price per litre because SSIP don't have access to subsidies and SSIP are unable to benefit from economies of scale. Nor can they assure the quality of their water. Externalities are not be taken into consideration by a private operator and hence the price is not reflecting the real cost. Price differences with the publicly supplied water range from 1.5 to 2.5 times the official public utility price, and may increase in times of scarcity. However, the key advantage of SSIP is that they deliver the water at home.

WUP (2003: 53) considers intermediate and independent service providers are filling the gap between the public suppliers and no supply. They suggest working with the local sub-network providers and water carriers and tankers to improve services. The small-scale providers have the potential to become local private operators in small towns. Over time they can play a more important role in medium and large towns.

There are strong reasons to try to increase the role of the private sector operators as a percentage of total turnover in the water and sanitation sector and to encourage their development. The government could impose a specific status for these operators in the water and sanitation sector, giving them for example fiscal incentives and asking a certain quality of water in return. For quality and environmental reasons governments may not want to promote private operators and on site solutions in drinking water. However, given the size of the sanitation problem and the ambitious MDG in this respect, given the difficulty to recover the cost of sanitation and the huge amounts that need to be invested for onsite public solutions involving a network and centralized waste water treatment governments may still be inclined to promote the existing on site private sanitation solutions. Then it should provide more space to SSIP in the water and sanitation sector.

If drinking water resale initiatives and private sanitation solutions are encouraged it is important to raise the public awareness of health and hygiene issues and to clarify the respective roles of public and private players in the water and sanitation market. The OECD global forum on sustainable development also concluded that policies are necessary to enable the private sector to play a greater role in helping to achieve the MDGs.[7]

7. Multi-governance issues related to sanitation

Sanitation is embedded in governance structures and different countries have selected different solutions. Table 6 summarizes the major institutional arrangements. The question is: which structures work and why?

	Institutional arrangement	Drinking water	Sanitation
Solutions on the plot	Public	Piped connection	Sewerage network
	Private	Well or bore hole Home delivery	Not connected latrine, but f.ex. pit latrines
Not on site solutions	Public	Standpipes Public wells Uncontrolled sources	Government supported community toilets
	Private	Autonomous water kiosks	NGO supported community toilets Private paying toilets Uncontrolled

Table 6. Private versus public on site and collective solutions

How does the official municipal utility cooperate with the authorities dealing with the 'informal' solutions? As an example: in Uganda NWSC is responsible for piped sanitation, while the City council is involved in toilet projects with all kinds of NGOs. Is this working and what can we learn from such experiences?

---

[7] 'Public-private partnerships in water supply and sanitation: trends and new opportunities' (www.oecd.org).

8. Consider the economics of different technological options for sanitation

Different toilets are available and different technologies have been suggested for waste water treatment. Which ones are doing better and why? Eco san solutions would also be considered. The different options in the water and sanitation sector are illustrated in table 2. Private sector involvement in sanitation has a double meaning. It means on the one hand that individual households need to find a solution for their sanitary problems. On the other hand the households may involve private firms for the construction or maintenance of the facilities. In the table we distinguish private versus public and on-site versus not on-site solutions. In the latter case these would be collective or community solutions (WSP, 1998).

For different reasons (see box 2) large-scale network sewerage solutions are too expensive to introduce on a large scale in developing countries. To achieve the relevant MDG a different approach will have to be taken.[8] Since in a number of countries there is no separate institutional structure in place to manage sewer systems, it is often left to the drinking water companies to take care of this issue.[9] In fact if the water companies can put a surcharge for sewerage on the water fee their financing problem will be partly solved (Pagiola and Platais, 2002). Otherwise we may have to rely more on small-scale independent providers in the case of sanitation.

1.	The necessary steel and concrete often needs to be imported requiring foreign exchange and risking huge debts in foreign currency
2.	The long term investments (50 to 100 years) are difficult to finance because no capital market for long term finance exists in most developing countries
3.	It is more difficult to recover the cost in the case of sanitation than in the case of drinking water
4.	Use need to be made of expensive consultants to design the system
5.	Technicians tend to overdo the dimensions of the system to be able to deal with future extension and one time disasters
6.	There is sometimes no sewerage system in place, or repairing the existing system would be very expensive because built under the ground fifty years ago
7.	Network sewerage solutions need a lot of maintenance, which is often not budgeted for
8.	There may be no institutional structure in place to manage sewer systems
9.	The tendering system may not always be competitive and transparent
10.	International contractors may be required given the scale of the projects

Box 2. Why large scale network sewerage solutions may be too expensive

[8] In the framework of the European Water Initiative (EUWI) efforts have been made to start a dialogue in a number of African countries on how to achieve these MDGs with the involvement of all actors: local governments, non governmental organizations (NGOs) and the private sector. After organizing a dialogue, a road map, or sector plan would have to specify the minimum acceptable level of access to water supply and sanitation. Subsequently a Financing strategy is developed to indicate how these objectives can be achieved. The objective of such a Financing strategy is achieving MDGs through private sector involvement in water and sanitation, or by tapping as many sources of finance as possible.

[9] Tunisia is an example of a separate National Sanitation Utility, which seems to work very well.

In sanitation competition also exists between informal and formal suppliers. The sanitation and drinking water sector in developing countries can be described at least as a dual system if 85 to 90 percent of the people in developing countries depend on private sanitary solutions and 65 percent on private water suppliers as is the case for example in urban Benin. Private water vendors play indeed an important role in supplying at the average 20 percent or more of the urban population in developing countries (World Bank, 1988). This implies that the role of the private sector is much more important than generally admitted. It competes with the relevant public utilities. The major mechanisms to achieve more efficiency in service delivery are the possibility of unbundling, technological progress and more competition. The factors influencing the choice of an appropriate sanitation solution are depicted in figure 2.

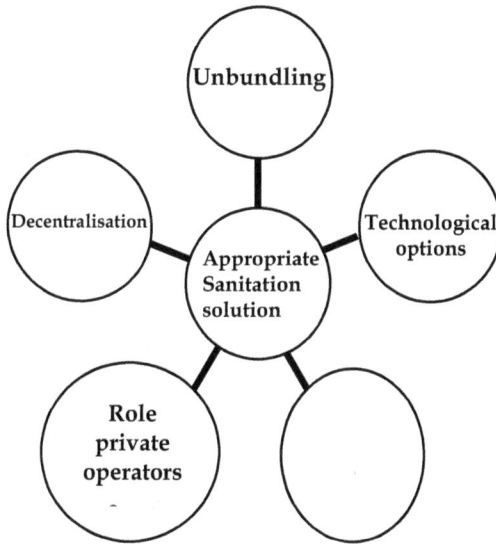

Fig. 2. Elements of an appropriate sanitation solution

9. Incorporating informality in the sanitation sector

Fransen et al. (eds. 2010) argue that if local governments cannot provide certain services they could facilitate other actors who do take up that challenge. The role of these non state providers is elaborated by Van Dijk (2008).

Liberalization is a process by which competition is introduced in situations or sectors hitherto characterized by exclusive or special rights, or monopoly, granted to historical operators. We argued that more competition is possible in sanitation and related activities than in the drinking water sector and hence different policies can be pursued. In fact competition is easier in sanitation than in drinking water and private solutions and PSI are more likely because there are no economies of scale, such as the ones existing in drinking water systems. Secondly, drinking water supply is more often characterized as a natural monopoly in the distribution system, which is not the case in sanitation, where there is a real dual system since often 85 to 90 percent of the urban population in developing countries depends on private solutions. The natural monopoly can be overcome with common carriage[10] and inset arrangements[11], which can create real competition and exists in several European countries (Van Dijk, 2003). In practice common carriage and inset arrangements make up only between 5 and 10 percent of the drinking water in countries like the Netherlands and England and Wales. However, through the possibility of inset arrangements regional monopolies do not work anymore. Even if currently its use is limited, the possibility to compete already limits monopolistic behaviour. The taste and quality of the water may be different in the case of sharing arrangements, and the responsibility for negative health consequences may become more diffuse in the case of inset arrangements. In drinking water it is not as easy as in for example the power sector to break natural monopolies by linking different networks.

10. Alternative ways of financing sanitation

The MDG for sanitation is to halve, by 2015, the proportion of people who have no access to basic sanitation. The estimated funding requirements range from US$ 2.1 to 23 billion and when going beyond the more basic definition of urban service provision will cost even more. Already the Camdesus report in 2003 (Winpenny, 2005) had suggested an additional US$ 32 billion a year and if a broader definition including treatment of all municipal and industrial wastewater and solid waste would be used US$ 100 billion a year would be necessary. An overview of the progress with household sanitation in South Asia is provided by Sijbesma (2008).

Research should contribute to the development of an innovative approach to water and sanitation in African slums, which is not only cheaper, but also institutionally, environmentally and financially sustainable. It starts with identifying the institutional network for providing water and sanitation to the slums. What is the role of the government, of the private sector (small scale providers) and of NGOs? If their impact is limited, private solutions will dominate and the willingness to go for collective solutions needs to be studied.

We noted that large scale sanitation activities are difficult to finance, given the large amounts needed and the lack of cost recovery mechanisms. The private sector can get in if the projects are really conceived as economic investments with a return. This requires an emphasis on ways and means to finance sanitation services and recover the cost. Cost recovery is possible through:

1.   Contributions from the people benefiting from the system, possibly in kind
2.   Linking sanitation to drinking water
3.   Charging connection fees
4.   Asking small contributions to the necessary investments

---

[10] The common carriage principle is also used for telephones and electricity, and tested for water in England and Wales. It means sharing the use of pipes (for raw water) and implies mixing different qualities of water.

[11] Inset appointments would allow in the UK under the original privatisation scheme, a new water or sewerage 'undertaker' to penetrate the area of an existing undertaker via this appointment (putting clean water in the network of another company).

5. Using private construction firms, and local small enterprises for building, O&M and for emptying and finally small enterprises for recycling the liquid waste products

Many alternative financial solutions have been suggested, ranging from cross subsidies to using micro loans to pay for connection fees (Winpenny, 2005). The bottom line is that some subsidy can be provided (for example cross subsidies for the poor) and the first 200 litres of drinking water can be provided for free (the life line approach), but if there is no money in the system, it will run dry. Sustainability involves not only environmental, but also institutional and financial sustainability. The issues of low household income, low social status of the customers and a limited degree of organization of poor people are linked with measurable consequences in terms of surface water pollution, poor health and a large number of children dying under such circumstances.

Decentralization requires more local revenues. However, the need for financial reform at the municipal level should be mentioned as a condition if municipalities want to qualify for loans to finance their infrastructure. Decentralization is the trend, but the financial means also need to be available at lower levels of government to carry out the tasks assigned to them. These means are often lacking.

More traditional sources of finance	More alternative sources of finance
Higher levels of government, financed out of tax revenues Project finance, with loans or bonds State Level Finance Institutions, or Municipal Infrastructure Development Funds: investment, capital funds, trust funds, or endowment funds Hedging, using futures and options to cover risks	Private sector involvement, for example through Public Private Partnerships & joint ventures Concessions, BOT (Build Operate Transfer), Design, Finance, Build and Operate (DFBO) and ROT (Rehabilitation Operate Transfer) Microcredit to finance water and sewerage connections, or rotating savings and credit associations (ROSCAs) linking savings with credit

Source: Van Dijk (2006).

Table 7. Innovative ways of financing infrastructure: water and sanitation

A solid and sound financial management system should comprise an improved municipal accounting system, but also a better budgeting system and budgetary control, improved internal control systems, internal audit systems and modern data processing facilities. A number of reforms at the municipal level are necessary for example to qualify for support in the framework of an urban infrastructure project in India

Different drinking water and sanitation options are available for the inhabitants, which can be introduced in other low-income neighbourhoods through NGOs or CBOs and local small scale private sector entrepreneurs. These 'private' solutions that people have chosen have their cost and need support from new sources of finance, such as the ones mentioned in table 7.

11. The politics of sanitation

For the improvement of sanitary services in developing countries different actors are important. Each one has its own specific interests and objectives which will try to achieve them. The interests of the different actors: landlords, tenants, Ministries, NGOs, donors, international lending agencies, etc. They are likely to conflict and, as such, achieving these interests and objectives is subject to contestation. As different entities pursue their, possibly conflicting, interests the provision of sanitation services becomes an inherently political process. The dynamics of this political process underlying the provision of sanitation services, within a slum setting, has largely been ignored by researchers in this field. The sanitation crisis in slum areas has largely been perceived as either an issue of developing appropriate technologies or, in recent years, as an issue of creating demand for sanitation services. Once a sanitation coverage gap is established, efforts have to focus on raising resources to build appropriate facilities, coupled with sensitization and the job is done. This portrayal of slum areas is oversimplified and underestimates the inherent social complexities of providing sustainable sanitation services in slum areas. Very little is known about demographic and social processes within slum areas and how these may impact provision of sustainable sanitation services.

PSI can make a contribution to the achievement of the Millennium Development Goals (MDGs), giving the financial and skill bottlenecks for the fulfilment of the Millennium Development Goals in the water and sanitation sector in Africa, Latin America and Asia. There are some dilemmas concerning the role of the private sector in relation to the achieving the Millennium Development Goals. The role of the private sector can never be to take over the political responsibility of the government. They can also not take the decision to go for large scale centralized or for decentralized wastewater treatment. Once decisions like what will be solved collectively and what will be left to individuals are taken the private sector can execute the activities required and will probably become more efficient than the government in supplying these services.

12. Conclusions

Local governments and utilities share the responsibility for waste water. Too often they do not link the idea of collection, transport and treatment. It is expected that an integrated approach as practices in a limited number of Third world cities would produce better results. It would mean that different actors work with different technologies and alternative sources of finance to deal with sanitation in an integrated way. Using local enterprises more in the sanitation value chain will increase employment and contribute to local development.

Realistic prices for such services and involving the private sector where and when adequate is important. One way to achieve satisfactory results is to follow the methods suggested by the European Union Water Initiative (EUWI). It is suggested to involve as many parties as possible in the construction, operation and financing of the required facilities and to bring them together before actually starting to identify possible bottlenecks.

We have suggested different ways of financing water and sanitation to allow more poor people to gain access to these services. Subsidizing may make it unaffordable for most governments in the long run, hence designing appropriate schemes which would be self financing is much more the challenge. All kinds of statements have been made about sanitation. The Joint Africa-EU statement mainly repeats the commitment, but is very brief about the ways of financing a different approach to sanitation, which would really benefit the poor. Also NGO initiatives like the Sustainable sanitation alliance is brief on funding (mentioning mainly to include sustainable sanitation issues in to existing funding instruments and initialising of new funding mechanisms in the sanitation sector), without being very specific. More experience need to be gained with the different options mentioned in table 2. Further research could then identify the conditions in which these approaches may work successfully.

Governments should recognize the importance of what we called 'private solutions'. They can be recognized and supported, for example by introducing adequate financing systems. This is what is called incorporating informality (Van Dijk, 2010) and would lead to a dynamic small scale private sector of service providers in the water and sanitation sector. We have shown that technological development, unbundling and competition, which are often looked for in the drinking water sector in fact exist in the sanitation sector. Their effectiveness can be enhanced through a different approach to sanitation, more support for it and appropriate financing mechanisms.

13. References

ADB (2004) Small piped water networks, Helping local entrepreneurs to invest. Manila: Asian Development Bank.

ADB (2007) Smarter sanitation, How to clean up your sanitation and waste water mess. Manila: Asian Development Bank.

Collignon, B. and M. Vezina (2000) Independent water and sanitation providers in African cities, full report of a ten-country study. Washington: Water and Sanitation Program.

Dijk, M.P. van (2003) Liberalization of drinking water in Europe and developing countries. Delft: UNESCO-IHE Institute for Water education.

Dijk, M.P. van (2006) Incorporating informality. In: Shelter HSMI New Delhi, Vol. 9, No. 4, December, pp. 14-21.

Dijk, M.P. van (2008) Role of small-scale private operators in water and sanitation in: International Journal of Water Vol. 4, No. 3/4, 2008, pp. 275-290.

Dijk, M.P. van (2010) Incorporating informality: 35 years of research and policies on the urban informal sector. In: J. Fransen et al. (eds, 2010), pages 1-15.

Fransen, J., S. Kassahun and M.P. van Dijk (eds., 2010) Formalization and informalization processes in urban Ethiopia: Incorporating informality, Maastricht: Shaker.

Kariuki, M. and J. Schwartz (2005) Small-scale private service providers of water supply and electricity. Washington: World Bank.

Liang, X. and Dijk, M.P. van (2010), Financial and economic feasibility of decentralized waste water reuse systems in Beijing. In: Water science and technology, 61(8) pp. 1965-1974.

Mathew, N.B. (2003) Performance analysis of privately and publicly managed waste water utilities in Indiana and surrounding states. Delft: UNESCO-IHE, MSc thesis.

Pagiola, S. and G. Platais (2002) Payments for environmental services. Environmental strategy notes, Washington: IBRD.

Sijbesma, C. (2008) Sanitation and hygiene in South Asia: Progress and challenges. South Asian Sanitation and hygiene practitioners workshop in Dhaka: BRAC, WaterAid & IRC.

Tu Shan (2006) The use of BOT contracts in the water sector in the People's Republic of
    China. UNESCO-IHE Institute for Water education, MSc thesis.

Winpenny, J. (2005) Guaranteeing development? The impact of financial guarantees. OECD.

World Bank (1988) The role of public finance, World development report. Washington:
    World Bank.

WSP (1998) Community water supply and sanitation conference, Proceedings. Nairobi:
    Water and Sanitation Program.

WSP (2004) Sanitation is a business, Approaches for demand-oriented policies. Nairobi:
    Water and Sanitation Program.

WUP (2003) Better water and sanitation for the urban poor, Good practice from sub-Saharan
    Africa. Nairobi: Water Utility Partnership for Capacity Building (WUP) Africa.

Part 2

Assessment of Technological Innovation

iTech: An Interactive Virtual Assistant for Technical Communication

Dale-Marie Wilson[1], Aqueasha M. Martin[2] and Juan E. Gilbert[2]
[1]University of North Carolina at Charlotte
[2]Clemson University
USA

1. Introduction

A manual accompanies almost every product or device. Manuals are usually included with products or services to provide customer assistance and provide technical information to users. However, Thimbleby states, "User manuals are the scapegoat of bad system design." (Major, 1985; Thimbleby, 1996). Technical communications are provided through several mediums and manuals are one example of this. Other mediums range from interactive animation to virtual reality (Hailey, 2004), with each new medium attempting to improve upon the drawbacks of the previous one. The first medium introduced was the paper manual. However, issues with paper manuals have been widely documented, especially by technicians in the armed forces. Problems include lack of portability, inaccuracy, and increasing content and complexity (Ventura, 1988). To improve upon the drawbacks of paper manuals alternative mediums such as online manuals for technical communication emerged.

Although, alternative mediums have improved upon some of the drawbacks of paper manuals, traditional technical communication mediums still often entail a timely search through a large paper manual or rigorous cognitive processing to generate an appropriate query to search an online manual. As a result, consumers oftentimes spend much more time searching for an appropriate solution, have trouble finding an appropriate solution, or become frustrated and result to other means to find a solution. Research suggests that when choosing a technical communication medium one should consider the needs of the audience, the functionalities of the new information technology and how the functionalities are to be utilized, and the application of the new medium and if it will prove the concepts for introducing the new medium. Therefore, alternative mediums for technical communication should also be compared for their ability to communicate information to consumers effectively and with reduced frustration.

In this chapter, iTech is introduced. ITech is an interactive technical assistant that was designed to assist users in finding information about a product, in this case vi, a programming editor. In this paper, the motivation and design of iTech are discussed. Furthermore, the design and results of a research study conducted to examine the usability, effectiveness, and efficiency of iTech are presented.

2. Literature review

2.1 Technical communication

Technical communication refers to the process of delivering technical information to the user. Albing defines it as, "...the creation, control, delivery, and maintenance of distributed information across the enterprise and in a network that includes sources and users." (pp. 67) (Albing, 1996). An effective technical document is determined by the following factors (Zachary et al., 2001):

1. Is the analysis of the communication problem complete?
2. Is the goal/task to be explained clearly identified?
3. Is the vocabulary used to explain the goal comprehensive and does it follow conventional guidelines?

These factors are used for evaluating all mediums of technical communications. However, as the need for manuals has grown, little investment has been placed in the development of these manuals. Paper manuals often need updates; therefore, become outdated quickly, are hard to understand, inaccessible, erroneous, and difficult to search if the index is not designed properly. On the other hand, while online manuals provide additional benefits such decreased search time, smaller documentation, and better search techniques (Barnett, 1998), they also require increased query pre-processing for either the user or the search engine. With the introduction of web-based mediums such as animations and virtual reality has also come the concern of available technologies, end user expectations, and user-centered design (Zachary et al., 2001). Therefore, this suggests that mediums that provide higher degrees of user satisfaction and ease of use may be important to delivering effective technical communication.

2.2 Animated agents and interactive assistants

Interactive assistants aim to aid users in managing their environment (Kirste & Rapp, 2001). Because computers are continually becoming more ubiquitous, permeating aspects of people's daily lives, there is a need for an efficient interface between users and computers. Interactive assistants address this need by providing natural, intuitive, and effective interaction between people and computers (Oviatt et al., 2000). Interactive assistants typically contain multimodal features including speech input and output, gesture and handwriting recognition, and animated agents or avatars. These features provide users with interaction choices that can circumvent personal and/or environmental limitations, require little or no training. They also have great potential to promote new forms of computing and expand the accessibility of computing to a diverse group of users ((Lester et al., 1997).

Interactive assistants have been used in various types of user help systems including training, education, and marketing [11, 12]. Additionally, research has been conducted on how the inclusion of such agents impacts user's interactions with the system. Some of the earlier agents were designed in the domain of education. Rosis et al. designed the XDM-Agent, an animated character that aids in illustrating interface objects for software development in user-adapted interfaces and explain which tasks may be performed and how (Rosis et al., 1999). Steve an animated pedagogical agent was created to help students perform procedural tasks (Johnson & Rickel, 1997). Other early animated agents such as Adele, Herman the Bug, Cosmo and PPP-Persona were also introduced as providing user assistance (Johnson & Rickel, 1997).

However, with all the advances in animated agent help systems, research aimed at the usefulness of conversational interactive assistants for assisting users in searching technical communication in particular is limited. Additionally, minimal research has been conducted on how the introduction of such agents and their design impacts user's interactions with technical communication documentation. iTech is an interactive technical assistant that was designed and developed to address some of the limitations of current mediums for technical communication while improving user satisfaction and the user experience by taking advantage of the opportunities provided by interactive assistants.

3. Research methodology

3.1 Design

iTech is an interactive virtual assistant that was designed and developed to address the limitations of the current mediums of technical communication including paper-based and online systems. More specifically, iTech was designed to address the limitations associated with current technical documentation including understandability, portability, accessibility, accuracy, search time, and the ability to make updates. Although these limitations do not apply to all mediums of technical communication, there is an application limitation for each medium. In the process of designing iTech several additional limitations associated with automatic speech recognition (ASR) engines were encountered, i.e. population of the question-database and conversational questioning answering. In the design of iTech, the goal was to provide iTech with the ability to understand natural language queries from a variety of speakers as is without any additional training. To do so, is was necessary to eliminate the preprocessing step that is associated with many other techniques. In addition, it was desirable to have iTech be able to effectively answer (return an appropriate answer) even if the question asked did not appear in the database.

The accuracy of automated speech recognition (ASR) engines for speaker-independent systems has a higher word error rate (WER) than those that are trained. The WER can be reduced by a limited grammar, but natural language questions necessitate a larger grammar to account for the questions that may be asked. To allow for a large grammar, the database used to generate the grammar must be populated with all relevant questions that a user may ask. Each answer must then be mapped to a relevant question. Additionally, each answer is not restricted to a specific question. Because of this the database must be populated with a massive amount data. Furthermore, current techniques for conversational question answering require pre-processing (parts-of-speech tagging, semantic interpretation) of queries before execution and removing what the authors argue to be relevant information.

iTech utilizes the Answers First (A1) approach for conversational question answering (Wilson et al., 2010). In A1, unlike many other information retrieval or natural language processing techniques, requires no language processing before the query is executed. The users query once recognized is sent to a server and decomposed into bigrams (word pairs). The bigrams are matched to a repository of questions using a question resolution algorithm and the question with the highest concentration of matched terms is returned. This process continues, prompting the user appropriately until an appropriate solution is found.

iTech has a client-server architecture as shown in Figure 4. iTech's Architecture. The user initiates the conversation with iTech by pressing a button to speak and ask a question. The built-in speech recognition engine, Microsoft English ASR Version 5 Engine, recognizes the user's question and passes the recognized speech to the browser environment of the page where the Speech Application Language Tags (SALT) are hosted (Cisco Systems Inc. et al, 2002). Additional client-side scripts then manipulate the SALT elements. The resulting text of the recognized speech is then sent as a request to the server.

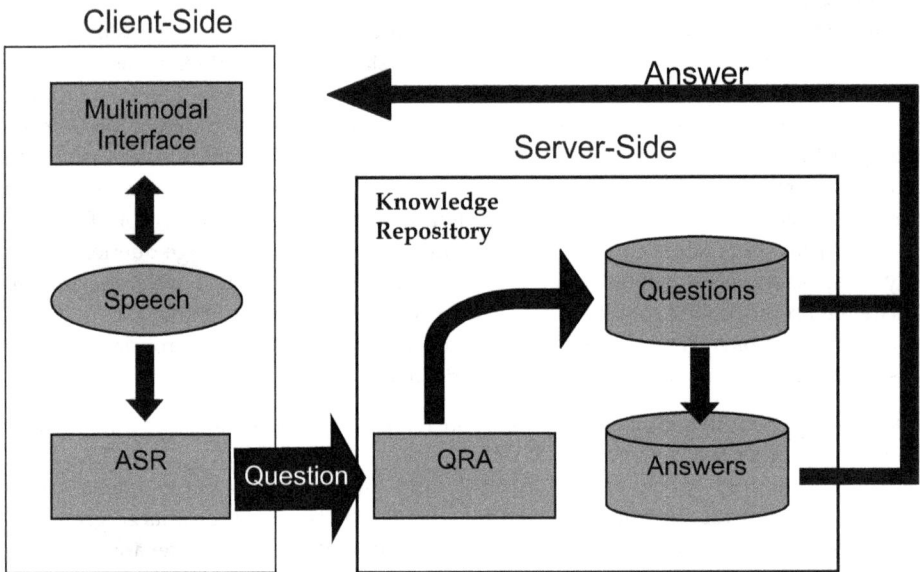

Fig. 1. iTech's architecture

The server side consists of the Knowledge Repository (KR) that is populated with question-answer pairs generated from the chosen manual (Wilson et al., 2010). The Question Resolution Algorithm (QRA) module resolves the recognized question with the KR, identifies the question-answer pair with the highest concentration of matched terms and retrieves the relevant answer (Wilson et al., 2010). The retrieved answer is then displayed to the user.

The system works in the following way. A user initiates the system by opening up the application's browser. Once loaded, iTech welcomes the user and tells them of his purpose and how to ask a question. The user presses the 'Push 2 Speak' button and asks their question. The browser interacts with the user and identifies the exact content of the question. The question is converted into text and sent to the QRA module (Wilson et al., 2010). The QRA module performs three tasks. First the users text is broken into bigrams or word pairs. Second, the QRA matches the question's terms against a corresponding table of word pairs residing in the KR. Third, the KR finds the question with the highest concentration of terms and the indexed answer to that question is returned to the iTech's interface with a link to the corresponding document. Finally, the answer is displayed for the user.

iTech's interface is multimodal and can be housed on any personal computing device with a microphone or the ability to add a microphone. The microphone is used to collect the user's speech. The graphical user interface (GUI) consists of two frames: the Navigation frame and the Content frame. The Navigation frame consists of an animated agent and the Speech Application Language Tags (SALT). The presence of a likeable animated pedagogical agent has been shown to improve student performance by enhancing the student's desire to learn (Baylor & Ryu, 2003). This desire is increased as the student forges a personal connection with the agent, thereby making the learning experience more enjoyable. However, the agent must possess certain characteristics for this to be effective. The agent must be engaging, person-like, and credible; promoting relationships with the learner requires the presence of these characteristics (Baylor & Ryu, 2003). iTech is male. This choice was deliberate and influenced because findings suggest that male pedagogical agents are perceived as more extraverted and agreeable resulting in a more satisfying experience by the learner (Baylor & Kim, 2003). The ethnicity of iTech was chosen as African-American. This choice was determined by study results that indicated African-Americans were more inclined to choose an agent of the same ethnicity than Caucasians (Baylor et al., 2003). The agent was generated using SitePal and embedded into a HTML file (Oddcast Inc., 2008). The SitePal application allows for greater developer control over the appearance of iTech. To enable the agent's perceived participation in conversations, SALT and JavaScript were used.

JavaScript was used to provide text-to-speech (TTS) capabilities to the agent. SALT is then used to enable iTech's hearing. SALT is embedded in a compliant browser and using Microsoft's recognition engine allows iTech to listen to user's questions. Once the question is recognized, the question resolution algorithm is applied, an answer is identified and retrieved and is displayed in the Content Frame.

When iTech is loaded for the first time, the Content frame displays the cover of the vi manual (See Figure 2. iTech's welcome screen and welcome instructions). Once interaction begins, the Content frame dynamically displays the solutions retrieved by the question resolution algorithm (QRA. The QRA is initiated by a PHP script that connects to a MySQL database that houses the KR.

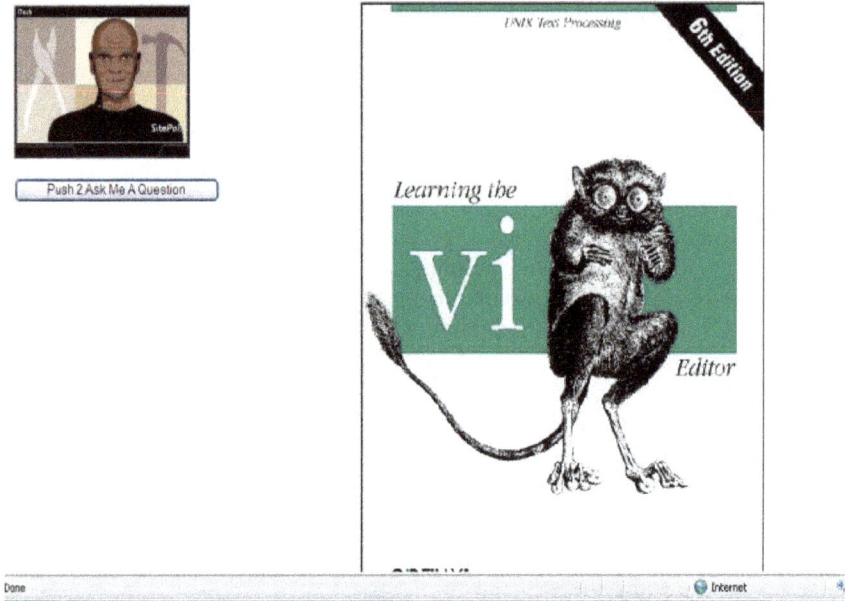

iTech: Hello, I am iTech ….
iTech: I am here to help you with the vi editor
iTech: When you need assistance. Just push the button to ask me a question

Fig. 2. iTech's welcome screen and welcome instructions

3.2 Equipment

To test the iTech design, a usability study to measure performance and user satisfaction was conducted. The authors set out to answer three research questions related to search time and user satisfaction:

- Does iTech improve search time compared to other technical communication mediums, namely the book and online mediums?
- Does iTech improve task completion time compared to other technical communication mediums, namely the book and online mediums?
- Does the introduction of an interactive agent, increase user satisfaction compared to other technical communication mediums?

Search time refers to the amount of time the participant spent referencing their assigned medium before the correct solution was found. After finding the correct solution, the user had to read the solution. The task completion time was the amount of time the participant spent referencing their assigned medium before the participant read the correct solution. User satisfaction refers to the effectiveness, efficiency and user's overall experience with the system.

The experiment was setup in a private room furnished with one large table and five chairs. All testing was conducted on a Gateway 2000e CPU running the Windows XP operating system and equipped with a 17″ Sony Monitor, a standard scroll mouse, and a Logitech USB headset. In addition, we downloaded Internet Explorer 6.x, the Microsoft Internet Speech Add-in 1.0., and the SecureCRT 4.07 software on the machine. All user interactions were recorded with a Sony 700x Digital Handy cam video recorder.

3.3 Participants

Seventy-four college level students were recruited as participants. Institutional Review Board (IRB) approval was granted by Auburn University and all participants signed informed consent forms before participating in the study. Participants had little or no exposure with the vi editor before participating in the study. The vi editor is a short hand editor used on Unix and Linux operating systems. Computer programmers often used the vi editor on these systems because it provided an efficient editing tool at the Unix/Linux command line; however, the vi editor has a huge learning curve. The vi editor is not a WYSIWYG (What You See Is What You Get) editor. It requires the user to know several keyboard shortcuts before using vi. As such, the vi editor is very difficult to use without proper training. Today, most users of Unix/Linux systems prefer WYSIWYG editors like pico; therefore, very few college students are familiar with vi. Many participants in the study had some experience using either Microsoft Office Word or Corel Word Perfect. To ensure that all participants had similar experience using a personal computer and editing text documents, our recruitment was focused on student enrolled in at least one course from the College of Engineering.

3.4 Experiment

The usability evaluation was designed as a controlled experiment. To reduce the casual effects of other factors, the following controls were applied:

- All participants sat in the same chair in the same room with the researcher.
- Each participant completed the same task in the same order. The only independent variable that changed was the medium of technical communication.
- Participants were randomly selected to use the book manual, online manual, or iTech.
- Participants who were assigned the iTech medium used the Logitech USB headset.
- The delay time before starting the survey was the same for each participant. The pre-experiment survey was started when the participant arrived in the experiment room. The post-experiment survey was started immediately after the participant finished his or her task.
- Participants were asked not to discuss the experiment with others to ensure that all participants had equal knowledge of the experiment.

The experiment was conducted for three different mediums. Medium I was a book manual entitled Learning the vi Editor published by O'Reilly. Medium II was the combination of a search engine and the electronic PDF version of the book manual used for Medium I. The combination of a search engine and the PDF was used to ensure that each medium being tested had the same content. To generate this medium, each section of the PDF was separated and saved as an individual file. Once the electronic manual was decomposed into individual sections, Google Desktop (Google Inc., 2009) was installed on the experiment computer. The preferences for Google Desktop were set to search a specific folder on the experiment computer's hard drive. This was to once again insure that the content of Medium II was the same as that of Mediums I and III. Medium II could be accessed through a floating desktop bar that was positioned in the top right corner of the monitor. When a participant entered a search query a list of all relevant documents was returned. Medium III was iTech. iTech was populated with information from the book manual. The answers indexed in iTech were the same electronic copies of the individual sections of the manual used in the online medium (Medium II). Consistency in content was maintained across all three mediums to reduce the probability that any difference in search and/or task completion were not due to any variable other than medium.

Twenty participants were assigned Medium I, twenty-four participants Medium II, and thirty participants Medium III at random. At the beginning of each experiment each participant was asked to fill out a pre-experiment questionnaire. Participants were then given an information sheet explaining the experiment and an instruction sheet that included tasks that the participants were asked to complete. Participants were assigned a medium and it was explained to the participant that they would be using the medium in completing the task. If the participant was assigned Medium I, they were given the book Learning the vi Editor. If the participant was assigned Medium II, the participant was directed to the floating desktop bar in the right hand corner of the screen and was instructed that he or she would be using a search engine linked to an online manual to assist in completing the task. If Medium III was assigned, the participant was instructed to put on and adjust the Logitech headset and iTech was launched. The participant was then directed to the SecureCRT terminal containing a file named example.txt to be edited. The participant was informed that they would be accessing the vi editor and the file from the current terminal. Lastly, the video recorder was started and the participant began his or her task.

The tasks were selected from the Exploring Microsoft Office 2003 textbook (Grauer, 2003). Participants were asked to figure out how to open the specified file and edit it. Editing included deleting individual words, changing words, changing characters, deleting and inserting sentences, and deleting and inserting paragraphs. When the participants completed the assigned tasks, they were then asked to fill out a post-experiment survey.

3.5 Data collection

During the course of the experiment, several approaches were used to collect data including video recordings and surveys. Table 1. Experimental Instruments and Measures provides an overview of the experimental measures and instruments used. Pre-experiment surveys were used to gather demographic information about participants and to determine whether they met the criteria established for classification as a vi editor novice. In addition, questions were asked about the participant's familiarity with computers such as how long they had used a computer, how often they use a computer, computer programming experience, and experience with specific software applications like word processors.

Performance data was collected using a video camera. Recordings were used to measure search, reading, and task completion times. Characteristics of spoken queries such as the average number of spoken queries per search per user, the number of recognition errors, and the total number of spoken terms per query were also derived from the participants' utterances. Informal and formal user observations were also employed to gather performance data.

Post-experiment questionnaires were used to gather user satisfaction data. Two post-experiment questionnaires were designed for the experiment. One was administered to participants that used iTech and the other was administered to all other participants. Part I of the questionnaire was identical in both versions of the questionnaire. It gathered overall participant ratings using six bi-polar rating scales. Part II of the questionnaire included a series of Likert-like scales where participants were asked to rate their reactions to the system. This part of the questionnaire included statements concerning the medium's ease of use and intuitiveness. The version of the questionnaire designed for iTech included statements concerning participant's reactions to the agent. Lastly, each questionnaire included a section where participants could share suggestions or comments regarding the medium assigned.

Instrument	Description
Pre-experiment Questionnaire	User background, demographics, computer literacy, etc.
Performance data	Time, QRA accuracy
User Observations	Qualitative and quantitative observations
Post-experiment Questionnaire	User satisfaction

Table 1. Experimental Instruments and Measures

4. Results

For the purposes of this paper, the authors will focus on results related to search time, task completion time and user satisfaction. The Jmp Statistical Software package was used to analyze the data collected for each of these measures (SAS Institute, 1984).

4.1 Participant analysis

An analysis of participant data shows that participants' ages ranged from 18 to 27 years with a mean age of 20 years (See Table 2). Of these, 71% were male and 29% were female. Participant's average number of years of computer use was 12 and the minimum number of years of computer use was 8. Therefore, the majority of the participants were comfortable using a computer.

Measurement	Medium I N = 20	Medium II N = 24	Medium III N = 30	Total
Average Age	19.15	19.22	22	20
% female	20%	37.5%	23.33%	26.67%
Avg. years of computer use	8.3	16.0	11.53	11.94
English as a 2nd Language	N/A	N/A	6.67%	N/A

Table 2. Participant Background Data

4.2 Performance results (search time and task completion)

To determine if iTech provided an improvement in search time compared to the book and online mediums, an analysis of the distribution of search times for each medium was performed. Additionally, the Shapiro-Wilk's test for normality was used for its resilience to the outliers present in the data. Table 3. *Mean Search Time by Medium* displays mean and standard deviation for each medium. This analysis suggests that Medium III (iTech) had the fastest average search time.

Measurement in secs	Book	Online	iTech
Mean	119.55	176.49	38.13
Standard Deviation	145.66	235.43	74.2

Table 3. Mean Search Times by Medium

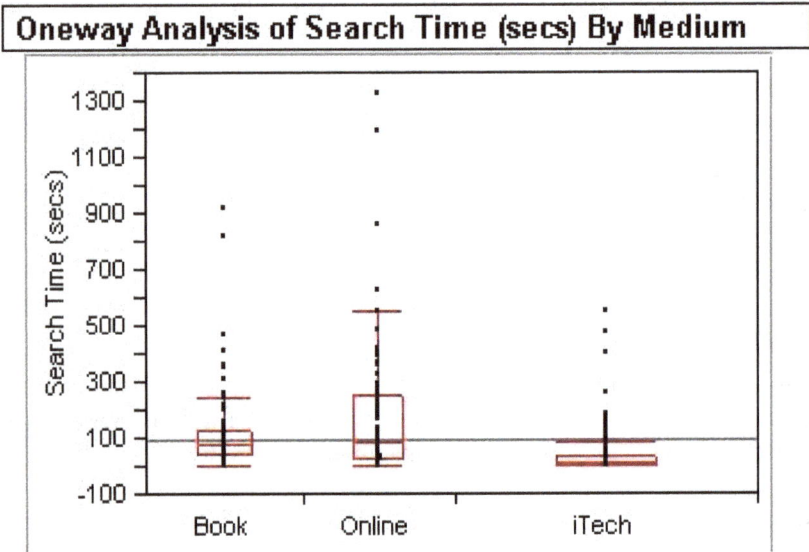

Fig. 3. Side-by-Side Box Plots of Medium Search Times

The Shapiro-Wilk's test provided very strong evidence to reject the null hypothesis that states that the means are normally distributed. With $\alpha = 0.05$ the Book medium [W = 0.6195, p = 0.0000], Online medium [W = 0.6811, p = 0.0000] and iTech medium [W = 0.5174, p = 0.0000] all strongly support this deduction. Because of this finding, the Kruskal-Wallis or Wilcoxon test was used to check for statistical significance.

The Kruskal-Wallis nonparametric analysis of variance provides a method for coping with data that contain extreme outliers and that have more than 2 independent variables. It does this by replacing the observation values by their ranks in a single sample and applying a one-way analysis of the F-test on the rank-transformed data (NIST, 2003). The result of this test [F (1,2) = 106.9946, p < .0001] is a Kruskal-Wallis test statistic of 106.9946 with a p-value < .0001 from a chi-square distribution with 2 degrees of freedom. The null hypothesis for this test states that the search time means for the mediums are equal. The Kruskal-Wallis test provided strong evidence to reject this null hypothesis. Thereby, there is statistical significance that supports that the search time means are different. Thus, there is strong evidence to reject the null hypothesis that states that the means were equal. The Kruskal-Wallis test allows for the comparison between three or more unpaired groups, however it does not allow for deductions between specific pairs or means. The resulting p-value, which is very small, indicates that the deduction can be made that the difference in the group means is not a coincidence. However, this does not mean that every group differs from every other group. The Kruskal-Wallis test only determines that at least one group differs from one of the others. Thus, a post-test was applied to determine which groups differed from the other groups.

The Tukey-Kramer test analyzes data of unequal sample sizes and determines whether the differences between all existing pairs are due to coincidence (NIST, 2003). The results of the Tukey-Kramer test provided very strong evidence that the differences in the pairs of means were statistically significant (See Table 4). The positive values between each pair of means indicate that their differences are significantly different. Thus, there is sufficient evidence to deduce that the independent variable of medium type had a statistically significant effect on the search times, with the search times for the iTech medium being the most expeditious. Next an analysis was performed to determine the effects of search time on task completion time.

	Online	Book	iTech
Online	-	3.368	91.582
Book	3.698	-	35.528
iTech	91.582	35.528	-

Table 4. Tukey-Kramer Test Results

The same tests were applied to medium search times to task completion times. The mean task completion times are displayed in Table 5. *Mean Task Completion Time by* and the normality spreads are shown in Figure 4. *Side-by-Side Box Plot of Medium Task Completion Time.* Preliminary observations indicate that Medium I (Book medium) had the fastest average task completion time. The Shapiro-Wilk's test did not however provide sufficient evidence to either strongly reject or accept the null hypothesis that states that the means are normally distributed. With $\alpha = 0.05$ the Book medium [W = 0.9395, p = 0.229], Online medium [W = 0.9664, p = 0.582] and iTech medium [W = 0.9501, p = 0.199] all results recommend the failure to reject the null hypothesis, indicating that the distributions are fairly normal. As a result, the Kruskal-Wallis was used to check for statistical significance.

Measurement	Book (secs)	Online (secs)	iTech (secs)
Mean	1360.58	1666.63	1377.87
Standard Deviation	290.95	500.1	420.87

Table 5. Mean Task Completion Time by Medium

An application of the Kruskal-Wallis yielded no significant differences. The result of this test [F (1,2) = 5.7065, p = 0.0577] suggests a failure to reject the null hypothesis that states that the differences in the mean task completion times is due to coincidence. Therefore, there is no statistical significance that indicates the task completion time means are different. An investigation as to the cause of this effect, led to the effect of reading times. Reading time was recorded as the time the participant spent reading and understanding the solution once it was presented to the user by the respective medium. This recorded time represented the time from the appearance of the solution on the monitor to the time the participant touched the keyboard. Results are shown in Table 6. *Mean Read Times by Medium* and Figure 5. *Side-by-Side Box Plots of Medium Read Times.*

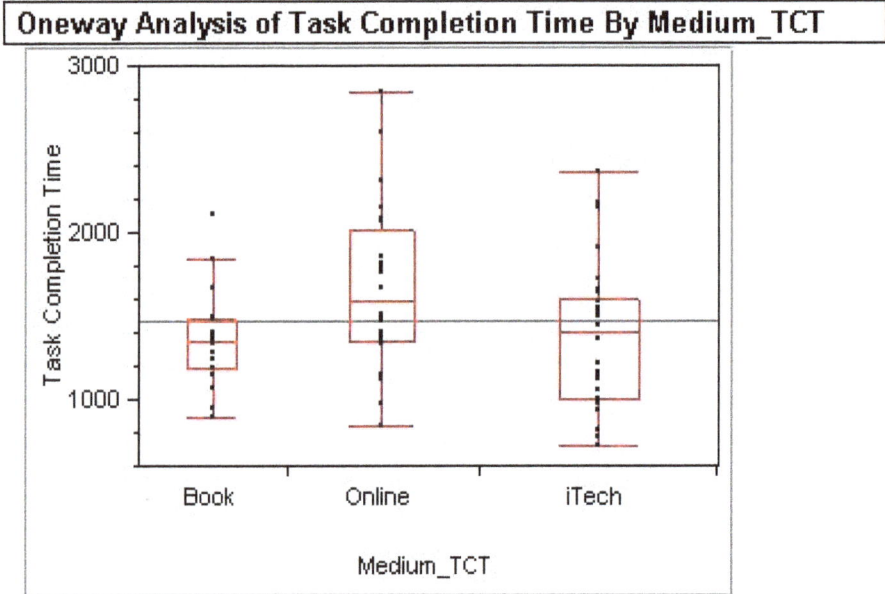

Fig. 4. Side-by-Side Box Plots of Medium Task Completion Time

Measurement	Book (secs)	Online (secs)	iTech (secs)
Mean	42.36	33.09	47.77
Standard Deviation	38.34	33.08	45.34

Table 6. Mean Read Times by Medium

Preliminary observations show that there is very little difference between the mean read times for the different mediums. Application of the Shapiro-Wilk's test for normality yielded these results; Medium I (book medium) [W=0.7854, p<.0000], Medium II (online medium) [W = 0.7166, p < .0000] and Medium III (iTech medium) [W = 0.7694, p < .0000] and provides strong evidence that the means are not normally distributed. As the Kruskal-Wallis test is just as effective on normal distributions and allows for consistency of applied tests, it was applied to the read time data. On application of the test, the result [F(1,2) = 9.1906, p = 0.0101] indicates that there is evidence that the difference between the means is statistically significant. However, further investigation as to which pairs were significantly different was required. The Tukey-Kramer test was applied and it demonstrated that only the difference between the online and iTech mediums was statistically significant. With respect to task completion time, it was observed that participants spent more time reading online than with the Book medium. When a solution was found, several participants' encountered difficulties

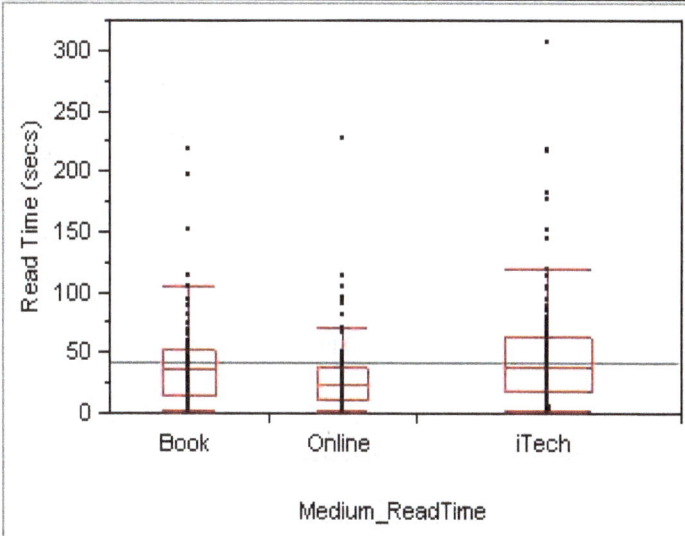

Fig. 5. Side-by-Side Box Plots of Medium Read Times

understanding the text, therefore although iTech found solutions faster, participants read the solutions longer than with the Book medium. It should be noted that the solutions were all identical regardless of the medium used. Also, a significant proportion of participants did not read the solution carefully and as a result either had to return to the solution several times, or implemented an incorrect action that led them further away from the correct action. These results suggest improvements in the content and understandability of technical communications may increase the improvements in search time provided by the iTech medium. Lastly, performance analysis of task success showed that nearly all participants were able to successfully complete the assigned tasks. Task success was determined by comparing the file updated by each participant to a correct version of the updated file. 95% of all participants successfully completed the task using one of the three mediums provided.

4.3 User satisfaction

To get a better idea of users reactions to iTech compared to the book and online mediums a post-experiment questionnaire was used to collect data using two rating scales. The first rating scale included a five-point bi-polar scale. This scale presented several qualities that might influence usability. The rating means are shown in Table 7. *Bi-polar Rating Scales assessing General Usability*. For each of these scales a higher rating indicates a number closer to the positive side with the exception of the anchor usable to not usable. For this anchor a higher rating indicates a number closer to the negative side. A quick review suggests that the participants' reactions to iTech were generally more favorable than the other two mediums. However, investigation of just the means does not provide a complete picture of the users' evaluations. For example, although iTech's rating of the Terrible-Wonderful anchor is lower than the Online medium, iTech received 19 ratings at levels 4 - 5 while the Online medium received only 16. Therefore, an analysis of the entire distribution for each rating was conducted.

Bi-Polar Scale Anchors	Book Ratings (Mean)	Online Ratings (Mean)	iTech Ratings (Mean)
Terrible – Wonderful	3.17	3.55	3.29
Frustrating – Satisfying	3.23	2.90	3.0
Dull – Stimulating	2.93	3.62	3.79
Usable – Not Usable	2.4	2.31	2.57
Boring – Fun	2.9	3.38	3.64

Table 7. Bi-polar Rating Scales Assessing General Usability

Three scales were used to examine the usability of iTech: 1) terrible – wonderful, 2) dull – stimulating and 3) boring – fun. The five-point rating inherently assigns the score of 3 a neutral rating, with scores 1 and 2 being negative and scores 3 and 4 positive. For the book medium, 33.33% of the participants rated that medium with a score of 4 or higher on the terrible to wonderful scale. The online medium received 53.33% and iTech 65.52% for the same score values (See Figure 6. Terrible Wonderful Distributions).

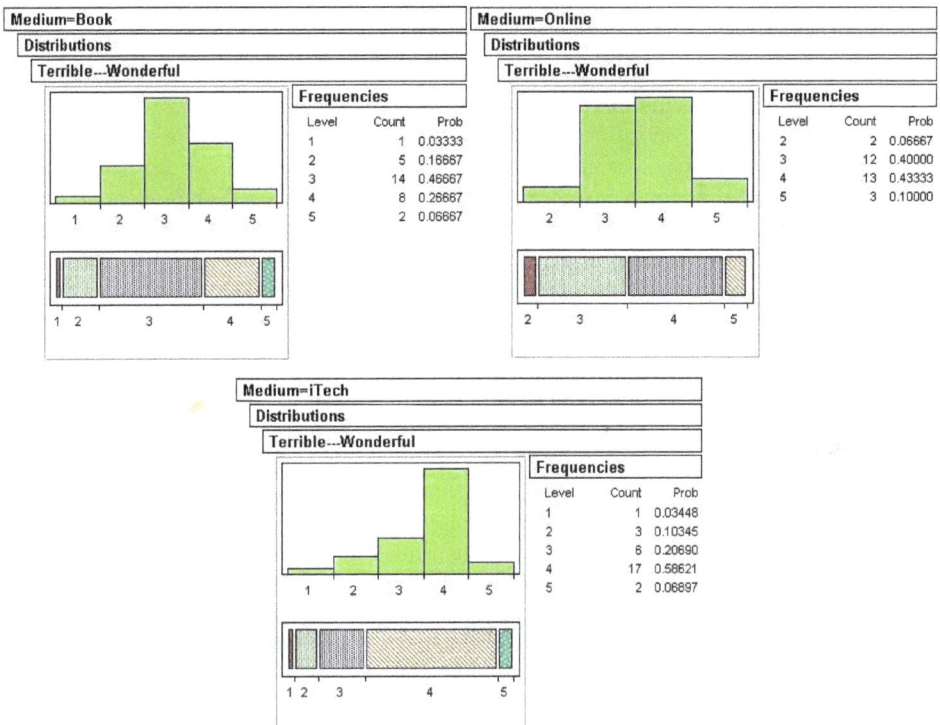

Fig. 6. Terrible - Wonderful Bi-polar Distribution

On the scales of dull to stimulating and boring to fun, iTech received the highest scores with respect to the other two mediums. For these scales, a much larger disparity in the distribution of scores is observed with respect to the book medium versus the online and iTech mediums. (See Figure 7. Dull-Simulating Distributions Distribution and Figures 8. Boring-Fun Bi-polar Distributions)

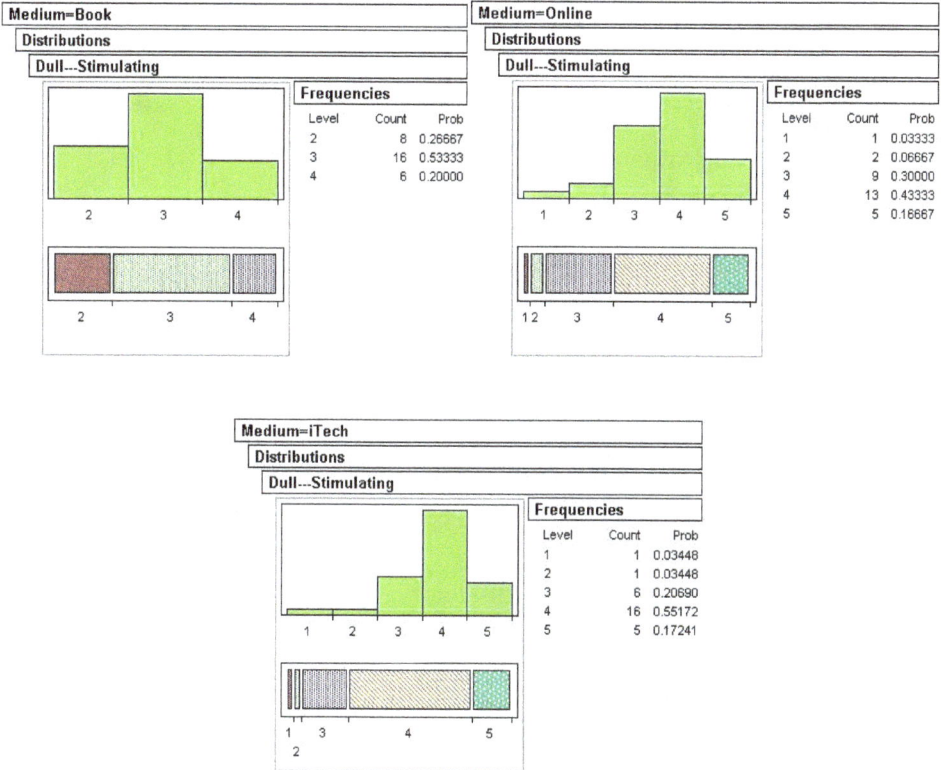

Fig. 7. Dull – Stimulating Bi-polar Distribution

The second set of rating scales consisted of items designed to examine reactions to specific aspects of the participants' interaction experience. These scales each contained an assertion e.g. 'The medium was easy to use', to which the participants responded using a five-point scale. This scale contained the following ratings: Strongly Agree, Agree, Neutral, Disagree and Strongly Disagree. We assigned each rating a weight. This weight was used for statistical analysis. The Strongly Agree was assigned a rating of 5, Agree a rating of 4, Neutral a rating of 3, Disagree a rating of 2, and Strongly Disagree a rating of 1.

Version I of the post-experiment survey contained 10 Likert-like ratings and Version II of our post-experiment survey contained 22 ratings. The first 9 ratings for each questionnaire were identical and as a result were compared across all three mediums.

Medium=Book

Distributions

Boring---Fun

Frequencies

Level	Count	Prob
1	3	0.10000
2	6	0.20000
3	13	0.43333
4	7	0.23333
5	1	0.03333

Medium=Online

Distributions

Boring---Fun

Frequencies

Level	Count	Prob
2	3	0.10000
3	15	0.50000
4	9	0.30000
5	3	0.10000

Medium=iTech

Distributions

Boring---Fun

Frequencies

Level	Count	Prob
1	1	0.03448
2	1	0.03448
3	9	0.31034
4	11	0.37931
5	7	0.24138

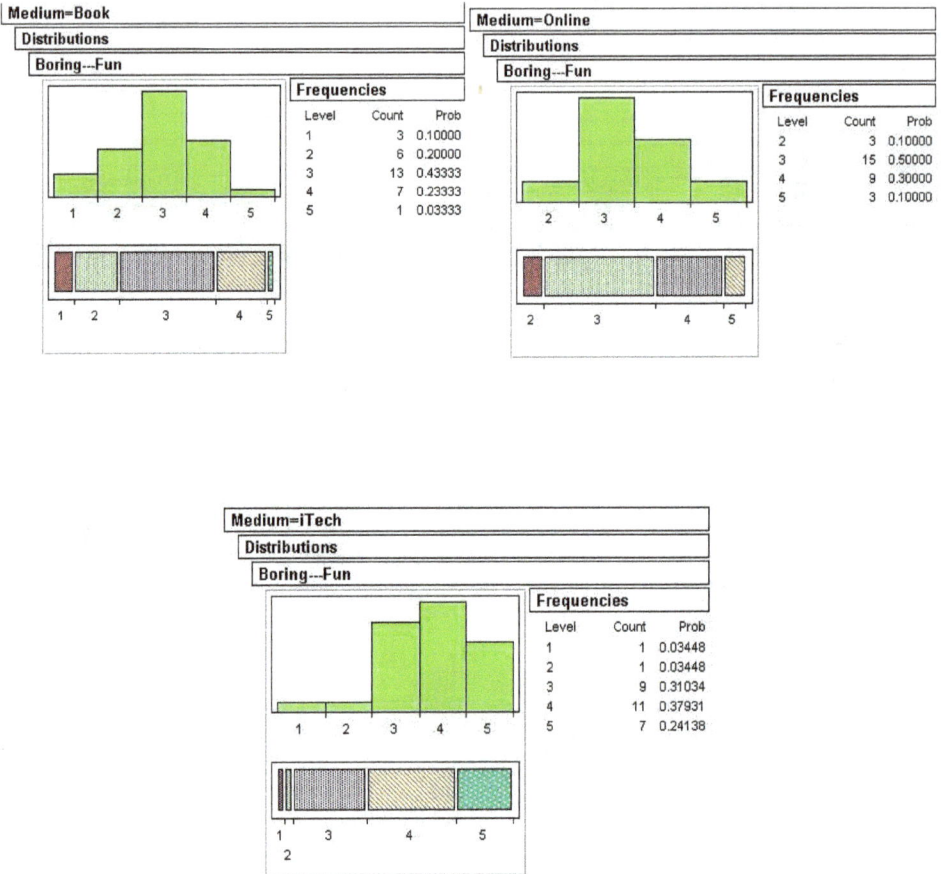

Fig. 8. Boring - Fun Bi-polar Distribution

The first property analyzed was the affordance of the mediums. This property was derived from the question, "It was easy to get started". Results show that iTech received a score of 4 or higher from 60.7% of the participants, while the book and online mediums received 46.67% and 30.0% respectively (See Figure 9. Affordance Distributions). This data is in agreement with the trends found in the mediums' search times. The online medium had the worst average search time with iTech having the best, suggesting that an application's affordance is an important feature of the application's success.

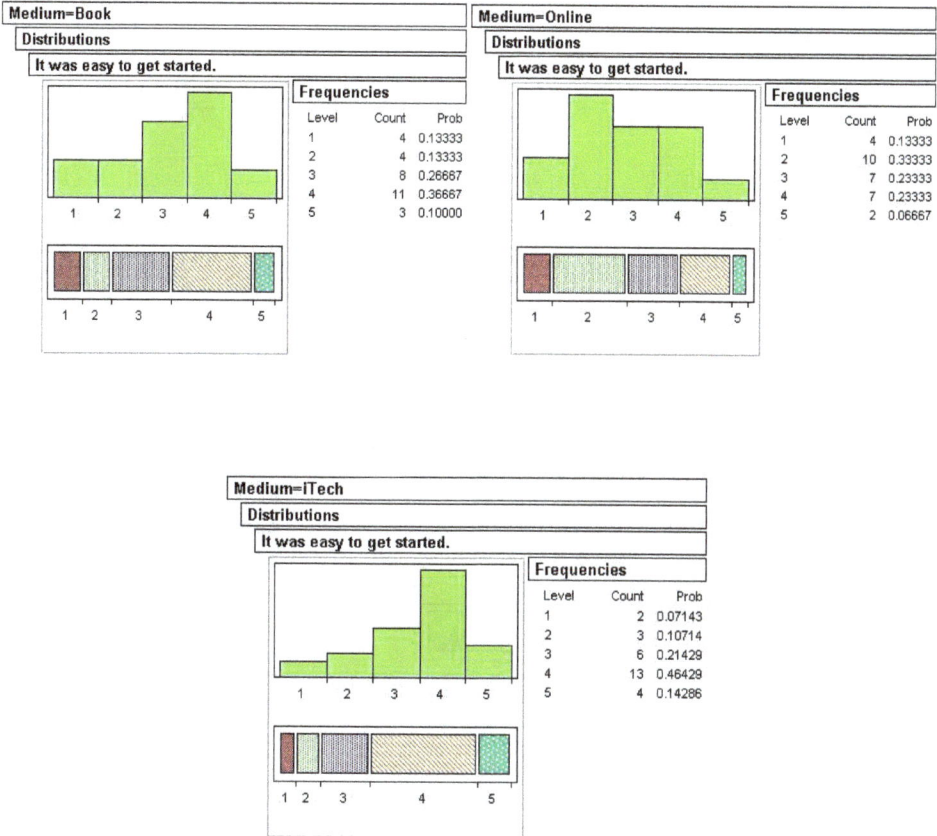

Fig. 9. Affordance Distributions

The scores for "understanding document updates" were over 80.0% for all mediums suggesting that we selected tasks that needed little training to get started. The results for the property of 'ease of use' reflect the problems with speech recognition accuracy. There were problems with recognition accuracy due to heavy southern accents and incorrect usage of the recording box. Subsequently, though the range for the medium averages is small, the scores for the iTech medium are the lowest in response to the statement, "It was easy retrieving an answer". The results are as follows: book medium – 63.3%, online medium – 50.0% and iTech medium – 48.27% for scores of 4 or higher. In spite of the recognition accuracy issues, iTech received the highest ratings with respect to knowing how to use the medium (See Figure 10. Getting Started Distributions). Next, we analyzed the user's reactions to the iTech medium.

Medium=Book

Distributions

It was easy to get started.

Frequencies		
Level	Count	Prob
1	4	0.13333
2	4	0.13333
3	8	0.26667
4	11	0.36667
5	3	0.10000

Medium=Online

Distributions

It was easy to get started.

Frequencies		
Level	Count	Prob
1	4	0.13333
2	10	0.33333
3	7	0.23333
4	7	0.23333
5	2	0.06667

Medium=iTech

Distributions

It was easy to get started.

Frequencies		
Level	Count	Prob
1	2	0.07143
2	3	0.10714
3	6	0.21429
4	13	0.46429
5	4	0.14286

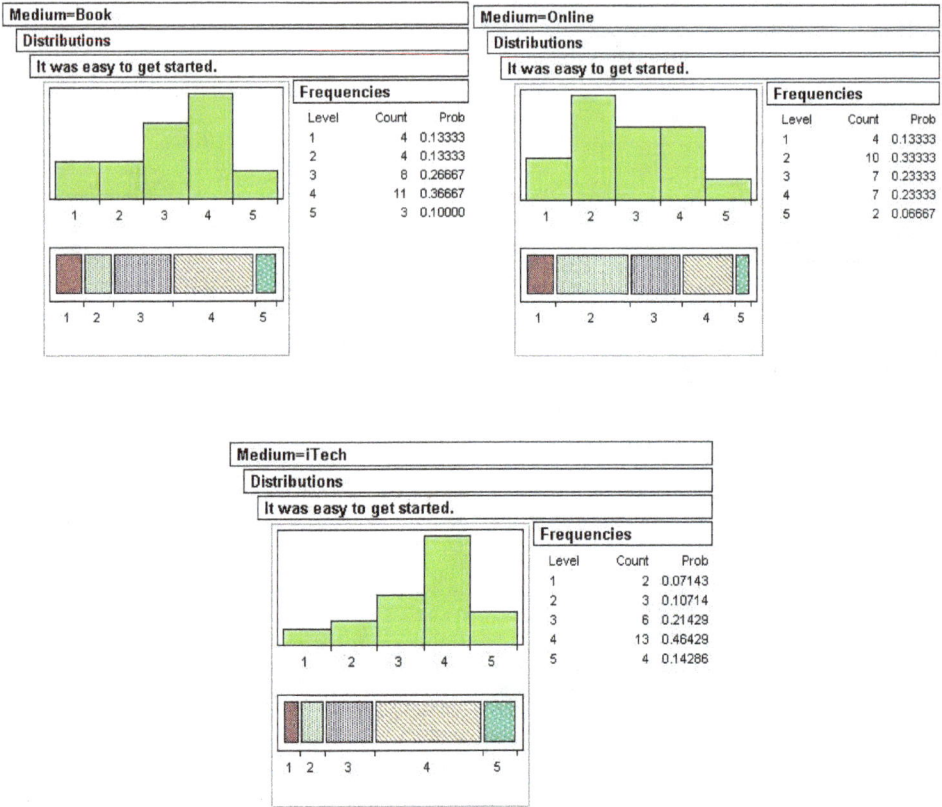

Fig. 10. Getting Started Distribution

Before beginning the analysis of user's reactions to iTech, statements unique to iTech were placed into one of six possible categories. These categories represented the six factors investigated user's attitudes towards speech systems (Hone, 2003). Results are shown in Table 8.

Participants liked the appearance of iTech and results suggest that they would reuse the application. Participants were able to understand iTech and thought that the application retrieved their answers in an expedient fashion. In addition, they agreed that computer novices would be able to use the application. The high user satisfaction ratings were solidified by additional comments.

"Worked greater than expectations based on previous speech help programs…"

"Pretty easy to use. User friendly"

"I really enjoyed iTech …the layout and technology used was great"

"It was overall very helpful and would be useful for people whom are computer literate".

Likert-like Scale Item	Mean	% with Score of 4 or higher
Cognitive Modeling		
iTech worked as I expected it during the task.	3.41	55.1
I was confident that iTech would be able to help me.	3.86	72.4
Perceived System Response Accuracy		
iTech gave me the correct answers.	3.69	72.41
iTech had problems understanding me.	3.52	68.97
Cognitive Demand		
I had problems understanding iTech.	2.32	71.14
Likeability		
I would use iTech again.	3.82	78.57
I liked the appearance of iTech.	4.18	89.29
I would have preferred a female technician.	2.97	13.79
I would have preferred iTech having no face, just a voice.	2.10	7.9
Habitability		
iTech would be easy to use by people who don't know a lot about computers.	3.62	75.86
Speed		
iTech was fast enough in response to my question.	4.03	86.21

Table 8. Analysis of User's Reactions to iTech

5. Discussion and implications

The results suggest that overall; iTech is a viable technology for use in the area of technical communications. The introduction of an animated agent that allows users to speak questions and return an appropriate solution through our research has been shown to decrease search time and task completion time, as well as overall user satisfaction compared to both book manuals and online searchable manuals. Therefore, such systems may provide some improvement over traditional technical communication mediums such as books and online search systems. More generally, because iTech is a speaker-independent system that employs conversational questioning answering techniques suggests additional advantages. Because iTech is speaker independent, there is no need for training. Additionally, because iTech allows users to answer spoken questions, query-preprocessing time is eliminated.

Another advantage of iTech like systems is that it improves on many of the limitations of online and paper manuals including portability and frequent updates. iTech is computer-based and therefore there is no need for a bulky paper manual, only a computer, cell-phone, or other Internet connected device is needed. Also, iTech transfers the search task from the user to the computer, removing the need for users to understand indexes. Because, iTech's content resides on a server in a database, the ability to make frequent updates is less time consuming and does not require an entire re-print and shipment of manuals to users. Users therefore will have access to the most recent version of the manual at all times. The study results present a new opportunity for professional communicators to incorporate the best of these two mediums, search engines and manuals. Furthermore, iTech has the potential to change the way technical information is communicated across numerous domains. For example, automobile manuals have significantly grown in size. At the same time, these manuals have found their way online in the form of Adobe Acrobat documents that can be easily searched by drivers. iTech has the potential to be integrated within the vehicle to provide instant access to manual information using the driver or passenger's voice. Another compelling domain is the military. When military personnel travel, they carry a great deal of equipment, including a laptop and manuals. iTech has the potential to consolidate their manuals into a single laptop with a natural language interface, e.g. typed or spoken text.

The need for effective technical communication to provide user assistance will continue to be an issue of importance as long as new products and devices are introduced into society. As these devices become more complex so will the documentation accompanying them. Therefore, usability and user satisfaction will continue to become an important factor in creating documentation that is among others easy to search, easy to understand, and easy to use. iTech addresses the limitations of paper and online manuals by providing technical communication through a personable virtual interactive technical assistant.

6. Conclusions

The results of this study show that iTech yielded faster search times than its paper and online counterparts. In addition, iTech had favorable usability results. Overall, the use of iTech was favorable and provided evidence that such a tool would be a viable option for providing technical assistance. In addition positive user comments show that users of iTech were satisfied with their experience with the tool. In addition, this research suggests that the application of interactive virtual assistants to technical communication is a viable research area for increasing usability and user satisfaction.

7. References

Albing, B. (1996). Process constraints in the management of technical documentation, in Proceedings of the 14th Annual international Conference on Systems Documentation: Marshaling New Technological Forces: Building A Corporate, Academic, and User-Oriented Triangle (Research Triangle Park, North Carolina, United States, October 19 - 22, 1996). SIGDOC '96. ACM, New York, NY, 67-74, 2006, Available: DOI= http://doi.acm.org/10.1145/238215.238257.

Barnett, M. (1998). Testing a digital library of technical manuals. Professional Communication, IEEE Transactions on, vol.41, no.2,1998, pp.116-122. Available: DOI: 10.1109/47.678553 URL: http://ieeexplore.ieee.org/stamp/stamp.jsp?tp=&arnumber=678553&isnumber=1 4943.

Baylor, A., & Ryu, J. (2003). The effect of image and animation in enhancing pedagogical agent persona. Journal of Educational Computing Research, 28(4), pp. 373-395.

Baylor, A. & Kim, Y. (2003). The Role of Gender and Ethnicity in Pedagogical Agent Perception. G. Richards (Ed.), Proceedings of World Conference on E-Learning in Corporate, Government, Healthcare, and Higher Education 2003. Chesapeake, VA: AACE, 2003, pp. 1503-1506.

Baylor, A., Shen, E. & Huang, X. (2003). Which Pedagogical Agent do Learners Choose? The Effects of Gender and Ethnicity. In G. Richards (Ed.), Proceedings of World

Conference on E-Learning in Corporate, Government, Healthcare, and Higher Education 2003. Chesapeake, VA: AACE, 2003, pp. 1507-1510.

Cisco Systems Inc., Comverse Inc., Interl Corporation, Microsoft Corporation, Philips Electronics N.V., & Speech Works International Inc. (2002). SALT Speech Application Language Tags (SALT) 1.0 Specification.

Google. (2009). Google Desktop. [Online] Available: http://desktop.google.com/.

Grauer, R. (2003). Exploring MS Office XP and Exploring FrontPage 2003 plus the Train and Assess IT Generation, Prentice Hall Publishing Co. ISBN: 0-536-83155-6.

Hailey, D.E. (2004). A Next Generation of Digital Genres: Expanding Documentation into Animation and Virtual Reality. Proceedings of the 22nd Annual International Conference on Design of Communication, Memphis, TN, 2004, pp.19-26.

Hone, K.S. & Graham, R. (2001). Subjective Assessment of Speech-System Interface Usability. Eurospeech .

Johnson, W. L. & Rickel, J. (1997). Steve: an animated pedagogical agent for procedural training in virtual environments. SIGART Bull. 8, 1-4 (Dec. 1997), 16-21. DOI= http://doi.acm.org/10.1145/272874.272877.

Kirste, T., & Rapp, S. (2001). Architecture for Multimodal Interactive Assistant Systems. Statu- stagung der Leitprojekte "Mensch-Technik-Interaktion", Saarbrucken, Germany, 2001, pp. 1-5.

Lester, J. C., Converse, S. A., Kahler, S. E., Barlow, S. T., Stone, B. A., & Bhogal, R. S. (1997). The persona effect: affective impact of animated pedagogical agents. In Proceedings of the SIGCHI Conference on Human Factors in Computing Systems (Atlanta, Georgia, United States, March 22 - 27, 1997). S. Pemberton, Ed. CHI '97. ACM, New York, NY, 359-366. DOI= http://doi.acm.org/10.1145/258549.258797.

Major, J. H. (1985). Pulling it all together: a well-designed user's manual. Proceedings of the 13th Annual ACM SIGUCCS Conference on User services: pulling it all together, Toledo, OH, 1996, pp 69 -76.

NIST. (2003). NIST/SEMATECH e-Handbook of Statistical Methods, http://www.itl.nist.gov/div898/handbook/.

Oddcast Inc. (2008). SitePal: Now you're talking business. [Online] Available: http://www.sitepal.com/.

Oviatt, S. L., Cohen, P., Vergo, J., Duncan, L., Suhn, B., Bers, J., Holzman, T., Winograd, T., Landay, J., Larson, J. & Ferro, D. (2000). Designing the User Interface for Multimodal Speech and Pen-based Gesture. Applications: State-of-the-Art Systems and Future Research Directions, Human Computer Interaction 15, 4, pp. 263-322.

Rosis, F., Carolis, B. & Pizzutilo, S. (1999). Software Documentation with Animated Agents.

SAS Institute. (1989). JMP IN: Software for statistical visualization on the Apple Macintosh Cary, NC.

Thimbleby, H. (1996). Creating user manuals for using in collaborative design. Conference Companion on Human Factors in Computing Systems: Common Ground (Vancouver, British Columbia, Canada, April 13 - 18, 1996. M. J. Tauber, Ed. CHI '96. ACM Press, New York, NY, pp. 279-280.

Ventura, C. A. (1988). Why Switch from Paper to Electronic Manuals?. Proceedings of ACM Conference on Document Processing Systems, ACM, New York, Santa Fe, New Mexico, 1988, pp. 111-116.

Wilson, D.M., Martin, A. & Gilbert, J.E. (2010). 'How May I Help You'-Spoken Queries for Technical Assistance. ACM Southeast Conference, Oxford, MS, April 15-17, 2010.

Zachary, C., Cargile-Cook, K., Faber, B., & Zachary, M. (2001). The Changing Face of Technical Communication: New Directions for the Field in a New Millennium. Proceedings of the 19th Annual International Conference on Systems Documentation, 2001, pp. 248 – 260.

6

Risk Assessment of Innovations in the Biopharmaceutical Industry

David Domonkos[1] and Imre Hronszky[2]
[1]TMTT Doctoral School of BME, Gedeon Richter Plc.
[2]Faculty of Economic and Social Sciences,
Budapest University of Technology and Economics (BME)
Hungary

1. Introduction

First, the chapter summarizes the specialties, which are appeared in the red (medical) biotechnology in the occurrent risks/uncertainties point of view. Then it draws attention to the fact that part of the literature about risks/uncertainties (for example in the environmental literature) serve as a broad basis for the analysis and evaluation of uncertainty. This seems also useful for the examinations of the uncertainties in the medical biotechnology, but as far as we know, it is not applied. Finally, the third part of the chapter follows a new analysis and it introduces that there is an uncertainty dilemma in the research of the medical biotechnology, which can be reduced, but can not be eliminated.

2. Biotechnological innovations and trends

Biotechnology is spreading rapidly in the pharmaceutical, environmental protection, agricultural, and other industrial environments. The number of molecules produced by biotechnological methods is growing rapidly, thanks to new methods and an almost exponentially increasing knowledge base.

The appearance of novelties is very fast. There is a significant technological leap, from time to time. These radical innovations are aimed at solving complex problems by implementing and integrating new technologies. Radical innovations that lead to disruptive technological development in biotechnology based industry and especially in red biotechnology are usually result of long term research. These innovations provide a broad platform for a new regime in technology, from time to time [1]. At the same time, disruptive innovation is not necessarily radical. Small innovations can also have great disruptive economical influence, provided they are introduced in a new milieu. Just think about the turning to containers in oversee ship cargos, for example, when containers had already been much earlier utilised in other areas of transport.

Companies were forced to cooperate due to the high risk associated with biotechnology, the complexity of strategic management rules and the unusually high amount of needed funds. First of all, the necessary monetary tools are available only at the largest companies. Second, the necessary competencies are often missing with smaller companies. For example, a smaller company, a market leader in R&D, most probably does not have the necessary experience either of the capability needed to clinical testing or production. Cooperation is necessary to fill these gaps. With this sharing of different sorts of risks will be realised. These risks - actually non-calculable uncertainties several times, or at least the calculations can not be serve as reliable planning tools - may be technology, market, regulatory or competition related. The competition related one reflects on the segments of all the other risks, since the rapid development of China, South Korea and India. The only comparative advantages can only be quality and knowledge the traditional pharma producing countries have. But precisely these are areas where China and India are developing rapidly, while maintaining the seemingly natural price advantage. Europe and the USA can only compete with these products if they do not count on price advantage, but on therapeutic advantage. This means producing a newer, better molecule, first of all. However this larger added intellectual value brings larger risks, uncertainties on behalf of technological, market and registration. These tendencies are also catalysts of cooperation, for cooperation means some risk sharing.

It is precisely these different, but interrelated risks that make pharmaceutical biotechnology complex. To successfully manage complex processes and instability necessitates cooperation. Instabilities are cross-linked. They can even strengthen or weaken each other. An example of mutual strengthening of uncertainty is the technological uncertainty of producing a new molecule, and the registration and legalization which follow. Registration gives the same molecule an added economical value and can, if it is registered already, decrease market instability, since it can become a market leader, a so called "blockbuster"[1], with multi-million dollar yearly turnover.

Drug manufacture is a multinational phenomenon, with an active global trade in intermediates (specialty chemicals), active pharmaceutical ingredients, and finished products. R&D, by contrast, is much more geographically concentrated; the bulk of all R&D expenditure occurs in the United States, a handful of European countries, and Japan. The pharmaceutical value chain encompasses many activities, ranging from basic scientific research to marketing and distribution. Innovation in the industry is tightly linked to basic biomedical science, and many companies participate actively in basic scientific research that generates new fundamental knowledge, data, and methods.

Drug discovery includes basic science and research on disease physiology, identification and validation of "druggable targets" in the body where therapeutic molecules may affect disease processes, identification and optimization of drug candidates, and preclinical testing. The development phase of research focuses on testing in humans, from the first small-scale trials directed at establishing basic physiological data in healthy volunteers through to large-scale trials on patients having the disease, which are designed to provide data on safety and efficacy to support applications for regulatory approval of the drug. Following marketing approval, research often continues to develop improved formulations of the product and to establish safety and efficacy in treatment of additional diseases or patient populations. Reflecting extraordinary advances in biology and biochemistry since the 1970s, the industry has become progressively more science intensive, relying closely on

---

[1] A blockbuster drug is a drug generating more than $1 billion of revenue for its owner each year.

fundamental advances in physiology, biochemistry, and molecular biology rather than "brute force" application of large-scale resources. If anything, this process has accelerated over the past decade as the industry has focused on complex and systemic diseases such as cancer, autoimmune diseases, and psychiatric conditions. Particularly in drug discovery, industrial and publicly funded research efforts are deeply intertwined·

Rapid growth in technological capabilities in low-cost emerging economies is presenting new opportunities and challenges for pharmaceutical companies. Some geographic redistribution of R&D activity does appear to be taking place. On the one hand, companies located in countries such as India and China are performing more in-house R&D oriented toward developing new drugs, rather than reverse-engineering existing products or improving production efficiency. On the other hand, reflecting the general trend of the industry toward greater specialization and external sourcing of R&D services, OECD-based companies are beginning to look to low-cost countries as suppliers of contract research services, and growing numbers of clinical trials are being conducted in emerging economies. India and China are the two countries most frequently mentioned in this regard; however, by some indicators significant growth in activity also appears to be taking place in some Eastern European countries, Argentina, Brazil, Taiwan, South Africa, and Israel. Over the past decade, the biotechnology industry has been the focus of increasing academic and policy interest as a potential source of regional and national economic development [2] [3].

Historical development of biotechnology can be divided into several large eras [4]:

- The period started with the first conscientious use of biotechnology. This process started in the second half of the XIX century (about 1865), when Pasteur discovered that fermentation is caused by microorganisms. After understanding the essence of this process through microbiology, its industrial application became feasible. The beer and alcohol industries developed, vinegar and lactic acid production began. The production of ethanol, butanol, acetone, glycerin, citric acid etc. through fermentation began.
- The discovery of antibiotics provided the momentum for the second great leap around 1940. The productivity of microorganisms was increased by biological, genetic and biochemical methods (mutation, selection). Building on these opportunities and the rapid development of fermentation techniques, the result was a veritable technological revolution. The most important results were the large scale production technologies of antibiotics, amino acids and enzymes.
- The next phase started in the first half of the 1970-s. The essence of this new biotechnology is that by altering the heritable material of living beings, through a conscious and planned manner, results in the development of new characteristics. Through the use of recombinant DNA and cell fusion, humans begun to alter the characteristics and functions of living organisms to suit their needs.
- The fourth era is linked to the first commercial sale of human insulin (1982). This is the first member of the rDNA pharmaceutical products, meaning the large scale distribution of the products of the previous era. Thus growth gained even more momentum.
- The fifth era can be marked by the latest great innovation, on one hand, the cloning of animals by the use of a cell nucleus, the creation of "Dolly" (1997) and later other cloned animals and on the other, the completion of the "Human Genome Project" (2000).

To make expectations essentially belongs to the development of new technologies. Many people see biotechnology as the industry of decisive strategic importance, following informatics at the start of the XXI century. According to optimistic forecasts, by the middle of the next decade, the pharmaceutical and biotechnology industries will become the leading industrial branches in the world, surpassing information technology and telecommunication. (In this article we will only concentrate on red (medical) biotechnology.) Expectations regularly realise some cyclic dynamic. One can argue that elements in that dynamic follow each other by necessity. Following the first hope even deep disillusion can be the next step as it was with the ecommerce bubble around the turn of the century. While with biotechnology the hope phase is still very strong, with time delay, unexpectedly raising costs, etc. in comparison to optimisticforecasts, the hope phase can partially turn into disillusion urging to change the earlier expected enthusiasm into more "rational" thinking.

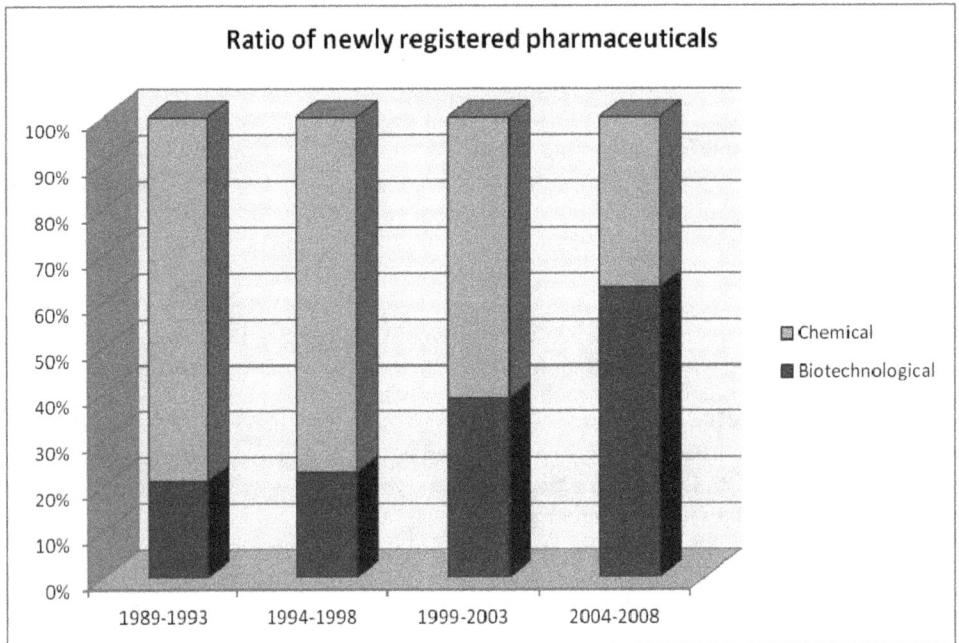

Fig. 1. [1] based on [7]: Ratio of newly registered pharmaceuticals: chemical and biotech entities

The EU considers the development of the biotech industry in view of the pharmaceutical industry exceptionally important and is rather optimistic about its future and the role the EU can play in it. This means strengthening collaboration between the two sectors [5]. Biotechnology plays an increasingly important role in pharmaceutical development, by preventing the onset of and curing previously un-curable diseases through the implementation of new diagnostic methods and treatments. Pharmaceuticals produced through biotech methods, such as proteins, antibodies, enzymes comprised 25% of pharmaceutical sales in 2003, already [6]. But most of the pharmaceuticals currently undergoing clinical trials are biotechnological in origin. The percentage of pharmaceuticals produced using biotechnological methods, is growing rapidly. Of all registered small molecules, significantly more are produced by biotechnological methods, than by synthetic methods. (Figure 1.)

These were all changes that radically altered the perspectives and tasks of biotechnology. Main directions of research were shifted, and the map of biotechnology was rearranged by economic factors as well. Therefore these are definitely disruptive innovations. Figure 2. shows growth of biotechnological knowledge, plotted against a timeline:

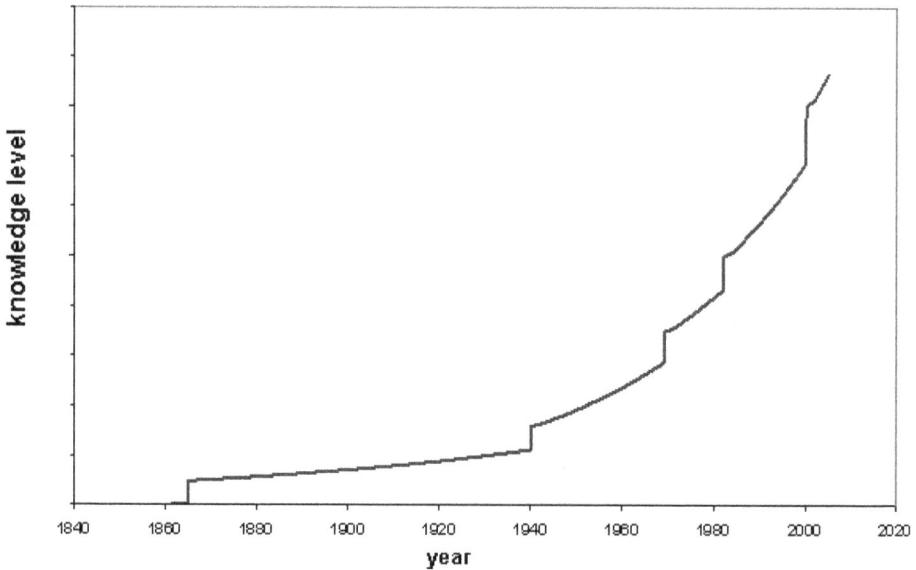

Fig. 2. [the authors]. Changes of the knowledge level in biotechnology

Usually, there is a complex, multidimensional, non-linearly correlated uncertainty surrounding disruptive, especially radical innovations, the solution of which often requires cross linked steps. It is important to state, that in terms of management, in opposition to small innovations, the management of radical innovations includes the ability to navigate in sight of unforeseeable events [1].

Biotechnology and pharmaceutical companies

Medical biotechnology is realised by two main types of companies. They are either large companies drawing on a long history in the given field and developing into more and more innovative biotechnology users, such as large pharmaceutical companies ("big pharma"). Or, modern biotechnological companies emerge, which the previously stated large companies purchase knowledge, projects or services from. Mainly the large companies control the biotechnology industry with regard to revenue. However this does not lead to strict adherence to traditions and the conservation of states of power. This is because, in terms of knowledge and the number of innovative projects, altogether small biotech companies have the comparative advantage.

Many biotech companies were founded in the 80-s. (e.g. Idec, SAFC Pharma, Enzyme Bio-Systems,Novagen). First they sought to become completely vertically integrated companies, encompassing everything from R&D to production and sales. They realised "closed innovation", only. Gradually, these companies brought new trends in their innovation strategies. At first the companies lacked two things that kept them from reaching their goals: the lack of funds, and experienced managers. However, these two things are essential (in addition to technology) for a company to grow from a spin-off enterprise to a large pharmaceutical company. The classic pharmaceutical companies, being on the top that time, already possessed these resources. Thus some of them purchased biotech companies while others however were not open to biotechnology in terms of investment and cooperation [8].

The volume and complexity of biotech and pharmaceutical projects grew in relation to the amount of available information and acquired knowledge in an environment of steadily growing needs for new knowledge. This placed further emphasis on cooperation, the sharing of costs and risks of producing new R&D results, because an industry of high risk-high benefit type emerged. This led to problems, but opportunities as well. Concerning the problems it was asked: Who will finance the costs of research? Will investors think that the industry is too risky? Naturally the significance of professional investors and specific tenders? increased with this.

Companies were forced to cooperate due to the high risk associated with biotechnology, the complexity of strategic management rules and the unusually high amount of needed funds. First of all, the necessary monetary tools are available only at the largest companies. Second, the necessary competencies are often missing with smaller companies. For example, a smaller company, a market leader in R&D, does not have the necessary experience either of the capability needed to clinical testing or production. Cooperation is necessary to fill these gaps. With this sharing of different sorts of risks will be realised. These risks, actually non-calculable uncertainties several times, may be technology, market, regulatory or competition related. The latter reflects on the segments of all the other risks, since the rapid development of China, South Korea and India. The only advantages can only be quality and knowledge for the traditional pharma producing countries. But precisely these are areas where China and India are developing rapidly, while maintaining the seemingly natural price advantage. Europe and the USA can only compete with these products if they do not count on price advantage, but on therapeutic advantage. This means producing a newer, better molecule, first of all. However this larger added intellectual value brings larger risks on behalf of technological, market and registration. These tendencies are also catalysts of cooperation.

It is precisely these different, yet interrelated risks that make pharmaceutical biotechnology complex. To successfully manage complex processes and instability necessitates cooperation. Instabilities are cross-linked, they can even strengthen or weaken each other. An example of mutual strengthening is the technological uncertainty of producing a new molecule, and the registration and legalization which follow. Registration gives the same molecule an added economical value and can, if it is registered already, decrease market instability, since it can become a market leader, a so called "blockbuster" with multi-million dollar yearly turnover.

Thus instabilities constitute a kind of synergic system. Instabilities are difficult to predict individually, their interrelations are even more so.

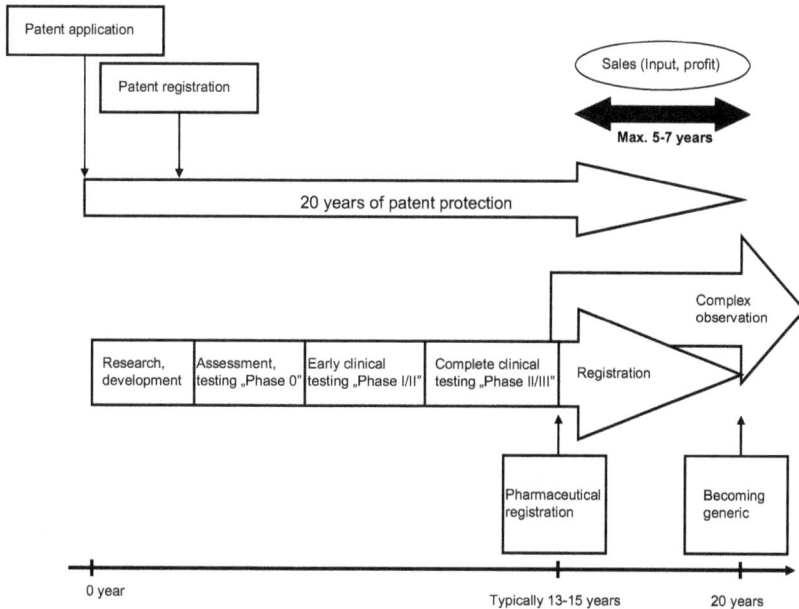

Fig. 3. [the authors]: a typical time-schedule of a new biotech identity

Necessity of cooperation can be explained from another point of view as well (figure 3). Validity period of a patent is 20 years from the date of application, which, in case of pharmaceuticals can be extended by at most 5 years (SPC). According to Figure 3, the product generally appears on the market 13-15 years after the patent application. With the end of the patent period, one must also count with the appearance of generic and biosimilar products.[2] Thus there is, at most 10, but more often only 5 years to cover the entire costs of R&D and clinical costs and make some revenue. Thus everyone seeks to make the time needed for R&D as short as possible. One method could be some sort of open innovation, which supports cooperation and outsourcing instead of solving everything in-house. /On "open innovation" you find more details in [9] and 10] /. There are numerous factors which make a part of the R&D earlier fully integrated in the vertical control target of outsourcing. To shorten the needed time to find a molecule and make it a drug, the steeply growing costs of keeping all the needed expertise within the firm, the decreasing costs of reaching the needed expertise outside, together the transaction costs arguments and the abundance of expertise outside are all for giving advantage to trust R&D tasks to outsiders who are already experts in the given field. This method definitely saves time and possibly costs as well and systematically open access to better solutions than those available in a "closed innovation" method.

---

[2] A **generic drug** (generic drugs, short: generics) is a drug which is produced and distributed without patent protection. The generic drug may still have a patent on the formulation but not on the active ingredient.
Biosimilars or follow-on biologics are terms used to describe officially approved new versions of innovator biopharmaceutical products, following patent expiry.

The consequences of uncertainty

Not all of the multinational drug companies had enough courage to apply the newest discoveries of biotechnology from their own budget. Obviously they regarded to these projects as they are too risky and for multinational and successful companies they didn't provide enough motivation to jeopardize their convenient state. But smaller enterprises, spin-off companies and biotechnological organizations must had to apply the new and more risky technology, which is based on new paradigm, because this was the only competitive edge for them against the big companies.

Until recently biotechnological companies have limited themselves to the early phase of the research and they sold their products, ideas and research results to drug companies. These biotechnological companies were quite small, and they had no possibilities to develop their own product as drog, only the "knowledge-import" were really achievable for them. Most of them became bankrupt, did not become successful, only some of them stayed alive after the initial phase.

Multinational drug companies often bought ready molecules from small biotechnology enterprises before or after the clinical phase II. With these purchase they could reduce their non-calculable risk attached to the uncertainties of R&D – although it stayed significant in this phase, too – but at the same time it caused success for small biotechnology enterprises. As a result new types of organizations appeared: e.g. contract manufacturing organizations (CMO) sites (contractual plant), contract research organizations CRO organizations (contractual research site), advisory and supply companies. This process can also interpret, which says that organizations share their risks similarly to their work and revenue.

So risks can be reduced thanks to cooperation and when risks are non-calculable, the precaution provides the other possible solution. The above mentioned risk reducing mechanisms suggest to apply the "open innovation" method, which can be read in details in the final document.

The four elements of the required framework highlight the key resources and dynamics associated with the emergence and sustainability of leading clusters in all segments of the biotechnology industry. First, as mentioned earlier, the development of biotechnology innovation requires access to specialized inputs, including researchers, risk capital, biological materials, and even intellectual property. By and large, accessing these resources is most easily accomplished within a regional context, rather than across long distances or political boundaries. For example, the development of the agricultural biotechnology cluster surrounding St. Louis depended on the ability of companies such as Monsanto to draw upon and reinforce the significant expertise and research capabilities of Washington University in St. Louis.

Second, a key driver of effective clustering in the biotechnology sector seems to be competition among locally based biotechnology companies. These companies compete on the basis of attracting talent, publishing high-quality scientific research, and attracting investment and interest from venture capitalists and downstream commercial partners, many of whom are located outside the cluster. This is perhaps most apparent in some of the clusters associated with health-oriented biotechnology; for example, the Massachusetts biotechnology cluster includes more than 400 different firms, 235 of which are developing therapeutic drugs [11].

Third, most leading biotechnology clusters are located not only near sources of high-quality basic research but also around areas with significant capacity in clinical innovation. For example, the pressures on the Massachusetts biotechnology cluster arise as much from the presence of demanding clinicians in the leading hospitals as from that of specialized genetics researchers. Similarly, the medical device cluster in Minneapolis is pushed by demanding consumers at the Mayo Clinic and related institutions, and industrial biotechnology innovation in Scandanavia depends in part on demanding customers in the chemical industry [12].

Finally, the biotechnology cluster depends on the presence of related and supporting industries, most notably an active venture capital industry to supply managerial expertise, risk capital, and relationship experience with downstream partners as well as key pieces of infrastructure (e.g., biological resource centers, specialized seed banks and agricultural research stations, specialized equipment and tools). Each of these factors encourages the investment of sunken assets and the development of specialized capabilities that reinforce the strength and ultimately the international competitiveness of that cluster environment.

While the United States remains the largest single national home for biotechnology activity, it is useful to note that the EU actually accounts for a greater number of companies than the United States. [13]. Along with the earlier employment statistics, this suggests that individual EU biotechnology companies have fewer employees (on average) than their U.S. counterparts. Simply put, this means that the scale of operations for a typical EU biotechnology firm is smaller than that of a biotechnology firm in the United States.

Furthermore, the European biotechnology companies seem to grow more slowly than their U.S. counterparts. By and large, young European firms are often overtaken by international competitors and even some of the oldest European biotechnology companies have been acquired by U.S. companies that have better access to financial and commercialization resources [8]. As in the employment statistics, this concentration of small companies seems to reflect the international distribution of employment activities.

This central insight—an increase in the number of regional innovation clusters, rather than a simple dispersion of biotechnology activity—holds several important implications for (1) evaluating the global biotechnology industry going forward and (2) developing effective policy to ensure continued U.S. leadership in this area.

First, some analysis suggests that the impact of globalization on biotechnology innovation seems to be different than that of traditional manufacturing sectors, such as the automobile industry or the IT sector. Specifically, the globalization of other industries reflects the increasing availability of low-cost locations to conduct activities that previously had been done in the United States. In contrast, the globalization of biotechnology reflects a "catching up" process by a small number of regions around the world that seek to compete head-to-head with leading regions in the United States.

Second, it is important to account for the range of activities now included within the biotechnology industry, including diverse applications in the life sciences, agriculture, and industry. Although most discussion focuses on life sciences—which remains the largest single segment of biotechnology in terms of employment, enterprises, investment, and patenting—the globalization of biotechnology is occurring most rapidly in industrial applications. Moreover, although the United States continues its historical advantage in agricultural applications, this may be due to political resistance in Europe and other regions rather than the presence of strong agglomeration economies within the United States. For example, the presence of extremely strong clusters with a high level of entrepreneurship that characterizes life sciences biotechnology seems to be a bit less salient for agricultural applications. The presence of multiple industrial segments—each of which is associated with distinct locational dynamics—raises the possibility that, even as individual clusters become more important within each application area, the total number of global clusters may increase with the range of applications.

Third, at least in terms of the available data, the United States maintains a very strong, even dominant, position within biotechnology. While some conceptual frameworks (e.g., the convergence effect) would suggest that early leadership by the United States would have been followed by a more even global distribution of biotechnology innovation, the "gap" between the United States and the rest of the world has remained relatively constant over the past decade or so. Indeed, itis likely that the United States has a historic opportunity to establish a long-term position as a global hub for biotechnology innovation, particularly in the life sciences and agricultural areas. In contrast to traditional debates about outsourcing, it is possible that increased global activity in biotechnology can complement rather than substitute for U.S. investment, employment, and innovation.

Finally, our analysis highlights the small size (in terms of absolute levels of employment) of the biotechnology industry. While industries such as IT may plausibly be associated with a large impact on the total workforces of individual states and regions, total employment in biotechnology is very small, although associated with very high average wages. The simple fact is that, if the biotechnology industry remains at roughly the same scale that it has achieved over the past decade or so, it is unlikely to be a major driver of employment patterns and overall job growth, either in the United States or abroad.

Trends in the Pharmaceutical industry[3]

In terms of individual pharmaceutical trends, there are many cited novel commercial models and the rising importance of emerging markets as the most promising. Further consolidation through **Mergers, Acquisitions** and alliances and partnerships made it a close third. Difficult market access and reimbursement were named as the biggest risks, along with pricing pressure and general cost containment. Novel commercial models have been an industry issue for a while. Not only have new stakeholders, such as payors and patients, gained influence. On a different note, many managers believe that by rethinking traditional models, corporations could improve their image.

The rising importance of emerging markets is reflected by a number of developments. For example, in 2008, GlaxoSmithKline established an Emerging Markets region and appointed its President to the Corporate Executive Team. Contrary to mature markets, the middle class in such regions has increased its purchasing power. Furthermore, the public provision of healthcare is improving. Yet some stumbling blocks remain, such as the issue of liberalization in Russia or the need for better protection of intellectual property in India. However, in summary, there is no doubt that emerging markets will provide a key growth

---

[3] We rely in this chaper on the [14] heavily.

engine in the midto long-term. In the meantime, the BRIC countries allow the industry to learn in a "non-traditional", much more consumer-driven environment.

The challenges ahead[4]

Market access and reimbursement have emerged as top management issues. According to our survey, generating demand with physicians is no longer sufficient. This can be traced back to the increasing hurdles related to reimbursement. Hence, many managers expect that the trend, not to launch products in certain markets, will accelerate – as could already be witnessed in the UK, Germany and France.

As with reimbursement, general cost containment has also resulted in significant pricing pressure across most markets. A major driver of this development is the discounts which are granted to payors. Moreover, due to fragmented budgets and decision-making, total costs are not relevant enough: It is the price of the product which counts.

Most executives concur that R&D productivity remains a key challenge. Some are even convinced that this is the underlying issue for all of the industry's problems. For one, costs are on a steady rise. Yet, due to poor clinical trial results and higher regulatory hurdles – which have increased costs by 50% and more – the number of approvals cannot keep pace.

"The industry needs to apply a model which is less fragmented and much more entrepreneurial", said one top manager. The issue of insufficient intellectual property protection earned mixed reactions. While some managers believe that it could challenge the existence of the entire industry, others are not as pessimistic. Some pointed out that, should patent protection fail, R&D expenses could be reduced by two thirds. Others see the matter as a call for action: "The industry should stop fighting for patent protection and learn to create protected market situations using different instruments, such as brands or customer loyalty."

The industry is reviewing its commercial model. While corporations are driven by the wish to better cope with changing customer structure and become more cost-effective, they are simultaneously investing in services. At this point in time, however, this is seen as an effort to maintain customer access and loyalty, rather than as a contribution to revenue and profit.

Even in today's challenging economic environment, the pharmaceutical industry can still be considered an industry with good long-term prospects. Expanding aging populations, increased wealth in emerging markets and unmet medical needs, accompanied by rapid technological progress, are fueling the demand for innovative drugs – and will continue to do so in the future.

Product innovation and patents – formerly the driving forces behind the industry – have lost momentum. After years of high growth for shareholders in the 1980s and 90s, significant value has been destroyed since the turn of the century. The old blockbuster business model has lost its appeal. The pipeline has dried up and the number of commercially viable candidates is down. Companies show limited willingness to have a large share of their sales depend on just a few products. Health systems are challenging the highmargin business

---

[4] We rely in this chaper on the [15] heavily.

model of the industry, primarily by questioning the value contribution of "pseudo-innovations" and "me-too" products.

Fierce competition from generics and a growing focus on price in tender business. The first step for executives in pharmaceutical companies is to review which therapeutic areas (TAs) they are currently active in and decide whether these are really the most promising ones. What are the TAs with the highest potential in terms of revenue growth and profitability in the next five years:

- Oncology
- CNS
- Cardiovascular
- Vaccines
- Diabetes
- Immunology

For many years the typical product manufactured by the pharmaceutical industry was the pill or capsule containing small molecules surrounded by galenic technology. This has changed dramatically in recent years and more changes are on the way. We asked participants in the survey what type of physical pharma product they thought would gain most in importance in the coming five years. Here's what they said:

- Biologicals will be the strongest growth drivers in the next five years (49% of respondents)
- Combinations of pharmaceuticals and diagnostics were ranked second (30%)
- Small molecules came third (9%) – a dramatic drop in the ranking from previous years
- Cell-based therapeutics came a close fourth (7%)

This shift in product types is expected to have a major impact on the pharmaceutical value chain. Thus, pharmaceutical companies seeking approval for expensive biologicals will need to clearly demonstrate the additional benefit of their drugs in order to ensure reimbursement. Indeed, in the survey, 74% of respondents considered reimbursement and market access the biggest challenges faced in the pharma value chain. Demographic change and technological advance are driving the demand for pharmaceutical products. But most pharmaceutical companies operate in markets which are not liberalized. In such markets, prices are not the result of supply and demand but of restrictive governmental healthcare systems that limit market growth. In light of the financial crisis and the resulting large fiscal debts, growth is set to slow even further. We therefore asked the participants in our survey to name the financial source that they thought would fuel the growth of the R&D-based pharma business model in the coming five years.

The majority of respondents in the survey (78%) said that the first step would be to improve the personal or "soft" skills of their employees. Specialist expertise is also seen as major challenge by respondents working in the area of R&D. One such respondent commented as follows: "It is not only about those soft skills. You need the top people with the top specialist know-how for those TAs you want to play in." To achieve cultural excellence, pharma executives intend to focus on three levers: (1) gaining access to the best talent; (2) fostering entrepreneurship rather than bureaucracy; and (3) focusing on the scientific culture in R&D, incentivizing employees accordingly. We will discuss each of these levers in turn below.

3. Risks and uncertainties

Calculating risk needs on the most basic level knowledge of the issues, the variables, damages or benefits, and knowledge of the occurrence of their probability. Andrew Stirling [16] elegantly demonstrates how with lessening of knowledge of the variables, that can occur as impossibility of setting just one model of the issues, because of value differences in indication what the important problem to choose is, and parallel with this, lessening knowledge of probabilities separates different areas of uncertainty.

Figure 4 shows the risk sharing process. By their nature, completely new biotechnological projects, aiming at radical innovation, originally fall into the "suicide box". They are characterized by high market and technological uncertainty. In a given situation, a small biotech company, since it has no other choice, working out the right technology, sells it to the larger pharmaceutical company. From the point of view of the big company, the technological uncertainty is reduced considerably, since it is purchasing a technology that has been proven to work. (The technology is over the proof of the concept phase) The market uncertainty remains now to solve, which can be assessed and estimated by the purchaser. Another extreme case is when a small innovative company tries to become a supplier for one of the large market players. Trying to meet its needs, perhaps even relocating closer to the purchaser, is thus reducing market uncertainty for both parties. Thus the reason for cooperation is to decrease at least one, but preferably both (marked by dashed arrow) uncertainties. By sharing the associated risks, the organizations will not be able to reach the small innovation level, as this is not the goal of the cooperation. But at least they can decrease risk somewhat.

The not always appropriate knowledge of the events causes an enormous problem for biotechnology. For example in drug manufacturing with genetically modified organisms, in comparison to traditional pharmaceuticals new problems appear, such as the social and environmental acceptance of the technology that can be doubtful. This generalized the judgement of the work with recombinant organisms, independently of the fact that the organisms are isolated when they are in use. However those "classic" events, which have influence on the behaviour of the pharmaceutical manufacturing, for example the cartel of the competitors and the status of the industrial property, have played a role yet. The appearance of biogeneric or biosimilar molecules, which generate further uncertainty (n-dimension uncertainty) in the field of regulation and licensing, is a new and not predicted problem. All this indicates that the actors in the red biotechnology are often unable to set risk calculations, first because unpredictable variables emerge, crop up.

The probabilities are the other problem. There are fields, where the probabilities can be estimated relatively easily (e.g. industrial property, technician feasibility), but elsewhere it is a really hard work (e.g. modification of regulation, variability of marketability, price or supply and demand). The latter cases cannot be generalized from the classical examples of the pharmaceutical or chemical industry, because they are biotech-specific. On the whole, there are also probabilities which can be regarded as unknowns. The most difficult thing is to take into consideration the hazard of the unknown processes before decision. So in the case of a radical biotechnological innovation we can talk about ignorance, the field of the real surprise in the figure.

Technology uncertainty

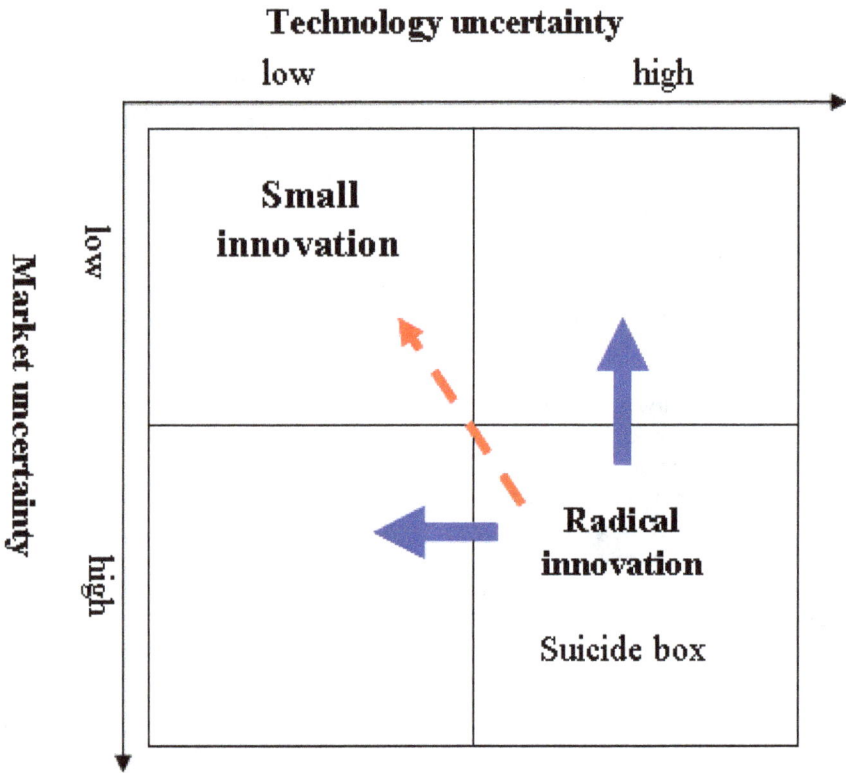

Fig. 4. The risk sharing process, based on[16] (modified by the author)

Hence we can regard that sooner or later in the field of the drastic decrease of the knowledge we can not only talk about uncertainty but its extreme case, ignorance, because we don't know or cannot know about neither the probability of the occurrence, nor the existence of the forthcoming events [17].

Breakthrough, radical innovations are created under circumstances which lead to genuine surprise. They necessarily imply essential previous ignorance, and result in genuine surprise. To different extents, this is the definition of radical innovations. That is the reason why managing radical innovation, the uncertainty and the unknown, the sphere of ignorance, has a consequence. That is that some sort of trial and error approach is a key issue, so any recognition of some, even very weak paternisation, regularity can be enormous comparative advantage. With this we acquire some plausible knowledge about a part of the "previously unknowable" while taking certain interrelations into account. It can be stated that during the evaluation of an uncertain situation partly the "I know that I know" problems should be handled. In this case a deterministic, at least a probability based answer can be provided for. Other problems are of an "I know that I don't know" nature (I know the phenomenon but I don't know the possibility of the occurrence). In this case we can provide plausible answers. The third type is the "I don't know that I don't know" problem, meaning ignorance, already a challenge that both even events and effects are unknown. These are extreme situations e.g. when extreme security requirements are to be realised or outstandingly high profit is searched for. In permanently turbulent environments these questions become the natural questions. [18]

Two types of problems emerge in very uncertain situations. First, we literally do not know what can happen (for example by the synergistic effects of known factors and what can the effect be) and second, we do not know the frequency of what happens either. It is most important to see that the main problem with these types of issues is not the calculability with prognostic aim. The problem is the lack of knowledge what can occur at all, the so called lack of ability of modelling.

In terms of biotechnology, uncertainty in the progressing realization of some innovation can be understood more as "ignorance" or "real surprise" for a while. With the accelerating development of biotech industry the domain of "we don't know what we don't know", the range of insufficiently known events and distributions, 'original surprises' is becoming increasingly important, in most cases also accented by irreversibility [19]. Fuzzy sets considerations can only be part of the solution for these types of issues. At their border "ignorance" is impossible not to take into account if there are reasons that the turbulence is very high.

The lack of the possibility to make reliable risk calculations poses a problem. For example, when looking at the production of pharmaceuticals through genetically modified organisms, producers have been faced by problems such of the societal acceptance of the questionable health and environmental consequences of the technology. These concerns result in an overall negative judgment of all technology using recombinant organisms, regardless of their isolation during use. At the same time, "classic" events influencing pharmaceutical production still play a role, such as the merging and cooperation of concurrent companies in the background, as well as the state of industrial rights protection. A new and non-foreseeable problem is the appearance of biogeneric / biosimilar molecules, which generate further uncertainty in the fields of regulation and registration (n-th dimension uncertainty).

Based on this we can ascertain that in case of the drastic reduction of knowledge, sooner or later we can begin talking about lack of knowledge, ignorance instead of uncertainty, since we do not have, in certain cases, we cannot have any information regarding the events to come, not just the frequency of their occurrence [20]. The trivial consequence for action is than that it is wortwhile to prepare to accommodate, as a very basic element of the strategy of firms, to new situations occurring as consequences of genuine surprises.

But to provide approapriate knowledge base for any risky situation, notwithstanding that they are calculable or not, is not enough to take into account risk/uncertainty facts, only. When dealing with the role of uncertainty for decision making mostly it is too much told about the risk facts, the analytical level. To make decision, action conclusions leads to empirical fallacy, a misbelief that facts alone can lead to decisions. But decision making unavoidably includes risk evaluation too. It is unfortunate that the risk assessment literature neglects this layer. Actually, it is a hidden assumption that everybody naturally makes the same evaluation of the same facts but this is a simple error. Following Schwartz and Thompson, and utilizing ideas from lectures of Imre Hronszky and the PhD Dissertation of Ágnes Fésüs we shortly demonstrate what with risk evaluation the issue is. Actually at risk evaluation for basic types of evaluating perspectives can be utilized. The Figure 5 below demonstrates them.

Types of nature represented by different potential curves accepted by the four different types of agents in society, according to Schwarz and Thompson:

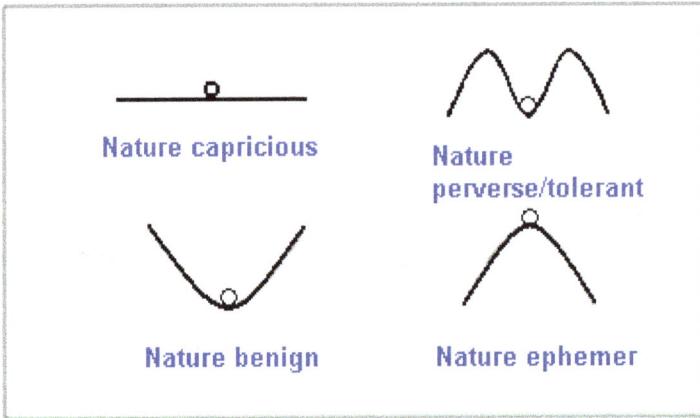

Fig. 5. Four different types of agents in society, by Schwarz and Thompson based on [17]

On risks and uncertainties

The assessment of risks (uncertainties) in the technological development is becoming an increasingly difficult task to solve. This is especially true in a rapidly changing turbulent environment, where environment and its knowledge changes from day to day, where in addition to small ones, radical innovations are typical as well. Understanding the necessary parameters is becoming more and more uncertain, thus also becoming limited. There is a huge literature on technological uncertainty just as there is on risks of financial issues.

Two types of problems emerge in very uncertain situations. First, we literally do not know what can happen (for example by the synergistic effects of known factors and what can the effect be) and second, we do not know the frequency of what happens either. It is most important to see that the main problem with these types of issues is not the calculability with prognostic aim. The problem is the lack of knowledge what can occur at all, the so called lack of ability of modelling.

The classical scientific assessment of uncertainty is the quantitative risk assessment (qRA). If you know the damaging events and their frequencies you certainly can make prognostic calculations, too. QRA has a quite long success story in modernity. But, for its reductive nature, that we have to know these very basic preconditions, it is with ignorance ('deep uncertainty') hopelessly challenged in strongly turbulent issues and with basic lacks in knowledge. These types of problems will be more and more often.

Understanding uncertainty in modernity has developed as a progressing capability of calculation of risks, the quantitative risk assessment (qRA). As its essential it is calculation of the probabilities of occurrence of some known set of events and multiplying these probabilities with the possible damages the set of events may cause. This way the qRA aims at the best possible prediction of risks. This approach includes into the (probabilistic) deterministic planning and command and control regulatory approach. All these issues, qRA, deterministic planning, command and control approaches require the existence of quite strong preconditions (as, concerning qRA I indicated above). These strong preconditions can up to a grade be weakened. For example fuzzy set techniques can be included concerning the estimations of events.

It is important to see at least in outline how uncertainty of events and likelihoods and the plurality of the values as a societal fact for democracies have challenged the modernist approach to uncertainty. Andy Stirling, in line with some other authors, elegantly summarizes the basic problematic of quantitative risk approach as providing for a reductive-aggregative way to interpreting uncertainty [20]. One can speak about reductivity in the meaning that the classical risk research reduces its interest in calculable risk. This could be done, concerning the production dimension, in time of mass production. Then long periods were stable and made the prognostic effort rather successful. We can make rational suppositions on the basic preconditions of risk calculation in such periods. Additionally, a lot of efforts have been made by practitioners in risk research to find methods to successfully assess situations when quantitative risk calculation ceases to work exactly. This applies both for making conjectures about the existence of events as well as their probabilities.

It is just a platitude to say that knowledge in any real case is incomplete in most decision situations. Important is what sorts of incompleteness are or/and are to be recognized. From the quantitative risk perspective, to be able to function, we should be able to identify the events that should be taken into account and should be able to attribute values to probabilities at least as subjective guesses. But one can also consider a further case when this does not work with either the events or the probabilities. Instead the earlier cases even strong 'surprises' may occur. History helps us to learn on empirical base that these sorts of cases, issues and effects of 'unknown unknowns' are real cases. These can occur without any human interaction or as results of them. It is possible to argue that high complexity of the issues provides for frequent occurrence of 'surprises'. All the radical innovations are in their realisation process for a while 'surprises', too.

On the precautionary approach

All this is connected to some recognition of the nature of uncertainty of the processes and the appropriate management efforts. The new permanent and decisive challenges are decision making and action under the pressure of comprehensive and irreparable information uncertainty in a world of 'ontological uncertainties', as this is typically formulated.

Risk management based on quantitative risk assessment does not help much in these issues. The reason is that both the needed type of assessment and the controllability of the process too are realisable in a limited way only. Risk assessment has to draw back to the second line, to the efforts made for isolation and control of special issues. But this should not open the way to nihilism. Instead, some sort of precautionary approach is possible to unify as we try to emphasize the commitment of the actor both for utilising suddenly appearing chances and avoiding hazards. While risk calculation, the extension of the calculative rationality attitude to uncertain situations, in principle promised the certainty of exact calculation of the one best solution (in an ideal world where neither the unavoidable multiplicity of values, the 'combinatorial complexity' neither real complexity disturbs) rational management of complex processes has to be satisfied with calculations unavoidably leaving essential place for uncertainty. This is where value choices have their structural place. With this we already want to call attention to the issue that it is reasonable to think of a continuum of approaches, made of different combinations of courage and drawing back. This continuum of behaviour reflects the variability of human agents' relation to the uncertainty as a world of chance and hazard.

Modernity first had success with mathematically handling deterministic issues. Then it went further and had immense success in handling probabilistically deterministic issues. As its pair in management caution and prevention may be seen as key categories developed by modernity in relation to mastering negative effects. This is mastering by and preventing based on (in principle exact) understanding of the probability of causal mechanisms. In this latter case it is risk that comprises the relation of modernity to uncertainty: uncertainty can in principle be bounded in exact calculus. As for exchange there is no place for surprise in issues where quantitative risk assessment is valid, as Frank Knight sharply recognized in 1916. This is to set against a 'post'-modern 'world' in which interactions and 'deep uncertainty' assessments get the supreme position. The realised rationality by modernity is calculation of isolated issues as exactly as it can be done, with the result of possible exact prediction of the effects. Its action part is to come to term with the outcomes the probabilities of which are calculated. This rationality of more and more exact calculation is the basis for acting through deterministically planning that is based on evaluating the realisability of probabilistically predicted positive and avoidability of the negative effects. It is oriented toward exactness in quantity. That means that the methodological effort can concentrate on identifying the degree of risk. Concerning the future, this open space of unknown, risk calculation commits itself to understand future through extrapolation based on some continuity. Estimated uncertainties in the future are compared with the known risks recently and these are hypothetically extended to trends into the future. (With this there is a, not always conscientious reductive presumption made on the type and 'measure' of novelty. Because the induction problem is unsolvable in principle trend extrapolation has to be decisionistically accepted. It will or won't be rejected. At some point we do not reject anymore as Bernstein [20] correctly recognized the extrapolative guess as measure for what can be novel in the future. There are two different basic types of practices to realise this non-rejection. We may make it because it is acceptable for modelling or for practical reasons in the real practice.)

Risks appear this way as if they were quasi-natural variables, objective, repeatable, and measurable in standardised situations. But we know that risks are social-natural variables, damages are damages in relation to some values, only. So, even when they seem to be quasi-natural variables, a justified multiplicity of risks can be identified around the same issue, expressing the (often conflicting) relations of interested groups to the risky issue. This is the mentioned ambiguity of the valuing relation. Life is obviously more complex than any of the leading ideas on which actions can be based on their purified forms. Modernity also experienced that values are irreducibly multiple in democratic societies and that uncertainties around the data and the models definitely hinder the unambiguity quantitative exactness risk management in its ideal form requires. The unceasable value plurality, as characteristic for the manifoldness of the social processes and a democratic society together is a strong challenge to classical risk assessment. It challenges at least its claimed punctual exactness, its capability to provide for any unique 'objective standard' for decision making. Risk assessment is unavoidably based on some 'subjective' framing. From the endless many perspectives one will be chosen to serve for realising a qRA process. One has further to see that risk is not only social-natural variable because it expresses a special evaluative relation (damage) to some object but risk is also constructed in reality: one can rightfully, strictly speak of quantitative risks of an action in standardized systems, only. So, either the situation is really standardised as far as possible in the real practice or is identified so as it really would be standardised.

4. Conclusion

Pharmaceuticals is a highly globalized industry, dominated by multinational companies that engage in significant business activity in many countries and whose products are distributed and marketed worldwide. Historically, the industry has been dominated by vertically integrated firms performing almost all of the activities in the value chain by the firm itself, from basic research through to sales and marketing. They realised some sort of "closed innovation", as Henry Chesbrough introduced the term. [12]

In recent decades the industry has undergone dramatic structural changes, with the rise of the biotechnology sector, substantial growth in demand driven by demographics, substitution away from other therapeutic modalities such as surgery, and increased competition from globally active generic manufacturers

Pharmaceutical biotechnology, just like other dynamically growing branches of industry, has been very rapidly changing. Disruptive innovations arise from time to time. Since this is a very high risk - high benefit industry, and already R&D phases often require several hundred millions of dollars, the participants seek to minimize and share risk, more precisely the uncertainties, not always calculable as quantitative risks.

In case of biotechnology the assessment of risks (uncertainties) in the technological development is becoming an increasingly difficult task to solve. This is especially true in a rapidly changing turbulent societal, economic, political, ideological environment, where that environment and knowledge of it changes from day to day, where, in addition to small ones, radical innovations are not seldom but typical as well. Understanding the necessary parameters is becoming more and more uncertain, thus also becoming profoundly limited, just as with the whole dynamics.

5. References

[1] Domonkos, D.: *The opportunities of the spread of innovation in the biotechnology industry,* Technical Chemical Days, 2006, Veszprém. Presentation (In Hungarian: *Az*

innováció terjedésének lehetőségei a biotechnológiai iparban, Műszaki Kémiai Napok, 2006, Veszprém. Előadás)

[2] Cortright, J., and H. Mayer. (2002). *Signs of Life: The Growth of Biotechnology Centers in the U.S.* Washington, D.C.: Brookings Institution Press.

[3] Feldman, M., and J. Francis. (2003). *Fortune favours the prepared region: The case of entrepreneurship and the capitol region biotechnology cluster.* European Planning Studies *11(7):765-788.*

[4] Frigyesi V.: *The economic preconditions for the development of biotechnology in Hungary.* Research Institute of Industrial Economics of the Hungarian Academy of Sciences, (1988) – 52p

[5] Critical, I. (2006). *Biotechnology in Europe: 2006 Comparative Study.* EuropaBio, The European Association for Bioindustries.

[6] Schmidt, F.R.: *Recombinant expression systems in the pharmaceutical industry*, Appl. Microbiol. Biotechnol. (2004) 65: 363-372

[7] Tuft Center for the Study of Drug Development (TCSDD): *Number of mAbs entering clinival study nearly tripled in last decad.* March/April 2008, Vol.10, No.2

[8] Murray, F.: *Innovation as co-evolution of scientific and technological networks: exploring tissue engineering*, Research Policy 31 (2002) 1389-1403

[9] Chesbrough, H.: *Open Innovation*, Harvard Business School Press, Boston 2003

[10] Chesbrough, H: *Open Services Innovation, Rethinking Your Business to Grow and Compete in a New Era*, Jossey-Bass, San Francisco 2011

[11] Massachusetts Biotechnology Council. (2007). *Massachusetts Biotechnology Company Directory.* Available at
http://massbio.org/directory/statistics/stats_comp_yrfound.html.
Accessed December 19, 2007.

[12] Hermans, R., M. Kulvik, and A.-J. Tahvanainen. (2006). The biotechnology industry in Finland. In *Sustainable Biotechnology Development – New Insights into Finland*, R. Hermans and M. Kulvik, eds. ETLA, The Research Institute of the Finnish Economy, Series B 217.

[13] The BIO report: *State Bioscience Initiatives (2004)*, www.bio.org/local/battelle2004/

[14] Macher, J.T- Mowery, D.C.: Jeffrey, T. Macher and David C. Mowery (Editors), (2008): *Innovation in Global Industries: U.S. Firms Competing in a New World (Collected Studies) Chapter 6 and 7.* The National Academies Press, Washington, D.C.

[15] Danner, S., Ruzicic, A., Biecheler, P. (2008): *Pharma at the crossroads.* Roland Berger Startegy consultants, www.rolandberger.com

[16] Stirling, A.: Precaution, Foresight and Sustainability: *Reflection and Reflexivity in the Governance of Science and Technology. Reflexive Governance for Sustainable Development.* J.-P. Voß, D. Bauknecht and R. Kemp. Cheltenham, Edward Elgar Publ: 225-272., 2006.

[17] Schwarz, M., Thompson M. *Divided we stand : redefining politics, technology, and social choice.* Philadelphia, University of Pennsylvania Press, 1990.

[18] Hronszky I. - Várkonyi L.: *Managing Radical Innovation, Harvard Business Manager* Review, 2006. 10. szám, 28-41./*In Hungarian: Radikális innovációk menedzselése /*

[19] Cooke, P: Biotechnology clusters, *'Big Pharma' and the knowledge-driven economy*, Int. J. Technology Management, Vol. 25, Nos. 1/2, 2003

[20] Bernstein, P. (1996): *Against the gods: the remarkable story of risk.* New York; Chichester, Wiley.

7

Performance Evaluation for Knowledge Transfer Organizations: Best European Practices and a Conceptual Framework

Anna Comacchio and Sara Bonesso
Ca' Foscari University of Venice
Italy

1. Introduction

The importance of Knowledge Transfer Organizations (KTOs) for boosting innovative performance both at regional and firm level has been highlighted by literature and empirical research (Kodama, 2008; Laranja, 2009; Muller & Zenker, 2001; Muscio, 2010; Tether &Tajar, 2008). KTOs encompass a set of diversified institutions, both public and private in nature, such as science parks, incubators, business innovation centres (BICs), industrial liaison offices (European Commission, 2004; Reisman, 2005). Their mission is to be providers of knowledge intensive services to firms-receivers in the different phases of their innovation process (Howells, 2006) as well as to be part of a Knowledge Transfer (KT) infrastructure which promotes and facilitates networking activities between companies and public or private research institutions. Due to the increasing diffusion of KTOs operating in a regional innovation system and the variety of services provided, performance evaluation of these organizations is becoming paramount from different viewpoints. First, a measurement system by which different actors may gather performance information could help to overcome one of the main difficulties of creating a market for technological knowledge (Arora et al., 2001; Arora & Gambardella, 2010; Decter et al., 2007; Dosi et al., 2006; Lichtenthaler & Ernst, 2007), which is information asymmetry. Second, from the demand side, firms-receivers require a univocal method to compare and evaluate the offer of the different KTOs. Third, even KTOs need a performance measurement system on which they can rely to define their product/service portfolio and craft their competitive strategy at regional, national and international level. Finally, also local and regional institutions need to assess KTOs in order to define innovation policies and to allocate resources effectively. Despite the increasing need for measuring the effectiveness of KTOs, a still limited effort has been made by research to develop a performance measurement system based on a robust methodological framework. Approaches implemented by institutions and KTO associations like the IASP (International Association of Science Parks) or the European BIC Network, are based on multiple measures encompassing financial and economic metrics (for example the amount of investment made, the turnover generated, return on asset and return on equity), output indicators of the technology transfer process (for example the number of collaborative research agreements stipulated, the number of licenses executed or the number of spin-offs established) and input measures (such as physical space available, amount of staff expenses for HR development and number of research partners). Most of these approaches are not fully developed and show several drawbacks. For example, while measurement approaches allow the evaluation of a specific KTO in comparison with those of the same type (e.g. science park, BICs, incubators, industrial liaison offices), they do not provide stakeholders (promoters, political institutions, clients) with a commensurable 'benchmarking' of a same service provided by different KTOs (for instance the incubation service provided by a science park or by a BIC). Moreover, methods are still fragmented and based on different approaches.

The aim of this study is twofold. First, we wish to provide an in-depth review of the extant literature on knowledge transfer evaluation and a comparison among different measurement approaches adopted in outstanding European KTOs. Second, we want to elaborate an analytical and integrated model that makes it possible to monitor and compare the performance of a single KTO over time and against other KTOs.

The chapter is structured as follows. The next section highlights the growing relevance of KTOs for the competitiveness of regions and firms. The third section discusses the causes of the complexity of the evaluation of KTOs' performance (Bigliardi et al., 2006; Gardner et al., 2010) and presents a review of the different approaches to KT measurement diffused in the European context (Autio & Laamanen, 1995; European Commission, 2009; Guy, 1996; Hogan, 1996; Samtani et al., 2008). In the fourth section we develop an integrated analytical model for KT performance measurement bridging two main perspectives: the literature on Balanced Scorecard (BSC) management system (Kaplan & Norton, 1992; 2004; 2007; Kaplan et al., 2010) and the studies on the innovation value chain (Hansen & Birkinshaw, 2007; Roper et al., 2008). For the first time, the proposed model combines the different approaches and metrics for KT transfer measurement, analysed in the third section, within the comprehensive framework of BSC tailored according to the complexity and the specificities of KTO management and performance. Moreover, the model, through the proposed perspective of KT processes, makes it possible to position and assess a KTO according to the different phases of the innovation value chain covered by its services. Lastly, the chapter concludes with policy and managerial implications concerning the implementation of the proposed model.

The contribution of the chapter is twofold. It addresses a key issue of technological innovation management in a time of open innovation, that is knowledge transfer at both micro and macro level, and it provides an original analytical framework of performance measurement, grounded on extensive literature review and case analysis.

2. The growing market for research and knowledge transfer services

2.1 Introduction

The knowledge transfer process has been defined as the "intentional, goal oriented interaction between two or more persons, groups or organizations in order to exchange technological knowledge and/or artefacts and rights" (Amesse & Cohendet, 2001: 1459-1460).

The inherent difficulty of this type of process was the main driver of the birth of KT providers, such as science parks, whose main aim was to facilitate the access and exploitation of scientific knowledge by companies.

The key role played by science parks at the inception of knowledge transfer from university to industry was that of a catalyst, providing a location for an integrated interaction and cross fertilization (Bigliardi et al., 2006). Besides science park, the supply side of market of technology has been enriched by the birth and the entrance of a large set of different institutions (European Commission, 2004; Geuna & Muscio, 2009).

Their role has been changing over the last few decades and the set of services provided nowadays has widen due to related processes such as the quick pace of university research, its increasing convergence, the diffusion of an open innovation approach that intensified the complexity of the innovation value chain (Hansen & Birkinshaw, 2007) of the companies-receivers and their need for external competences and resources. The following two sections will briefly analyse the market of KT services and its recent evolutions, focusing on both the demand side of the firms-receivers and the supply side of the providers. In Section 2.2 we will discuss firms' emerging needs for KTO services at each stage of the innovation value chain; Section 2.3 will present the broadening set of services offered by KTOs and the wide range of organizations operating as knowledge providers. In the conclusions some preliminary implications for the performance evaluation of Knowledge Transfer Organizations will be drawn.

2.2 The emerging demand for KT services: Firm's needs and the innovation value chain

The evolution of technology transfer activities and of the role of KTOs can be understood by looking at the demand side of the market for KT services and considering the emerging challenges that innovative firms face as a driver of the quest for outside specialized partners-providers. KTOs act by complementing and stimulating firms' internal innovation processes and capabilities, which nowadays are not sufficient, in small as well as in large companies, to deal with scientific, technological and market changes and opportunities.

The expanding demand for KT services is related to macro challenges already highlighted by research on open innovation (Chesbrough, 2003; Gassman, 2006): globalization and its consequences in terms of economies of scale and time-to-market shrinking; technology intensity and the difficulties that even large companies have in coping with it; technology fusion and more interdisciplinary cross-border research; new business opportunities and the benefits of complementary partnerships, and finally the relevance of knowledge leveraging and market for ideas that promotes knowledge brokers.

At firm level, organizations are challenged to build up stronger and more efficient innovative processes, fostering not only the central activity of project development but also improving their overall innovation chain, from idea generation to its delivery to clients. This end-to-end approach to manage innovation processes, based on the comprehensive framework of innovation value chain, should guide not only managers, as already suggested by scholars (Hansen & Birkinshaw, 2007; Roper et al., 2008), but also KT providers willing to assist them in fostering innovation.

Thus, emerging needs for KTO services can be analysed, according to the innovation value chain model, by breaking down the innovation process into the following phases: knowledge generation, transformation and exploitation.

In the early stage of *knowledge generation*, firms are engaged in activities of monitoring scientific advancement and scouting for new ideas that are increasingly complex and risky.

First, according to the OECD, scientific progress is driven by the convergence of research fields. The interaction between some research disciplines, such as physics and chemistry, may lead to new research areas; moreover, where this interaction is not yet strong enough "space between fields may become the ground for a new area" OECD (2010). Second, due to technological convergence (Daim et al., 2009; Mendonça, 2009; OECD, 2010) that more recently affects not only high-tech companies but also medium and low-tech ones, new product or service concepts are likely to emerge at the intersection of different sectors. Consequently, firms need KTOs' assistance to analyse in-depth their own and other industries' state-of-the-art and trends, to forecast technological scenarios and scan patents data sets in search of ideas that are carrier of radical innovations and thus far from their in-house prior knowledge. Indeed, the value of new and distant fields is difficult to evaluate in itself and in relation to a firm's technological requirements; consequently there might be a need for complementary services of demand articulation and of semantic translation of domain specific knowledge into a language closer to firms' communication codes (Bessant & Rush, 1995; Carlile, 2004; Gassmann et al., 2011; Hagardon & Sutton, 1997; Howells, 2006; McEvily & Zaheer, 1999). Moreover, new business opportunities might arise outside organizational boundaries; therefore there is an increasing call for activities of business intelligence and validation of new business initiatives, especially in case of the start-up of a new company.

The second phase of the innovation value chain, *knowledge transformation*, is the stage during which "ideas must be turned into revenue-generating products, services, and processes" (Hansen & Birkinshaw, 2007: 125). It encompasses, for instance, processes of concept development, prototyping and validation, field test and launch, post-launch review (Howells, 2006). Fierce competition on prices and time-to-market spurs firms to focus their in-house knowledge transformation activities and complement them with external services (Tether and Tajar, 2008) of providers, that can deliver them better and faster and with whom firms can share innovation costs and the risks of flawed projects. When a firm decides to outsource to a KTO, its search ranges in terms of partners such as laboratories, research institutes or universities and in terms of geography. Today, this search might go well beyond the national boundaries. Consequently, firms might need a broker able to tap into a network of local and international providers, to support the activation and the coordination of multi-institutional projects (Corley at al., 2006; Fleming & Waguespack, 2007). Furthermore, considering the shortage of resources for R&D, firms might need assistance in implementing innovation projects financed by public institutions and programmes.

Finally, in the last stage of the innovation value chain, firms *exploit and disseminate* the newly developed knowledge and seek to profit from innovation: a new product, process or service is commercialized, patents are licenced. In this phase firms, especially those with limited resources, must find ways to ensure that innovative solutions, that are delivered to the marketplace, are protected by professional patent policy, are properly financed and commercialised.

Along with the different stages of the innovation value chain, firms can choose among a wide array of "ideas": from shopping for *raw ideas* through licensing (which implies contacting inventors directly) to buying *ready-to-market concepts* and competences (licencing enriched by

R&D collaboration agreements) to finally buying *ready-to-market products*, which are products
or services ready for launch (for instance, acquiring a company) (Nambisan & Sawhney, 2007).
Each type of "shopping" decision, given the different knowledge and contractual arrangement
features that it implies, requires a specific competent support from a KTO.

In contrast to this broad set of requirements at each stage of the innovation value chain,
firms still face several obstacles in finding suitable partners and in collaborating with them.
As highlighted recently, the decision about who to collaborate with to create an effective
network can be difficult, particularly for SMEs (Lee et al., 2010). Results of the Community
Innovation Survey confirm the propensity of firms to rely on traditional sources.
Institutional sources are less frequently consulted than internal or market sources; and
cooperation is easier with suppliers or customers than with consultants, commercial labs or
private R&D institutes and even than with universities or public research institutes (Parvan,
2007). This problem is even more complicated considering that firms should decide not only
with which partners to collaborate but also how these collaborations should be organised: as
a competitive market or as a collaborative community (Boudreau & Lakhani, 2009). As
recently shown, because the dynamics of communities and markets are inherently different,
a firm has to carefully understand which one is more coherent with its innovation value
chain and even with a specific project (Boudreau & Lakhani, 2009). Consequently,
innovative organizations might need external assistance for internally auditing their
objectives and innovation processes and for choosing the right set of interactions and
relationships with outside partners.

Firms' reluctance to use the market of KT activities points to additional KTO services whose
aim is building upon the inside capabilities of the firm. These services are for instance
training programmes that help organizations to reinforce their absorptive capacity, to
develop the culture of collaboration (Lee et al., 2010) for better cognitive and cultural
closeness to research institutes, factors that favour subsequent forms of face-to-face
collaboration (Balconi & Laboranti, 2006).

The positive impact of KTOs on the firms' innovation value chain can also be considered as
an intermediate effect of the role they play within a specific region or innovation cluster.
KTOs acting as intermediaries embedded within a geographical area can spur technological
spillovers and knowledge-sharing by increasing the network density among universities,
research centres and companies. Moreover, KTOs that hire qualified employees (some of
whom on temporary contracts) can also foster researchers' mobility, thus positively affecting
knowledge flows within a region (Breschi et al., 2005). Finally, KTOs might foster
entrepreneurship by incubating start-up projects and favouring spin-offs.

2.3 The differentiated supply side: Knowledge transfer services and providers

Triple helix initiatives, involving academia, government and industry, have spurred, since
the 1970s in the USA and the 1980s in Europe, the birth of a number of actors whose task it is
to facilitate the transfer of scientific knowledge from universities to firms-receivers. The
triple helix model ascribes to the university a third mission, in addition to research and
teaching, namely nurturing economic and industrial development (Etzkowitz &
Leydesdorff, 2000). In other words, universities are requested to implement actions in order
to favour the effective exploitation by companies of the scientific knowledge generated in

the academic labs. However, the knowledge transfer process between these two systems is characterised by high transaction costs and cognitive distance that hamper the direct contribution that universities can offer toward the commercialization of viable technologies (Gilsing et al., 2011; Kodama, 2008; Polt et al., 2001; Yusuf, 2008) (Figure 1). First, the establishment of university-industry linkages implies high search costs, since firms need to invest more time and resources in seeking and assessing academic partners than they do with those belonging to their supply chain. This can be ascribed, on one hand, to information asymmetry that makes it difficult for firms to stay up-to-date on the state-of-the-art projects and the related results carried out at the university and, on the other hand, to the uncertainty about the future output deriving from the application of general and theoretical knowledge. Moreover, the information asymmetry and the uncertainty which characterize the TT process also generate high bargaining costs in terms of negotiation and coordination of the parties. Indeed, "firms and universities are exposed to different incentive schemes that shape their interest in the transfer process" (Gilsing et al., 2011: 4). Scientists aim to contribute to the generation of public knowledge through dissemination; whereas firms want to appropriate the advantages deriving from the rapid commercialization of products/services that embody the new knowledge. This leads to the identification of complex solutions to overcome motivation problems and to increase the reliability among partners. Finally, the differences between universities and firms in terms of systems of perception, interpretation and evaluation, as well as of "shared meanings" linked to the organizational culture generate barriers to the combining of knowledge.

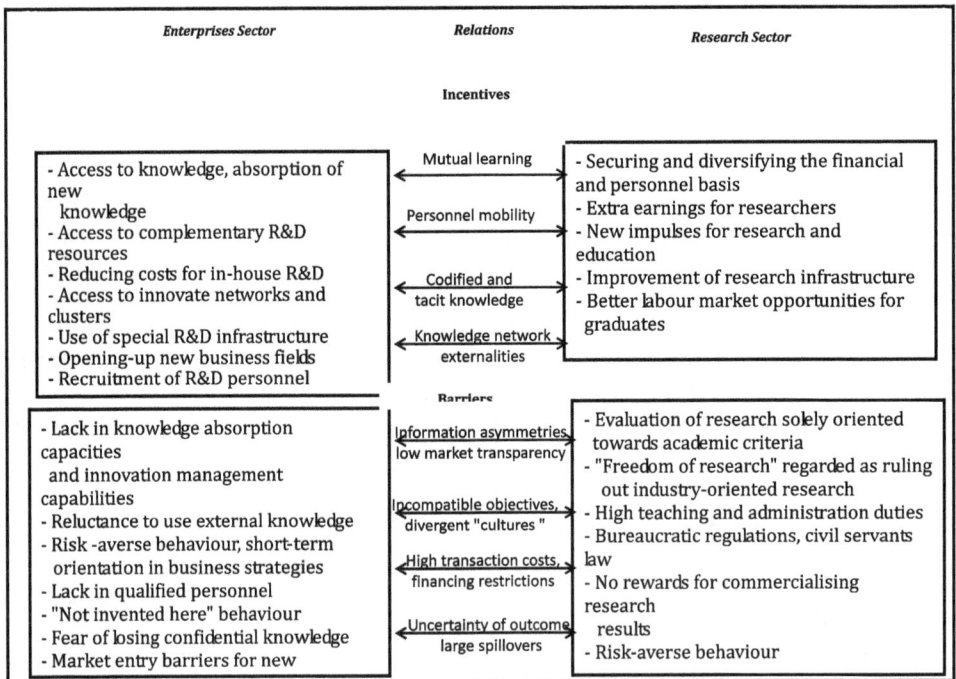

Fig. 1. Incentive and barriers on the relationship between firms and Public Research
Source: Polt et al., 2001

In order to increase the closeness between the different actors as well as to reduce the search and bargaining costs that this KT process implies, different types of organizations have been sponsored by public institutions over the last few decades. KTOs are knowledge-intensive providers which aim to facilitate the interaction between these heterogeneous actors and to support the exploitation of the research results of universities by industrial firms. For instance, science parks have been established in order to nurture innovative environments in which firms and academic institutions can interact via informal channels favouring knowledge spillover. Examples of well-known science parks include Stanford Research Park and Research Triangle Park of North Carolina in the US (Link & Link, 2003), Cambridge in the U.K. and Sophia-Antipolis in France (Longhi, 1999; Keeble et al., 1999). Another type of KTO, the academic incubator, has the mission to foster the creation of start-up firms based on university-owned or licensed technologies (Phan et al., 2005). Moreover, technology transfer offices or industrial liaison offices have been established in order to reduce the information asymmetry typically encountered in the scientific knowledge market through the management of intellectual property rights generated by scientists (Macho-Stadler et al., 2007).

As discussed in the previous sections, the complexity of applying scientific knowledge, the technological convergence trend which calls for multidisciplinary research, and the fierce competition on prices and time-to-market, spur firms to focus their in-house R&D and complement it with external scientific knowledge (Tether & Tajar, 2008). In this regard, in order to meet the firms' requirements, recent research highlights that KTOs are expanding rapidly and further typologies are emerging, such as business innovation centres, innovation agencies, R&D labs, technology consultants, technical testing and analysis labs (Comacchio et al., 2011; Consoli & Elche-Hortelano, 2010; Laranja, 2009). Moreover, studies show that KTOs are progressively evolving in terms of wider TT services portfolios and additional roles performed. Considering the three main phases of the innovation value chain, the offer of KTOs nowadays encompasses a broad range of services concerning (Howells, 2006; Lee et al., 2010; Muller & Zenker, 2001; Spithoven et al., 2010):

- *knowledge generation*: KTOs continuously scan the evolution of the scientific environment and describe the path to follow in order to integrate the technology to products and services (foresight, forecasting and technology roadmapping); they enable receivers to articulate their ideas and needs about new scientific and technological knowledge (demand articulation); they gather information and identify the potential collaborative partners (scanning and information processing);
- *knowledge transformation*: besides providing prototyping, testing and validation facilities, KTOs manage the process of knowledge re-engineering through in-house R&D and its recombination with the knowledge generated by heterogeneous partners;
- *knowledge exploitation*: KTOs support firms to assess and manage their inventions for Intellectual Property (IP) protection; they identify market opportunities, develop business plans, provide training and support for the commercialization of new products and services.

More recently, KTOs have started to fulfil a further activity throughout the innovation value chain of firms, namely the intermediation between firms-receivers and their scientific and technological partners. A recent research study describes this activity in terms of roles performed by KTOs, which can act as (Johnson, 2008): *mediators or arbitrators* in cases of

TYPOLOGIES	MISSION STATEMENT
Experimental station	Carrying out pre-competitive industrial research and development analysis, testing and experimentation of products, processes and new technologies, technical and scientific dissemination of knowledge and documentation in specific sectors at the national level.
Scientific park and technological hub	Promoting the economic development and competitiveness of regions and cities by creating new business opportunities and adding value to mature companies; fostering entrepreneurship and incubating new innovative companies; generating knowledge-based jobs; building attractive spaces for the emerging knowledge workers; enhancing the synergy between universities and companies.
Technological transfer office	Supporting the academic staff to identify and manage the organization's intellectual assets, including protecting intellectual property and transferring or licensing rights to other parties to enhance prospects for further development.
Incubator	Accelerating the growth and success of entrepreneurial companies through an array of business support resources and services that could include physical space, capital, coaching, common services, and networking connections.
Business Innovation Centre	Offering a range of integrated guidance and support services for projects carried out by innovative SMEs, thereby contributing to regional and local development.
Chamber of commerce special agency and laboratory	Furthering the development and expansion of technological innovation through the offer of services that meet the requirements of the firms associated to the Chamber of commerce.
Territorial development enterprise	Gathering and coordinating scientific, organizational and financial resources in the region in order to transfer acquired information on new production processes and research results to the entrepreneurial context.
Topic centre	Promoting a specific industry or a specific technological area inside a geographical context.
Multi-sector centre	Supplying of diversified services to firms operating in several sectors.
Public research organization	Performing research in its own Institutes, promoting innovation and competitiveness of the national industrial system, promoting the internationalization of the national research system, providing technologies and solutions to emerging public and private needs, offering advice to the Government and other public bodies, and contributing to the qualification of human resources.

TYPOLOGIES	MISSION STATEMENT
Laboratory of applied research and/or testing and analysis	Supplying qualified services of research and development, analysis and testing to client firms.

Table 1. Typologies of KTOs and mission statements
Source: Comacchio et al. 2011

dispute, thus facilitating contract negotiation for the accomplishment of a project; *sponsors and distributors* of funding for innovation efforts; *filters and legitimators* of projects that are worthy of support; *technology brokers* acting as a repository of information about technology experts and new technology opportunities and as a bridge between seekers and solvers of innovative solutions; *resource/management providers* in terms of project management practices that can be helpful in facilitating R&D collaboration.

The variety of KTO types (European Commission, 2004) and their differentiated offer (see Table 1) together with their relevance in promoting innovation and entrepreneurship at the firm level as well as at regional level raise the issue of identifying the appropriate criteria to assess the effectiveness of the knowledge transfer process performed by these organizations.

2.4 Summary: Why it is important to evaluate the KTOs' performance

The importance of evaluating the performance of KTOs has become paramount due to the increasing need for companies to rely on providers and intermediaries that are able to support them during the different phases of the innovation value chain. In order to nurture a market for what KTOs offer, there is a quest for more information about their characteristics in terms of services/products provided as well as for the performance they are able to achieve. The availability and the diffusion of this information may contribute to reduce the asymmetry, and thus increase the transparency of the transactions which occur in the market for knowledge. Promoting the disclosure of the capabilities and performance of KTOs may be beneficial from different points of view.

First, KTOs identifying their value proposition and assessing their performance are better able to define their product/service portfolio and craft their competitive strategy at regional, national and international level. Second, from the demand side, firms-receivers can use a univocal method to compare and evaluate the offer of the different KTOs, thus they may benefit from the reduction of transactional costs related to the search for and the selection of appropriate partners, as well as the monitoring of the results attained. Finally, local and regional stakeholders, such as universities, institutions and policy makers, may define their innovation policies and allocate resources effectively.

However, the variety of KTOs that have been set up over the last few decades together with the progressive broadening of their mission and the heterogeneity of stakeholders with different perspectives increase the complexity of the evaluation process. In the following section, we will present different approaches and attempts to KTO performance assessment, highlighting their strengths and weaknesses.

3. Knowledge transfer evaluation and a comparative analysis of European best practice

3.1 Introduction: The challenge of evaluating KTOs

During the last decade, European and local institutions have sponsored initiatives to identify methodologies in supporting KTOs to define their KT objectives, to assess the way these objectives are being fulfilled, and to set up permanent monitoring systems that can guarantee the evaluation of performance over time (European Commission, 2009; Guy, 1996). However, the attempts towards the definition of an evaluation system of KTOs have encountered several difficulties that can be ascribed to the characteristics of KTOs and the specific features of the knowledge transfer process.

First, KTOs present a high level of heterogeneity in terms of mission statements and business models adopted. Indeed, the institutional statements are often too broad, encompassing a wide range of aims that make it difficult to identify the KT objectives to set, monitor and evaluate. Moreover, each type of KTO has its own particular aims to pursue, increasing the difficulties of making comparisons across categories of KTOs (Hogan, 1996). Furthermore, the strategy and the subsequent business models implemented are determined by multiple stakeholders who show different expectations regarding the KTO performance measurement system. On the one hand, local and regional institutions aim to obtain an objective approach to measure allocated resources and to plan future investments consistently with the regional policy guidelines; on the other, firms want to obtain a performance appraisal tool that enables them to compare KTOs and select the effective partner to involve in their innovation process (Bigliardi et al., 2006). This makes designing a common performance measurement system, that conciliates these multiple interests, highly complex.

Second, the broad services portfolio provided nowadays by KTO increases the complexity of performance measurement due to the fact that each type of service should be assessed by a set of specific indicators. Furthermore, the knowledge contents transferred by KTO are characterized by different levels of tacitness and uncertainty that generate complexity in identifying their value. Indeed, a KTO may provide explicit and codified knowledge more easily valuable in a market for technological knowledge (such as the licensing of IP rights or the testing and prototyping of activities), but also knowledge characterized by a high degree of uncertainty and tacitness that makes it difficult to define appraisal metrics (for instance the promotion and coordination of research projects for new product development).

Third, the measurement of the impact of the KT process encompasses both the macro and the micro level of analysis, since the KTOs' activities have an effect on the single firm-receiver as well as on the local economic environment (region, industrial cluster) in which the KTO operates. This generates problems in terms of data availability and accessibility concerning the benefits attained by the local system and the receivers, but also in terms of isolating the results that can be ascribed to the specific intervention of KTOs (Gardner et al., 2010; Guy, 1996).

Finally, KTOs do not produce instant results. Indeed, the impact of the KT process on the firms-receivers and on the local environment takes time to emerge, whereas many of the costs are incurred when the service is provided (Gardner et al., 2010; Hogan, 1996), as in the case of the spin-off processes and research programs.

The aforementioned causes of complexity in achieving a uniform methodology for the evaluation of KTOs explain the wide diversity of approaches implemented by these organizations so far. This section provides a review of the extant methods used in assessing the performance of KTOs drawing on initiatives and best practices identified in the European context. The review aims to identify the set of core areas of evaluation and the related Key Performance Indicators (KPIs) which are likely to be common to all KTOs, in order to find convergence in appraisal methods. Drawing on institutional reports elaborated by European KTOs associations (such as the Association of European Science and Technology Transfer Professionals ASTP, the European BIC Network or the Pan-European Network of Knowledge Transfer Offices ProTon) as well as on appraisal tools elaborated by some European KTOs, the following sub-sections will present and discuss two main sets of KPIs related to two macro-areas of evaluation, which are the *output* of the KT process (in terms of research results, economic-financial returns and impact on firms-receivers and the local innovation system) and *input* (the structural, human and social capital of the KTOs which enable them to provide KT services).

3.2 Output evaluation

The evaluation approach which focuses on the results attained by the KTO may embrace three key areas.

The first area aims to capture the level of effectiveness of the KTO in achieving its KT goals, such as establishing high-tech firms in the region or promoting research collaborations between academia and firms. In this regard, metrics should assess both intermediate and final outputs of the knowledge transfer process. For instance, the effectiveness a KTO whose goal is to nurture the creation of new firms through incubation services can be assessed by monitoring the number of tenants hosted by the KTO (intermediate output) but also the enterprise survival rate following graduation (final output).

Table 2 shows in the first column the main aims included in the mission statement of KTOs: i) to establish channels of communication and collaboration between firms and research institutions; ii) to promote the commercialization of intellectual property rights generated by scientists; iii) to nurture entrepreneurship in the local innovation system; iv) to support the creation of new companies. For each area of evaluation the second column reports the Key Performance Indicators that are usually used to monitor the achievement of the KT goals.

Area of evaluation	Key Performance Indicators
Research activities	• Number of collaborative research agreements stipulated during the year where both firms, universities, institutions and other KTOs participate in the design of the research project, contribute to its implementation and share the project outputs • Number of collaborative research agreements stipulated during the year with other trans-regional or international KTOs

	• Number of contract research agreements stipulated during the year where all research is performed by the KTO • Number of consultancy agreements stipulated during the year where the KTO provides expert advice without performing new research • Number and type of pre-competitive research programs initiated during the year by established labs • Number and type of conventions stipulated during the year with established labs aiming at the concurrent development of products/processes • Number of scientific publications published during the year, with their relative impact factor • Number of patent researches carried out during the year
IP exploitation	• Total number of licenses executed during the year • Number of active spin-off companies with a formal license agreement with the KTO • Number of active spin-off companies with a formal equity agreement with the KTO • Number of spin-offs established during the year • Active spin-off companies' average during the year • Enterprise Survival Rate of spin-off companies
Entrepreneurship promotion	• Number of events organized during the year to promote entrepreneurship • Number of people that attended events to promote entrepreneurship • Number of training events/number of training hours organized during the year • Number of people that attended training events
Enterprise creation	• Average incubator space occupancy rate (%) • Average incubation time (years) • Number of tenants in incubators • Total employment by tenants • Number and type of new product prototypes launched by incubated firms • Number of start-ups created during the year • Number of jobs created in start-ups in the year • Enterprise Survival Rate (within the incubation period) • Enterprise Survival Rate (3 years following graduation) • Total numbers of contacts for enterprise creation • Number of feasibility studies created during the year • Number of enterprise creation projects during the year • Number of business plans produced during the year

Table 2. Key Performance Indicators of KTOs' knowledge transfer output
Sources: Area SciencePark, 2011; Arundel & Bordoy, 2010; Bigliardi et al., 2006; EBN, 2011; European Commission, 2009; Piccaluga et al., 2011.

Area of evaluation	Key Performance Indicators
Impact on the firms-receivers and the local innovation system	• Growth in turnover in companies attributed to KTO intervention • Number of jobs created in companies attributed to KTO intervention • Number of collaborative relationships and joint ventures among local, extra-regional and international firms favored by KTO • Number and type of environmental improvement carried out in collaboration with KTO • Investment flows installed by KTO from other regions or from foreign countries • Personnel flows installed by KTO from other regions or from foreign countries • Laboratories of extra-regional or foreign firms installed by the KTO

Table 3. Key Performance Indicators of KTOs' impact
Sources: Bigliardi et al., 2006; EBN, 2011; European Commission, 2009.

Area of evaluation	Key Performance Indicators
Financial returns	• Total turnover for services and its trend • Total turnover for area location and its trend • Total license revenue and its trend • Total financial value of all research agreements and its trend • Revenues generated during the year from profits and/or sales of equity in spin-offs in which the KTO holds equity • Financial autonomy (own current revenues / endowment)
Economic performance	• Total revenues per FTE member of KTO staff • ROA and its trend • ROE and its trend • Average number of start-ups per FTE member of KTO staff • Average number of jobs created per FTE member of KTO staff • Average number of business plans created per FTE member of KTO staff • Average number of companies assisted per FTE member of KTO staff • Cost per job created with the support of a KTO • Public financial contribution per job created • Average number of start-ups per 100K€ of KTO • Average number of jobs created per 100K€ of KTO income • Average number of business plans created per 100K€ of KTO expenditure • Average number of companies assisted per 100K€ of KTO income

Table 4. Key performance indicators of KTOs' financial and economic returns
Sources: Area SciencePark, 2011; Arundel & Bordoy, 2010; Bigliardi et al., 2006; EBN, 2011; European Commission, 2009; Piccaluga et al., 2011.

A second area of result concerns the evaluation of the success of the KT process, therefore its impact on the firms-receivers' performance and on the economic growth of the region (table 3). It is worth bearing in mind that these effects require a long period of time to emerge, and they are mediated by the innovation activities carried out by each receiver and by the different actors operating in the local innovation system.

The last area of result encompasses the financial and the economic performance attained by the KTO (Table 4). The related KPIs capture the ability of the KTO to generate financial returns from the supply of its services and to attain efficiency.

3.3 Evaluation of inputs

The measurement approach based on input relies on the Intellectual Capital model (Edvinsoon, 1997), often used by KTOs in their annual disclosure activities. Table 5 shows the Intellectual Capital components and the related KPIs tailored for the KTOs:

- *Structural capital*: this represents the "organizational value left when employees go home", and it is also defined as "the organizational capability, including the physical systems used to transmit and store intellectual material" (Edvinsoon, 1997). The structural capital of KTOs is related to research infrastructure that enables the knowledge transfer (research labs, equipment, etc.) as well as the portfolio of IP rights which represents the potential commercialization of public science.
- *Human capital:* this encompasses the competence, skills, and the relevant knowledge possessed by employees of KTOs. Research shows that the level of education and the continuous investment on employees' competency development not only enable KTOs to provide high-quality KT services but also to assume an intermediation role between firms and universities (Comacchio et al., 2011);
- *Relational capital*: this comprises the set of resources rooted in relationships that the KTOs establish with research and institutional partners (universities, research labs, policy makers) and with the firms-receivers.

Area of evaluation	Key Performance Indicators
Structural Capital	*R&D assets* • Number of years since establishment • Number of research laboratories hosted by the KTOs • Number of high-tech firms hosted by the KTOs • Total square meters available • Average square meters available for incubation activities of owned incubators • Average square meters available for research activities • Value of fixed assets (plant, machinery, etc.) • Total expenditures on infrastructure • Total expenditures on technology transfer activities (patent portfolio management costs, contract costs, etc.) • Total expenditures on basic and applied research (staff dedicated to research, capital expenditures on new equipment, etc.)

Area of evaluation	Key Performance Indicators
	IP assets • Number of invention disclosures received during the year • Total number of priority patent applications • Number of priority patent applications submitted in the year • Number of patents granted in the year
Human Capital	*Size, breakdown and level of education of staff* • Total number of KTO staff in full-time equivalents (FTEs) – including all professional, administrative and support staff for knowledge transfer activities • Number of permanent staff • Number of temporary staff • Number of contract staff • Number of professional staff (FTEs) • Number of graduates • Number of post-graduate degrees and doctorates *Competences development* • Number of training days • Number of participants in training courses • Total staff expenses for training • Total staff expenses - HR development *Retention* • Staff turnover • Average seniority of staff • Employee satisfaction (work environment), evaluation scale 1-5
Relational Capital	*Partners* • Number of partners • Number of new partnership established during the year • Number of foreign partners • Number of R&D partners • Number of institutional partners *Receiver satisfaction* • Number of KTO firms-receivers • Number of new KTO firms-receivers acquired during the year • Customer satisfaction: positive comments from participants in KT initiatives • Number of complaints from receivers

Table 5. Key performance indicators of KTOs' intellectual capital
Sources: Area SciencePark, 2011; Arundel & Bordoy, 2010; Bigliardi et al., 2006; EBN, 2011; European Commission, 2009; Piccaluga et al., 2011.

KTOs which have a structured, multidisciplinary, and international network of partners are able to obtain critical information more rapidly, to involve complementary partners in joint research programs and to more easily access the circuit of financed innovation projects (Adler & Kwon, 2002; Nahapiet & Ghoshal, 1998). Moreover, the activation of a network of external qualified relations increases the identity and the reputation of the KTO within the innovation system.

The relational capital also captures the customer relationships developed by the KTOs. In this regard, metrics rely on follow-up procedures or satisfaction surveys.

3.4 Summary

This section provided a review of the recent debate on the difficulties and opportunities of measuring KTOs' performance. Furthermore, the in-depth analysis carried out has shown the state-of-the-art of metrics and indicators used by KTOs and institutions to evaluate knowledge transfer benefits. This analysis has shed light on the fragmentation of practices implemented by KTOs, in terms of aims, areas of results and metrics.

According to a recent article, which provides a broad analysis of methods for quantifying and qualifying the benefits of KT around the world, four key issues have to be taken into account: inputs vs outputs, quality vs quantity, subjectivity vs objectivity and time series vs cross-sectional analysis (Gardner et al., 2010). We would contribute to this discussion, providing new implications for each issue, on the base of our in-depth analysis of evaluation practices and related indicators in Europe.

- *Inputs vs outputs:* Our analysis suggests that both inputs and output indicators are used. The first as a measure of stock of resources and of activities occurred, the second as evaluation of results achieved. However, they are rarely used in an integrated manner within the same measurement system.
- *Quality vs quantity:* While it is difficult to distinguish between the quality and quantity of results, an issue could be raised with accuracy of quantitative indicators, for instance the simple number of agreements does not inform about how many of them have been completed successfully. Also, our in-depth analysis shows that there is a first attempt to provide both measurements for a more accurate picture of results attained; however, especially in the area of research activity and IP exploitation, quantitative indicators overwhelm qualitative ones.
- *Subjectivity vs objectivity:* Objective evaluations should be preferred instead of subjective descriptions and measures. Approaches analyzed in this section rely on objective and measurable indicators. Beside this advantage, we maintain that objectives metrics should be related to the specific strategy and mission of a KTO, in order to signal effectively the coherence between results obtained and strategic aims pursued by a provider.
- *Time series vs cross-sectional analysis:* Evaluation could follow two methods: time series, namely the comparison within the same organization of results over time, and cross-sectional analysis, that is an inter-organization snapshot of results at one moment. Our analysis suggests that a comprehensive framework could help not only to investigate within the same organization trends and causal relationships among actions and

performance, but could also help to find an overall accurate framework for worthy comparisons among different types of KTOs.

In the following section we will discuss how the KTO performance management system could be improved by providing a preliminary comprehensive frame that draws on the Balanced Scorecard (BSC) approach, in order to bring within an overall integrated and coherent model different indicators of KTO performance.

4. An analytical model for the measurement of KTOs: A balanced scorecard approach

4.1 The balanced scorecard approach

Among the seminal performance measurement models, the Kaplan-Norton Balanced Scorecard continues to be the most referenced framework in the literature (Kaplan & Norton, 1992; 2004; 2007; Kaplan et al., 2010). This is due to its ability to support firms' strategic learning, namely "gathering feedback, testing the hypotheses on which strategy was based, and making the necessary adjustments" (Kaplan & Norton, 2007: 159). Indeed, the BSC helps firms to articulate their vision and to define in clear and operational terms the results that they aim to achieve. Moreover, the BSC is a comprehensive frame that enables companies to get a feedback system thanks to the specification of the causal relationships between performance drivers and objectives: "Companies build their strategy maps from the top down, starting with their long-term financial goals and then determining the value proposition that will deliver the revenue growth specified in those goals, identifying the processes most critical to creating and delivering that value proposition, and, finally, determining the human, information, and organization capital the processes require." (Kaplan & Norton, 2004: 55). Figure 2 provides a representation of the four perspectives (financial, customer, internal process, learning and growth) and the relationships with a firm's strategy.

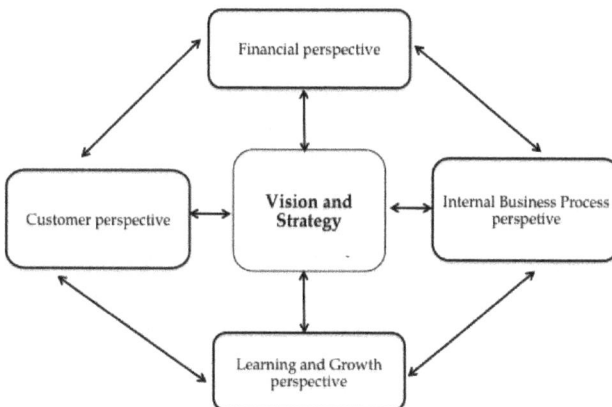

Fig. 2. The Balanced Scorecard Framework

The BSC framework has been widely applied in many manufacturing and service industries, but only recently has been extended to the research setting (Bremser & Barsky, 2004; Eilat et al., 2008; Sartorius et al., 2010; Wang et al., 2010). Drawing on the KT and BSC literature, we provide a preliminary contribution to this debate re-conceptualising the Balanced Scorecard coherently to the KTO context.

4.2 The balanced scorecard of KT

4.2.1 Introduction

The BSC strategy map (Fig.3) provides a framework for linking intangible assets to shareholder value creation through four interrelated perspectives: financial, customer, internal process, learning and growth. We suggest that this scheme is a coherent architecture by which the fragmented evaluation system of KTOs could be unified. Moreover, this framework provides a strategic way of thinking about performance assessment of KT services, useful for providers, receivers and policy-makers. Indeed a well-developed and comprehensive performance measurement systems could reduce the barriers that hamper the relationship between demand and supply of services (Polt et al., 2001).

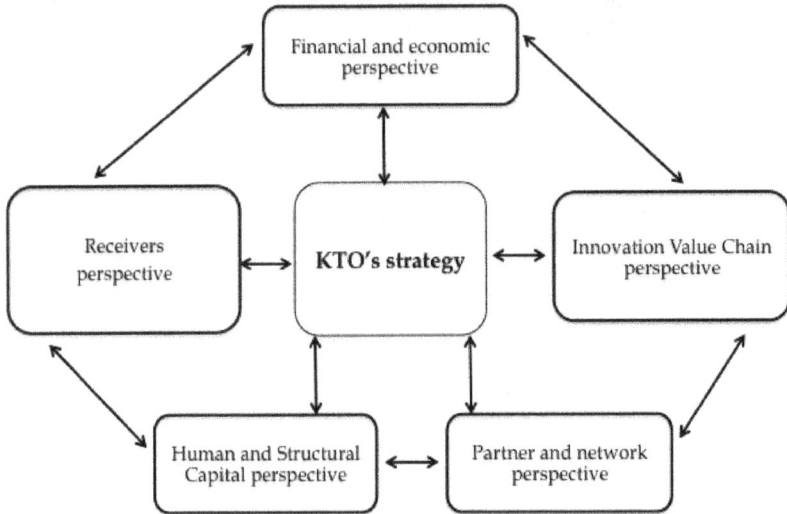

Fig. 3. The Balanced Scorecard Framework for KTO

4.2.2 The managerial perspectives of BSC applied to KTOs

The four managerial perspectives of the BSC provide KTOs' management with a comprehensive tool to measure performances along different but integrated dimensions. We propose that each dimension of the BSC could be adapted to the specific activities pursued by KT bodies. Accordingly in the following discussion each dimension of the original BSC frame is presented and is adjusted in order to be applied to the context of the market of knowledge transfer services. We finalize the discussion of each dimension by giving some examples of key indicators that can be used, coherently with metrics of KTO performance management, presented in Section 3.

The *Financial Perspective* describes the tangible outcomes of the strategy in financial terms, such as profitability, revenue growth and cost efficiency. It can be conceived as the final result of the cause-effect relationship among the other three dimensions of the scorecard (learning and growth performances affects internal business processes, which impact on customer outcomes, which finally drive financial results). This perspective applied to KTOs' performance measurement encompasses metrics related to the capability of the organization to gather external resources through its knowledge transfer services. It regards both the short-term profitability generated by commercialized services, the productivity and the efficiency of business processes as well as long-term financial sustainability of research and intermediation activities. Examples of financial performance indicators are the following: total turn-over from services, amount of public funded projects, ROA, costs. Moreover, within this perspective the impact of KTOs on a local innovation system is considered and measured by economic indicators, for instance personnel and investment flows attracted by KTOs in the region.

The *Customer Perspective* looks at customers' satisfaction with the product or services delivered by a firm and their loyalty. We suggest that as far as a KTO is concerned, customer perspective could be applied to acquisition of clients such as firms and to their retention. However, considering the brokerage role that KTOs play across firms and research institutes, this perspective could be extended to other clients such as Universities or private research bodies. Moreover, the relationships between a KTO and receivers could be characterised by different degrees of interdependences and duration. This dimension could be re-defined as *Receivers Perspective* in order to measure the performance of the overall receivers network. Beside the well known indicators of customer satisfaction and retention, other indicators are those that measure the network size, namely number of clients or agreements and network dynamics, such as partners development rate. Finally, indicators of social network analysis could be applied to assess the structural features of a KTO's ego network namely degree centrality.

The *Internal Perspective* regards the key internal processes that make up the value proposition to customers, thus those processes by which a firm is able to meet customers' needs. From a knowledge transfer point of view this dimension should consider client's requirements that are tightly related to the phases of the innovation value chain. Consequently, we suggest to reconceptualise this dimension as *Innovation Value Chain perspective*. The measures it includes assess whether a KTO is able to act as an external competent partner along the three main phases: knowledge generation, transformation and exploitation. Indicators for the knowledge generation phase are related to three main areas of results: i) basic research (number of scientific publications per year, number of patent researches carried out during the year); ii) technology scouting (number of technological audit provided); iii) brokerage (number of events organised to promote networking among innovators). Indicators of results in the second phase of knowledge transformation are the following: type of conventions stipulated during the year aiming at the concurrent development of products/processes and number of application for patent; number of research and development projects; number of training events organized; number of prototyping and testing services provided; etc. Indicators of results in the third phase of knowledge exploitation are: number of application for patent; number of business plans;

BSC perspectives	Area of results	Sample metrics
Financial Perspective	Profitability of KT services	• Total turnover for services and its trend; Total license revenue and its trend
	Financial sustainability	• Financial autonomy; amount of public founded research projects
	Productivity	• Total revenues per KTO employee; ROA and its trend
	Economic impact on local innovation system	• Personnel flows installed by KTO from other regions or from foreign countries
Receivers Perspective	Network size	• Number of clients • Number of agreements
	Network structure	• KTO degree centrality
	Network dynamics	• Customer retention rate • New clients development rate • Customer satisfaction • Number of complaints from customers
Innovation Value Chain Perspective	Knowledge Generation	• Basic research (number of scientific publications per year, number of patent researches carried out during the year) • Technology scouting (number of technological audit provided) • Brokerage (number of events organised to promote networking among innovators)
	Knowledge Transformation	• Number and type of conventions stipulated during the year aiming at the concurrent development of products/processes • Number of research and development project • Number of training events organized • Number of prototyping and testing services provided
	Knowledge	• Number of application for patent • Number of business plans • Number of spin-offs established during the year

BSC perspectives	Area of results	Sample metrics
	Exploitation	• Active spin-off companies' average during the year • Number of tenants in incubators
Human and Structural Capital Perspective	Human capital	*Size, breakdown and level of education of staff* • Total number of KTO staff in full-time equivalents (FTEs) • Number of graduates and post-graduate degrees and doctorates *Competences development* • Number of training days • Number of participants in training courses *Retention* • Staff turnover, average seniority of staff or satisfaction
	Structural capital	*R&D assets* • Total expenditures on infrastructure • Total expenditures on technology transfer activities • Total expenditures on basic and applied research *IP assets* • Number of invention disclosures received during the year • Number of patents granted in the year
Partner and Network Perspective	Network size and quality	• Number of partners • Number of new partnership established during the year • Number of external consultants; • Skill and capabilities provided by partners.
	Network dynamics	• Composition of partners and relationships over time

Table 6. The Balanced Scorecard of KTOs

number of spin-offs established during the year; active spin-off companies' average during the year; number of tenants in incubators, etc.

The *Learning and Growth Perspective* is the dimension that measures the intangible assets of an organization. As argued by Kaplan and Norton (2004: 54) "the objectives in this perspective identify which jobs (the human capital), which systems (the information capital), and what kind of climate (the organization capital) are required to support the value-creating internal processes". The same dimensions could be applied to KTO in which research activity relies both on human competencies and prior R&D experience, thus we suggest to name this aspect as *Human and Structural Capital Perspective*. Coherently, indicators are those traditionally used to measure KTO intangible input and already presented in Section 3, a sample of which is: indicators of *size, breakdown and level of education of staff* (total number of KTO staff in full-time equivalents (FTEs), number of graduates and post-graduate degrees and doctorates), indicators of *competences development* (number of training days or number of participants in training courses) and finally indicators of *retention and satisfaction* (staff turnover, average seniority of staff or satisfaction). Concerning *structural capital* some indicators are those of *R&D assets* (total expenditures on infrastructure/on technology transfer activities /on basic and applied research) and of *IP assets* (number of invention disclosures received during the year, number of patents granted in the year).

In addition to human and structural capital a further enabler of the competitive advantage of KTOs is the network of external contractors and organizational partners by which a KTO provides its services, such as consultants, researchers. Thus, accordingly with the most recent contributions on BCS in R&D settings (Lazzarotti et al., 2011) we enrich the original BCS map with an additional dimensions of KTO performance measurement: *Partner and Network Perspective*. Indicators are the following: number of partners, number of new partnership established during the year and number of external consultants; skills and capabilities provided by partners (network size and quality) and composition of partners and relationships over time (network dynamics).

Table 6 shows the BSC framework tailored for KTOs setting, illustrating the five complementary perspectives and the related area of results and indicators. The fragmented measurement systems presented in the Section 3 are therefore composed in a comprehensive framework.

5. Conclusion

In this chapter, after having illustrated the main drivers of the evolution of the market of knowledge transfer services, we maintained the relevance of the evaluation of KTO performances, notwithstanding its intrinsic difficulties. Grounding our discussion on an in-depth analysis of key indicators used in Europe to measure knowledge transfer performances, we highlighted the limits of the extant evaluation systems, specifically we argued that they are fragmented and disentangled from the strategy of KTOs. In line with these critiques we maintained that a performance measurement system based on an integrated frame such as the BSC has several advantages and positive effects.

First, as an assessment tool, it helps to overcome the main barriers between the demand and supply of KT services. From the point of view of firms-receivers, a BSC might increase transparency, lower information asymmetries reducing transaction costs (Polt et al., 2001).

Second, as a managerial tool, the BSC helps KTO to craft their strategy in terms of key services offered, targeted receivers served, core competences to mobilize internally and externally in order to achieve financial and economic returns (Bremser & Barsky, 2004). Moreover the description of the multiple strategic goals a KTO wants to accomplish through the five BSC perspectives might simplify the selection of the most appropriate measures (Kaplan, 2010).

Third, this comprehensive scheme could become a monitoring tool for policy makers willing to make cross-sectional comparisons of providers based on a univocal method, in order to manage resource allocation and support regional development.

In terms of future lines of research, we suggest that from a theoretical point of view the model could benefit from further conceptual development in two directions: by a more in-depth analysis of each perspective and related metrics and by an analysis of the causal relationships among perspectives within a KT organization. From an empirical point of view a field work by case studies could be a first test of the theoretical model to understand how much this framework is implemented by KTOs.

We also draw some preliminary managerial implications for KTO, considering how this framework could be applied by a KTO, for instance what information is easily available to implement the framework and how much detailed the frame should be to meet the different stakeholders expectations.

6. References

Adler, P.S. & Kwon, S.W. (2002). Social capital: Prospects for a new concept, *Academy of Management Review*, Vol.27, No.1, pp. 17–40.,

Amesse, F. & Cohendet, P. (2001). Technology transfer revisited from the perspective of the knowledge-based economy, *Research Policy*, Vol.30, pp.1459–1478.

Area SciencePark (2011), *Intellectual capital report 2009*. Trieste.

Arora, A. & Gambardella, A. (2010). Ideas for rent: an overview of markets for technology, *Industrial & Corporate Change*, Vol.19, No.3, pp. 775-803.

Arora, A., Fosfuri, A. & Gambardella, A. (2001). *Markets for technology: The economics of innovation and corporate strategy*, The MIT Press, Cambridge MA.

Arundel, A & Bordoy, C. (2010), Summary *Respondent Report: ASTP Survey for Fiscal Year 2008*, UNU-MERIT.

Autio, E. & Laamanen, T. (1995). Measurement and evaluation of knowledge transfer: Review of knowledge transfer mechanism and indicators, *International Journal of Technology Management*, Vol.10, No.7/8, pp. 643-664.

Balconi, M. & Laboranti, M. (2006). University–industry interactions in applied research: The case of microelectronics, *Research Policy*, Vol.35, pp.1616–1630.

Bessant, J. & Rush, H. (1995). Building bridges for innovation: The role of consultants in technology transfer, *Research Policy*, Vol.24, pp.97–114.

Bigliardi, B. Dormio, A.I., Nosella, A. & Petroni, G. (2006). Assessing science parks' performances: Direction from selected italian case studies, *Technovation*, Vol.26, pp. 489-505.

Boudreau, K.J. & Lakhani, K.R. (2009). How to manage outside innovation, MITSloan *Management Review*, Vol.50, N0.4, pp. 69-76.

Bremser, W.G. & Barsky N.P. (2004). Utilising the balanced scorecard for R&D performance measurement, *R&D Management*, Vol. 34, No.3, pp. 229-238.

Breschi, S., Lissoni, F. & Montobbio, F. (2005). The geography of knowledge spillovers: Conceptual issues and measurement problems. In S. Breschi & F. Malerba (Eds.), *Clusters, networks and innovation* (pp. 343-378). Oxford: Oxford University Press.

Carlile, P.R. (2004). Organization science transferring, translating, and transforming: An integrative framework for managing knowledge across boundaries, *Organization Science*, Vol.15, No.5, pp.555-568.

Chesbrough, H.W. (2003). *Open innovation: The new imperative for creating and profiting from technology*, Harvard Business School Press, Boston.

Comacchio, A., Bonesso, S. & Pizzi C. (2011). Boundary spanning between industry and university: the role of Technology Transfer Centres, *Journal of Technology Transfer*, DOI 10.1007/s10961-011-9227-6.

Consoli, D. & Elche-Hortelano D. (2010). Variety in the knowledge base of Knowledge Intensive Business Services, *Research Policy*, Vol.39, pp. 1303–1310.

Corley, P., Boardman, C. & Bozeman, B. (2006). Design and the management of multi-institutional research collaborations: Theoretical implications from two case studies, *Research Policy*, Vol.35, pp.975–993.

Daim, T.U., Kocaoglu, D.F. & Anderson, T.R. (2009). Knowledge driven planning tools for emerging and converging technologies, *Technological Forecasting & Social Change*, Vol.76, pp.1.

Decter, M., Bennett, D. & Leseure, M. (2007). University to business knowledge transfer – UK and USA comparisons, *Technovation*, Vol.27, pp. 145–155.

Dosi G., Llerena, P. & Sylos Labini, M. (2006). The relationships between science, technologies and their industrial exploitation: An illustration through the myths and realities of the so-called 'European Paradox', *Research Policy*, Vol.35, pp. 1450–1464.

EBN (2011), *BIC Observatory 2011 (Data 2010)*, European BIC Network.

Edvinsson, L. (1997). Developing intellectual capital at Skandia, *Long Range Planning*, Vol.30, No.3, pp. 266-373.

Eilat, H., Golany, B. & Shtub, A. (2008). R&D project evaluation: An integrated DEA and balanced scorecard approach, *Omega*, Vol.36, pp.895-912.

Etzkowitz, H. & Leydesdorff, L. (2000). The dynamics of innovation: from National Systems and "Mode 2" to a Triple Helix of university–industry–government relations, *Research Policy*, Vol.29, pp. 109-123.

European Commission (2004). *Knowledge transfer institutions in Europe: An overview*, Brussels.

European Commission (2009). Metrics for knowledge transfer from public research organizations in Europe, In: *Report from the European Commission's expert group on knowledge transfer metrics*, European Commission Directorate-General for Research Communication Unit, Brussel.

Fleming, L. & Waguespack, D. M. (2007). Brokerage, boundary spanning, and leadership in open innovation communities, *Organization Science*, Vol.18, No.2, pp.165–180.

Gardner, P.L., Fong, A.Y. & Huang, R.L., (2010). Measuring the impact of knowledge transfer from public research organisations: a comparison of metrics used around the world, *International Journal of Learning and Intellectual Capital*, Vol.7, No.3/4, pp. 318-327.

Gassmann, O. (2006). Editorial. Opening up the innovation process: towards an agenda. In *R&D Management*, Vol.36, pp.223-228.

Gassmann, O., Daiber, M. & Enkel, E. (2011). The role of intermediaries in cross-industry innovation processes, *R&D Management*, Vol.41, No.5, pp. 457-469.

Geuna, A. & Muscio, A. (2009). The governance of university knowledge transfer: A critical review of the literature, *Minerva*, Vol.47, pp.93–114.

Gilsing, V., Bekkers, R., Bodas Freitas, I.M. & van der Stehen, M. (2011). Differences in knowledge transfer between science-based and development-based industries: Transfer mechanisms and barriers, *Technovation*, Vol.31, No.12, pp.638-647.

Guy, K. (1996). Designing a science park evaluation, In: *The Science Park Evaluation Handbook*. European Innovation Monitoring System (EIMS) publication n. 61.

Hagardon, A. & Sutton, R. (1997). Technology brokering and innovation in a product development firm, *Administrative Science Quarterly*, Vol.42, pp.716–749.

Hansen, M. & Birkinshaw, J. (2007). The innovation value chain, *Harvard Business Review*, Vol.85, No.6, pp. 121-130.

Hogan, B. (1996). Evaluation of science and technology parks, In: *The Science Park Evaluation Handbook*, European Innovation Monitoring System (EIMS) publication n. 61.

Howells, J. (2006). Intermediation and the role of intermediaries in innovation, *Research Policy*, Vol.35, pp. 715–728.

Johnson, W.H.A. (2008). Roles, resources and benefits of intermediate organizations supporting triple helix collaborative R&D: The case of Precarn, *Technovation*, Vol.28, pp. 495–505.

Kaplan, R.S. (2010). Conceptual foundations of the balanced scorecard, *Harvard Business School Working Paper*, 10-074

Kaplan, R.S. & Norton, D.P. (1992). The balanced scorecard-measures that drive performance, *Harvard Business Review*, Vol.70, No.1, January-February, pp. 71-79.

Kaplan, R.S. & Norton, D.P. (2004). Measuring the strategic readiness of intangible assets, *Harvard Business Review*, January-February, Vol.82, No.2pp. 52-63.

Kaplan, R.S. & Norton, D.P. (2007). Using the balanced scorecard as a strategic management system, *Harvard Business Review*, July-August, Vol.85, No.7/8, pp. 150-161.

Kaplan, R.S., Norton, D.P. & Rugelsjoen, B. (2010). Managing alliances with the Balanced Scorecard, *Harvard Business Review*, Vol.88, No.1/2, January-February, pp. 114-120.

Keeble, D., Lawson, C., Moore B. & Wilkinson F. (1999). Collective learning processes, networking and 'institutional thickness' in the Cambridge Region, *Regional Studies*, Vol.33, No.4, pp. 319-332.

Kodama, T. (2008). The role of intermediation and absorptive capacity in facilitating university-industry linkages - An empirical study of TAMA in Japan, *Research Policy*, Vol.37, pp. 1224–1240.

Laranja, M. (2009). The development of technology infrastructure in Portugal and the need to pull innovation using proactive intermediation policies, *Technovation*, Vol.29, pp. 23–34.

Lazzarotti, V., Manzini, R. & Mari L. (2011). A model for R&D performance measurement, *International journal of production economics*, Vol.134, pp. 212-223

Lee, S., Park, G., Yoon, B. & Park, J (2010). Open innovation in SMEs — An intermediated network model, *Research Policy*, Vol.39, pp. 290–300.

Lichtenthaler, U. & Ernst, H. (2007). Developing reputation to overcome the imperfections in the markets for knowledge, *Research Policy*, Vol.36, pp. 37–55.

Link, A.N. & Link, K.R. (2003). On the growth of U.S. science parks, *Journal of Knowledge transfer*, Vol.28, pp. 81–85.

Longhi, C. (1999). Networks, collective learning and technology development in innovative high technology regions: The case of Sophia-Antipolis, *Regional Studies*, Vol.33, No.4, pp. 333-342.

Macho-Stadler, I., Pérez-Castrillo, D. & Veugelers, R. (2007). Licensing of university inventions: The role of a technology transfer office, *International Journal of Industrial Organization*, Vol.25, pp. 483–510.

McEvily, B. & Zaheer, A. (1999). Bridging the ties: A source of firms heterogeneity in competitive capabilities, *Strategic Management Journal*, Vol.20, pp.1133–1156.

Mendonça, S., (2009). Brave old world: Accounting for 'high-tech' knowledge in 'low-tech' industries., *Research Policy*, Vol.38, No.3, pp. 470-482.

Muller, E. & Zenker, A. (2001). Business services as actors of knowledge transformation: the role of KIBS in regional and national innovation systems, *Research Policy*, Vol.30, pp. 1501–1516.

Muscio, A. (2010). What drives the university use of knowledge transfer offices? Evidence from Italy, *Journal of Knowledge transfer*, Vol.35, No.2, pp. 181-202.

Nahapiet, J. & Ghoshal, S. (1998). Social capital, intellectual capital, and the organizational advantage, *Academy of Management Review*, Vol.23, No.2, pp. 242–266.

Nambisan, S. & Sawhney, M. (2007). A buyer's guide to the innovation bazaar, *Harvard Business Review*, Vol.86, No.6, pp.109-118.

OECD (2010). *Measuring innovation. A new perspective*. OECD publishing.

Parvan, S.V. (2007). *Community Innovation Statistics. Weak link between innovative enterprises and public research institutes/universities*, Statistics in focus, Science and Technology, 81/2007.

Phan, P.H., Siegel, D.S. & Wright M. (2005). Science parks and incubators: observations, synthesis and future research, *Journal of Business Venturing*, Vol.20, pp. 165–182.

Piccaluga, A., Balderi, C., Patrono, A. (2011). *The ProTon Europe. Seventh annual survey report (fiscal year 2009)*. Institute of Management Scuola Superiore Sant'Anna (Pisa, Italy), ProTon Europe.

Polt, W., Rammer, C., Schartinger, D., Gassler, H. & Schibany, A. (2001). Benchmarking industry–science relations: the role of framework conditions, *Science and Public Policy*, Vol.28, No.4, pp.247–258.

Reisman, A. (2005). Transfer of technologies: a cross-disciplinary taxonomy, *Omega*, Vol.33, pp. 189–202.

Roper, S., Dub, J. & Love, J.H. (2008). Modelling the innovation value chain, *Research Policy*, Vol.37, pp. 961-977.

Samtani, L.A., Mohannak, K. & Hughes, S.W. (2008). Knowledge transfer evaluation in the high technology industry: An interdisciplinary perspective, In Goa, J., Jay, L., Jun, N., Lin, M. & Joseph, M., Eds. Proceedings 3rd World Congress on Engineering Asset Management and Intelligent Maintenance Systems Conference (WCEAM-IMS 2008): *Engineering Asset Management – A Foundation for Sustainable Development*, pp. 1357-1365, Beijing, China.

Sartorius, K, Trollip, N. & Eitzen, C. (2010). Performance measurement frameworks in a state controlled research organization: Can the Balanced Scorecard (BSC) be modified? *South African Journal of Business Management*, Vol.41, No.(2), pp.51-63.

Spithoven, A., Clarysse, B. & Knockaert, M. (2010). Building absorptive capacity to organise inbound open innovation in traditional industries, *Technovation*, Vol.30, No.2, pp. 130–141.

Tether, B.S. & Tajar, A. (2008). Beyond industry–university links: Sourcing knowledge for innovation from consultants, private research organisations and the public science-base, *Research Policy*, Vol.37, pp. 1079–1095.

Wang, J., Lin, W., Huang, Y.H. (2010). A performance-oriented risk management framework for innovative R&D projects, *Technovation*, Vol.30, pp.601-611.

Yusuf, S. (2008). Intermediating knowledge exchange between universities and businesses, *Research Policy*, Vol.37, pp. 1167–1174.

8

Understanding Innovation Deployment and Evaluation in Healthcare: The Triality Framework

Urvashi Sharma, Julie Barnett and Malcolm Clarke

Brunel University, Department of Information Systems and Computing
United Kingdom

1. Introduction

This chapter encompasses two objectives. The first objective is to understand different conceptualisation of technology, context and the user according to various theoretical perspectives within information systems (IS) literature and in the field of healthcare/medical informatics; and in addition, outline perspectives that have expounded the nature of relationship between these three entities. The need to explore such notions stems from authors' interest to elicit how these theoretical perspectives and concepts have been applied in the field of healthcare to understand technological innovation deployment and evaluation processes by taking the tacit, ephemeral, and complicated nature of healthcare work practices into account.

The second objective involves drawing parallels between these two bodies of literature and conceptualising a framework that can be employed to explain how and why technological innovation deployment and evaluation processes in the field of healthcare cause many challenges to arise.

1.1 Chapter outline

The chapter starts by drawing on well known theories in the field of IS to study deployment processes. These include the social construction of technology (SCOT), actor network theory (ANT), diffusion of innovation, contextual approach and the work of sociologist, Anthony Giddens in structuration theory and high modernity, and its derivatives in the field of IS by authors such as Barley and Orlikowski. By drawing on these perspectives, the conceptual framework is presented in its skeletal form outlining the salience of considering technological innovation, context, and user as three entities interrelated to each other through recursive relationship.

In the following section of this chapter, theories which provide understanding on deployment and evaluation processes in the field of healthcare are presented. These include theories such as normalisation process theory (NPT), routinisation theory, the structure, process and outcome framework and, concepts such as virtual and invisible work. These theories and concepts provide an insight into the micro-dynamics that occur between the context, technology and the user, which in this research, are considered to be the attributes of recursive relationships between the three entities. This leads to the final iteration of the framework, called the Triality framework, according to which, the technology is conceptualised as technological innovation, the context is conceptualised as healthcare social system and the user is conceptualised as human agent. It further highlights that each recursive relationship has set of attributes associated with it and there are nine attributes in total.

By drawing on this framework, this chapter aims at eliciting the complexities associated with healthcare context and the dilemmas surrounding healthcare professionals when technological innovations such as telehealth and electronic records are deployed and evaluated.

2. Theoretical perspectives from the field of IS: Understanding conceptualisation of technological content, context and user, and the nature of interrelationship between them

Exploring theoretical perspectives from IS provides insight into how technological content, context and the user have been conceptualised, and what is the nature of relationship between the three entities. Thus, theoretical perspectives that are of essence are discussed next.

2.1 Social Construction of Technology (SCOT)

SCOT stems from science and technology studies, which concern themselves with the impact of technology on society and its dynamics. These studies argue that the technology design and its acceptance trajectory are affected by social dynamics within a given society (William & Edge, 1996). SCOT therefore, strives to explain why certain technological artefacts or "*variants*" rise in society at given point in time whereas the others "*die*" (Pinch & Bijker, 1984). This view was proposed in response to technology determinism that establishes an opposite perspective that technology impacts the surroundings and its users (William & Edge, 1996). SCOT presents specific conceptualisation of technology, context and the user. These are discussed below.

Technology according to SCOT is conceptualised as an artefact of which the design and use is the subject of various interpretations resulting due to enabling and constraining properties of the technology as experienced by the groups under a given social context (Avgerou, 2002; Bijker, 1995; Klein & Klienman, 2002; Pinch & Bijker, 1984; Jackson et al., 2002). This constitutes the concept of interpretive flexibility, which also ascertains that despite different interpretations of an artefact within a society, these interpretations can co-exist together as they are dependent on desired outcomes.

In addition, the degree to which a technological artefact is accepted within a society is defined by the concept of closure and stabilisation. Closure is achieved when "*all groups' problems have been addressed and groups achieve consensus on a particular design*" (Klein & Klienman, 2002). Stabilization on the other hand, occurs when an artefact no longer requires high specifications. Closure can be attained in two ways, through rhetorical closure where the relevant social group sees the technology as solving the problem that it was intended to address, and closure by redefinition of a problem that sees the use of technology as the solution to an entirely different problem than originally proposed. For example, when the prelimatic tyre was originally introduced to solve the vibration problems in cycles, it was rejected; however, when used instead in racing cycles, it was accepted. This illustrates the closure by redefinition of the problem.

Context according to SCOT includes the social and political dimensions. It is argued to shape the values and norms of the groups. It does not however provide detailed insight into this area and has been criticised for this weakness.

The user in SCOT, is conceptualised as a member belonging to a particular social group, who despite sharing a common interpretation of technological artefact as a member of the group also has an individual interpretation which can be subjected to the group's scrutiny over time. This notion is constituted within the concept of technology frames, and has been regarded as its contribution. Another contribution of SCOT includes its attention to the influence of the designer on design during the design process.

However, despite the vital contributions, SCOT has been argued to have drawbacks that include the lack of acknowledging that the design of an artefact might be embedded with a particular group and their intention, and therefore may be unsuitable for other contexts and groups. In addition it is further argued that SCOT does not consider user opinion and extrapolates findings by only following trends (Faulkner, 2009).

In the field of IS, SCOT is furthered by Orlikowski and Gash (1994), who introduced the concept of technology frames of reference (TFR) by drawing on Bijker's notion of the technology frame. According to TFR, the perceptions, expectations and experiences of an individual pertaining to a provided technology, constitute their view and attitude about its nature, value and use. Individuals are argued to have separate and different TFRs and these are often created when a change is introduced that causes disturbance to established routines and work practices (Ciborra & Lanzara, 1994).

In the case where the TFRs of individuals match, they are called to be in congruence such that "*alignment of frames on key elements or categories*" exists, but when there are differences, they lead to "*incongruence of technology frames*" (Orlikowski & Gash, 1994). Significant incongruence would result in technology being abandoned, while congruent technology frames would achieve use and become embedded in routines.

By drawing from SCOT and TFR, it can be argued that users of technology constitute perceptions as frames that are influenced by their interaction with their surrounding environment and group members (Davidson, 2006; Davidson, 2002; Davidson, 1997). These frames govern their reaction to any change introduced and if congruent then positive impacts such as use of provided technology can be achieved. On the other hand, if incongruent frames dominate, the use of technology will cease and efforts to deploy or evaluate would be wasted. The notion of frames discussed here and the way in which SCOT views the three entities of context, technology and the user, facilitates to draw holistic definition of each of the three entities in Triality Framework. The framework is presented later in the chapter.

The focus now moves to ANT which differs greatly from SCOT in terms of its analytical approach.

2.2 Actor Network Theory (ANT)

In actor network theory (ANT), there is no analytical difference between a technology artefact and a human user (Latour, 1999; Greenhalgh & Stones, 2010). Both are described as actants that pose as an agency, and together aim at achieving a goal within a heterogeneous network, while drawing on artefacts, text, conventions such as money, people or a hybrid of these intermediaries.

An agency that is described as the technical artefact and human user is argued to be semiotically equivalent; which requires thinking symmetrically about human and nonhuman agents. Both Pickering (1993) and Orlikoswki (2007, 2008) favour the use of this notion and have extended it. For example Pickering introduced the notion of the 'Mangle of practice', according to which, he argues that the human agency and the machine agency are temporally emergent in practice, with the goals of the human agency governing the outcome. In other words, human agency uses technology as its "*temporal extension*" and by which it can succeed in its endeavours. He further suggests that it is the interaction between the human agency and the machine agency which leads to resistance and accommodation.

More recently, using the underlying philosophy of ANT, Orlikowski (2007) introduced the concept of 'Cognitive entanglement', which "*presumes that there are no independently existing entities with inherent characteristics*". The entities in this statement pertain to the context, technology and the user. The notion suggests that there is no recursive relationship between the entities, and according to this approach the "*focus is on agencies that have so thoroughly saturated each other that previously taken for granted boundaries are dissolved*" (Orlikowski, 2008).

An approach encompassing such a perspective towards technology and human as a user is favoured because its argued to solves the problem of representation (Pickering, 1993; Orlikowski, 2007, 2008); which because of the subject object dichotomy employed within other perspectives, fails to acknowledge the issues arising due to power imbalance and shift in knowledge that occurs when a technological artefact is introduced within heterogeneous networks (Latour, 1999).

A heterogeneous network in ANT is a reference to the context in which human-machine interactions are situated (Law & Callon, 1992). The context is governed by sociology of translation. Translation here refers to the processes and actions that are involved in achieving a result due to the interaction between the actants, and it depends upon irreversibility and alignment of the interests of the actors with the overall network aims and goals. In addition, it is suggested that networks are successfully translated when they conform to the regulation, norms (local) and are constituted by legitimate actors. Such translations allow networks to converge and therefore, be effective and durable (Avgerou, 2001, 2002). This notion of sociology of translation is applied in the field of healthcare by Nicolini (2010) in his work to understand the adoption and assimilation of telemedicine innovation.

Among the main drawbacks of ANT is its analytical dimension in which human and machine agencies are treated as the same (Pickering, 1993; Avgerou, 2002). Labelled as the post-humanist approach, it is argued that the use of this analytical lens crumbles when the complexity and number of networks increases. In addition, it is debated that the context in which both human and nonhuman actors interact is only acknowledged to be politically influenced and the social and cultural avenues are ignored.

However, the very criticism of ANT is also seen as its main strength as some argue that by using such an analytical concept, ANT accounts for negotiating, redefining and appropriation of interest of human actors and those *"inscribed within a technical artefact"* (Orlikowski et al., 1996). One example that used the notion of negotiation in healthcare is study where the implementation of telehealth is studied, and it is argued that the interaction resulting between context, information communication technology (ICT), and the user due to ICT implementation is a social process. It requires negotiating between power, politics and meaning encompassed within ICT and its intended use (this concept uses the term ICT instead of IS innovation) (Constantinides & Barrett, 2006a, 2006b).

Another important concept that ANT provides is that of normalisation. This concept facilitates understanding about technology and its assimilation within networks. More importantly this concept highlights the longitudinal nature of technology assimilation by users in their routines (Avgerou, 2002).

In addition, ANT's concept of 'Unintended consequences' outlines that the outcomes of introducing technology in a network might not always be desired and expected, and that unintended ways of technology use increases when users and designers are spatially absent or distanced (Nicolini, 2007). One study using this concept in healthcare (Harrison et al., 2007) suggested that unintended consequences can be attributed to the change introduced by the healthcare information technology (HIT) implementation, and that the role of such unintended consequence can be assessed by evaluating five types of sociotechnical interactions, shown in the form of an interactive sociotechnical analysis (ISTA) model. The model allows examination of the actual use of HIT, the impact of the technical and physical settings of work on HIT, users' interpretation of HIT use, and the recursive dependence of these factors on each other.

The enhanced understanding that ANT provides on conceptualisation of technology, context and the user is drawn upon while defining these three entities within Triality Framework.

So far, we looked at the variants of Science and Technology Studies (STS), the focus now moves to another theoretical perspective, Diffusion Of Innovation (DOI). This perspective provides a valuable understanding on how technology diffuses within a society by perceiving technology, context and the user from different perspective.

2.3 Diffusion of innovation

In this perspective, technology is conceptualised as innovation where innovation is encompassed in a broad definition as *"an idea, practice, or object that is perceived as new by an individual or other unit of adoption"* (Rogers, 2003). An innovation is argued to have five attributes of: relative advantage, compatibility, complexity, trialability, and observability. The definition of each attribute is presented in table1.

Innovation as technology is also suggested to be a composite of hardware and software, and an enabler of accomplishing actions to achieve desired goals.

The context according to this perspective is a social system consisting of social structures and norms. The structures are a type of information that regulate the individual's behaviour and are defined as *"patterned arrangements of units in a system"*. Norms on the other hand are defined as regularised behaviour patterns that are acceptable. The notion of a social system is inclusive of an organisation or a network of organisations. When an innovation is proposed, an organisation is argued to go through a process of innovation which, involves defining organisational need, matching the need with an innovation, going through the process of redefining and restructuring the innovation and organisational routines to achieve fit, clarify if any gaps remain, and finally achieve routinization where innovation is accepted with activities such that it *"loses its identity"* over time (Rogers, 2003; Lee 2004).

Relative advantage   *"is the degree to which an innovation is perceived as better than the idea it supersedes"*
Compatibility   *"is the degree to which an innovation is perceived as being consistent with existing values, past experiences, and needs of the potential adopters"*
Complexity   *"is the degree to which an innovation is perceived as difficult to understand and use"*
Trialability   *"is the degree to which an innovation may be experimented with on a limited basis"*
Observalibility   *"is the degree to which the results of an innovation are visible to others"*

Table 1. Five attributes of innovation (taken from Rogers, 2003)

The user, according to this perspective is understood to be an individual who is a part of the innovation-decision process that involves mapping the journey of an individual through gathering knowledge on the innovation, developing a perception towards the innovation, evaluating and taking the decision to either accept or reject the innovation, use the innovation, and finally confirm the decision (to use it or not). The user is also conceptualised to communicate his/her understanding about the technological innovation to the other individuals within the social system, thus, impacting its diffusion.

One of the main contributions of this perspective is the concept of consonance and dissonance, which argues that an individual's behaviour is attributed to their constant effort towards eliminating or reducing the uncomforting feeling due to change. In IS, this notion is used to understand how use of IT changes work practices at a micro level, and how practices change due to perceived dissonance between the context, technology and the action entailed within practices. Overtime, individuals enact practice as routines to experience consonance (Vaast & Walsham, 2005).

This perspective also proposes that the change in social structures and norms due to the introduction of innovation leads to resistance (Greenhalgh et al., 2008). Using this perspective to understand the diffusion of electronic records in the NHS. U.K. resulted in a conceptual model that considers *"the determinants of diffusion, dissemination, and implementation of innovations in health service delivery and organization"* (Greenhalgh et al., 2004; Greenhalgh et al., 2008).

This perspective and how it had been used by other authors to understand the processes that impact diffusion of innovation provides a valuable contribution to the framework in this work to conceptualise the context, technology and user.

Having discussed theories that acknowledge the three entities of technology, context and the user as important yet do not discuss the nature of relationship between these three entities, we move our attention to the theories that provide such conceptualisation of relationship between the entities. These theories and concepts include the contextualist approach, the structuration theory and the technology-in-practice. We start with contextualist approach that elaborates on the recursive relationship between the (technological) content and the context.

2.4 Contextualist approach

The contextualist approach elucidates the importance of considering context when studying change and the processual nature of such a change. According to this approach, context is conceptualised to consist of outer and inner contexts. The outer context refers to inter-organisational conditions such as political, social, economical and competitive environment (Pettigrew, 1985, 1987, 1997). On the other hand, inner context accounts for intra-organisational aspects such as structure, organisational culture and political circumstances.

In IS, the contribution of the contextualist approach is considered important as defining the context can be often problematic (Avgerou & Madon, 2004); and by extrapolating this notion to IS, context can be defined as a set of variables that affect information systems and in turn are influenced by it. This is termed "*the environment*", and informs that context should be considered as emergent (Avgerou, 2001).

Conceptualisation of technology according to this perspective is that of a content that can be understood as "*the particular areas of transformation under investigation*" (Pettigrew, 1987). Apart from introducing new technology, when viewed from the perspective of organisational context, these areas of transformation can also include geographical positioning, change in organisational culture and work force. It may be further argued that content and context are bound in a mutual relationship, where one is continually shaped by and shapes the other. Outlining the recursive relationship between content and the context is one of the contributions of this approach (Pettigrew, 1985, 1987, 1997).

The relationship between context and content is enabled through process, where process encapsulates the actions, reactions and interactions of the various parties that are involved in changing of an organisation from one state to another (Markus & Daniel, 1988; Pettigrew , 1987). Thus, it represents continuity and interdependent sequence of actions and events related to a given phenomena and allows understanding of its (phenomena's) origin, continuance and resultant outcome. At an actor level, process is represented and described by verb forms such as interacting, acting, reacting, responding and adapting. Whereas, at a system level, emerging, elaborating, mobilizing, continuing, changing, dissolving and transforming describe process.

The strength of the contextualist approach lies in the provision of "*guiding assumptions*" to carry out the research appropriately (Pettigrew, 1997). Among these assumptions is the inclusion of vertical and horizontal analysis. It is argued that having multiple levels of analysis enables the emergent and situational nature of process under the given context to be captured. Analysis at the vertical dimension includes studying group dynamics and organisational issues, and the interrelationship between those levels. Horizontal analysis involves interconnecting longitudinally the phenomena of interest studied at the vertical level (Pettigrew 1987, 1990, 1997; Pettigrew et al., 2001; Walsham & Waema, 1994; Walsham 1993). Such form of analysis suggests that a theory or theories can be used as a motor to drive analysis, and that the processes under analysis should be linked to the outcome.

In IS, this approach is adhered to link content, context and process to the "'what', 'why' and 'how' of evaluation", where it is argued that "effective evaluation requires a thorough understanding of the interactions between these three elements" (Symons, 1991). In the field of healthcare, the contextualist approach is used to understand the adoption, use and diffusion of telehealth, and it is argued that the contextualist approach through its characteristic of facilitating multi-level analysis allows the researcher to "study network-level innovations involving multiple organizations and stakeholders" (Cho, 2007).

This perspective offers a valuable contribution to Triality framework as it confirms the recursive relationship between the entities of context and technology.

The focus now moves to structuration theory which introduces the concept of 'Duality of structure', emphasising that there exist a relationship between the context and the user.

2.5 Structuration theory

According to structuration theory (ST), the changes in social systems are not simply an outcome of either human action (subject to knowledgeability) or social structures, but a product of their interaction. This relationship of simultaneous mutual shaping is known as the 'Duality of structure', which is "that the structural properties of social systems are both the medium and the outcome of practices that constitute those systems" (Giddens, 1979). In other words, there exists a recursive relationship between the user (in this theory called human agent) and the context (in this theory termed social system), and thus, it offers a valuable contribution to Triality framework as it confirms the recursive relationship between context and the user. Understanding this reciprocal and recursive relationship involves three main concepts: the social system, structure and the human agency.

Social systems are the reservoir of recursive social practices that human agents enact daily. In modern social system, these practices are institutionalised and constitute deeply embedded routine work; changes to which are resisted. The rules and resources that human agents draw on, and simultaneously enact while accomplishing their daily routine through interaction within the social systems, are known as structures. Structures reside in human "memory traces" and are argued to be solely dependent on human engagement. Structures are further argued to the facilitators and the constrainers of the human agent's engagement (Giddens 1979, 1984).

It is suggested that the structure are produced and reproduced through human interaction and therefore, constitute the 'Duality of structure'. For example, structure of signification (structures of meaning) is enacted by human agents by drawing on their interpretive scheme, through communication, to evaluate the underlying motive of their actions and those around them. The structure of domination is enacted when human agents exercise power through facilities such as ability to locate resources (Walsham, 1997; Walsham & Han, 1991). The structure of legitimation is enacted when agents assess their actions and sanction them through norms (morality and ethics) (Giddens 1979, 1984, 1990, 1991; Sharma et al., 2011).

The above notion also implies that human agents have an inherent capacity to act and transform with an exercise of power (transformative capacity), and this is called the 'agency'. Furthermore, 'agency' is argued to be reflexive in nature such that human agents continually monitor their actions and that of others. In essence, it can be argued that in order to enact new routines, the agents utilise reflexivity and transformative capacity to enhance their knowledgeability which in turn reciprocates their decisions (Giddens 1979, 1984, 1990, 1991; Sharma et al., 2011).

The structuration theory has been widely acknowledged in various fields, and especially in IS (Jones et al., 2004; Sharma et al., 2011). However, there are challenges that need to be overcome when using this theory, and these are described next.

The first challenge faced when using structuration theory is that it encompasses number of concepts that are used to elucidate the 'Duality of structure'. This affects the applicability of ST in the field of IS, and presents a dilemma of whether to use the theory in its entirety (as a meta-theory) and risk abandoning salient concepts, or use specific concepts and risk losing its overarching perspective (Walsham & Han, 1991; Thompson, 2004; Jones & Karsten, 2008). However, one solution is to use specific concepts to facilitate detailed and meaningful exploration of a problem.

The second potential challenge of applying ST to the area of IS is its lack of focus on technology (DeSanctis & Poole, 1994; Pozzebon & Pinsonneault, 2005). This can be mapped onto adopting a nominal view about technology (Orlikowski & Icano, 2001). Many IS researchers have acknowledged this as a critical gap and have provided ways in which this can be addressed (Barley, 1986; Orlikowski 1992, 2000). These efforts are described next.

2.5.1 The scripts

One of the first examples of the use of ST to study the effects of technology introduction was to expound how the computed tomography (CT) scanners introduced in the radiology department of two hospitals enabled enactment of different structures (Barley, 1986). However, in order to overcome the lack of emphasis on technology in structuration theory and its effect on the interaction between agents and thus the structures enacted, the technology was defined as *"an intervention into the relationship between human agents and organisational structure, which potentially changes it"* (Orlikowski & Baroudi, 1991).

Moreover, by drawing on the duality of structure, other peripheral concepts within ST were diminished and the concept of 'scripts' was introduced, which was termed as *"observable, recurrent activities and patterns of interaction characteristic of a particular setting"* (Barley & Tolbert, 1997). This allows the link between action and institution to be explored using structuration theory and organisational theories, with a resultant sequential model of institutionalisation that has four *"transition states"* or moments based on scripts: encode, enact, replicate or revise and, externalise and objectify. These being a representative of practices that are enacted and in the light of changes, they are replicated or revised and eventually become part of daily routine or institutionalised.

The institutionalization model may be extended to explore the relationship between deployment of new technology and expertise (Black et al., 2004). This was achieved by applying mathematical modelling to ethnographic data, to determine how differences in the expertise of users affect their willingness to collaborate with others. It is suggested that enhancing users' knowledge on using technology does not necessarily guarantee its use and improved outcome. Instead, the success of technology is related to distribution of expertise amongst the users, where, the concept of relational distribution is explained as "*a relative balance in operational knowledge*" (Black et al., 2004).

2.5.2 Technology-in-practice

In 'Duality of technology', the technology is considered as an artefact embedded with structure, where structures are defined as the "*mental models*" that human agents enact while drawing on the rules and resources available to them and interacting with colleagues and the environment (Giddens 1979, 1984; Orlikowski 1992). An artefact is described as a tool that may be employed for intended use or other purposes, some of which may be completely unrelated to the original (Orlikowski & Iacono, 2001). Furthermore, it is the user's contextual surroundings that might influence the way in which the tool is used or allow for its use to be re-invented (Orlikowski, 2000; Constantinides & Barrett, 2006a, 2006b; Pettigrew, 1987).

However, the idea of an artefact embedded with structures contradicts the emergent use of technology and therefore, Orlikowski proposed an enhanced view in 2000 in which she proposed technology-in-practice lens that distinguishes between technology not in use and technology in use. According to this lens, it is argued that technology is to be considered as an artefact when not used but, when agents in their daily routines use technology, it becomes 'technology-in-practice', where structures of technology use are enacted by agents while they use and interact with a given technology in a given context (Orlikowski, 2000).

The phenomenon of emergent technology-in-practice was studied in three different organisations, all using the same technology (the Notes), and identified three types of enactment and six types of technology-in-practice (Orlikowski, 2000; Orlikowski & Barley, 2001; Orlikowski, 1993). It is observed that the types of enactments (structures enacted) are distinguished by the degree or extent of change in the process, technology or structure (context), which are termed as "*interactional, technological and institutional consequence*" respectively. Change in technology would result in the first type of enactment, "*inertia*", which is characterised by limited-use technology-in-practice. Change in technology and process would result in "*application*" as enactment and could involve collaboration, individual productivity, collective problem solving and process-support as technology-in-practice. The third type of enactment, "*change*", would involve improvisation technology-in-practice and is usually observed when there is a significant modification of the process, technology and structure.

This perspective offers a valuable contribution to Triality framework, as it confirms the recursive relationship between the technology and the user.

2.6 Conceptualising technological content, context and the user

Section one has presented theoretical perspectives to provide an understanding of the conceptualisation of technology, context, the user and the nature of relationship between them (See table2 below).

Theory	Conception of context	Conception of technology	Conception of the user
Social Construction of Technology	*Social context – society that is politically and economically influenced*	*Artefact (designed)*	*Social group comprising of individuals*
Actor Network Theory	*Politically driven-power play*	*Actant with semiotic agency*	*Actant with semiotic agency*
Diffusion of innovation	*Social system inherent to which are structures and norms*	*Innovation*	*Individual with goals to achieve and ability to learn and draw on experiences*
Contextualist approach	*Inner and outer*	*Content*	*Actor whose interacting, acting, reacting, responding and adapting*
Structuration Theory	*Social System*	*-*	*Human agent with an agency, and who is always involved in reflexivity of actions*

Table 2. Conception of context, technology and user according to different theoretical perspectives

Taking the essence of these perspectives into account, this research conceptualises technological content as IS innovation, context as healthcare social system, and the user as human agent. It further argues that a recursive relationship exists between each entity. This results in a conceptual framework that combines the congruent relationships, and enhances the common. This framework is called the Triality framework.

2.6.1 Conceptualising technological content as technological innovation

The term technological innovation is used to represent 'technology' in the Triality Framework as it ensures that technology is understood not just as an artefact, but rather a modality with emergent and dynamic use. Furthermore, although the innovation may be embedded within an overall vision, the way it is to be used is not fixed, and users may alter or modify it. This not only accounts for the attributes of innovation as outlined by Roger (2003) but also provides an explanation of the ways in which human agents in social systems enact practices, and how the process of enactment is facilitated through using innovation as a mediator, where its use can be improvised through the ability of the agents to learn and circumnavigate (Orlikowski, 2000, 2002).

This conceptualisation approach takes into account the ensemble view of technological artefact which focuses on *"dynamic interactions between people and technology whether during construction implementation or use in organizations or during the deployment of technology in society at large"* (Orlikowski & Icano, 2001). Moreover, ensemble view is highly apt for a complex context as healthcare (Chiasson & Davidson, 2004).

This notion also supports the view that where an innovation is considered to be a part of social change, it may be an enabler and a constrainer of actions due to various institutional, organisational, social, rational and idiosyncratic reasons (Avgerou, 2001; Avgerou & Madon, 2004).

The technological innovation term further includes the property of 'informate' where information technology possesses the duality whereby through its ability to automate the organisational transactions it also *"create(s) a vast overview of an organisation's operations, with many levels of data coordinated and accessible for a variety of analytical efforts"* (Zuboff, 1988)

2.6.2 Conceptualising context as healthcare social system

The introduction of technology in healthcare system introduce a number of changes to overall healthcare delivery, including the context where care is delivered and the way in which it is delivered (Boddy et al., 2009). This can result in factors contributing to conflict among its users and contradiction in overall goals (May, 1993; May et al., 2001, May & Ellis, 2001; May et al., 2003a; May et al., 2004; May et al., 2005; Whitten & Adams, 2003; Whitten & Mickus, 2007; Whitten & Mackert, 2005; Wootton et al., 2006). Studying and understanding such contextual changes can be facilitated by using ST and high modernity (Gammon et al., 2008; Hardcastle et al., 2005; Kouroubali, 2002; Lehoux et al., 2002; Peddle, 2007; Walsham, 1997; Whitten et al., 2007).

However, to apply this approach appropriately, IS innovation based services such as telehealth and electronic records have to be seen as an extension of a healthcare social system that spans across time and space. For this, Giddens's definition of social system and concept of disembedding mechanism will be called upon. As a healthcare system involves institutionalised practices, and by recalling the definition of a social system presented earlier (in SCOT, diffusion of innovation and ST), it is argued that a modern healthcare system can be perceived as a large social system that contains individuals interacting with each other, and whilst doing so, they enact various structures through continuous reflexive monitoring of their actions. Endogenous to these social systems are the disembedding mechanisms, to which, Giddens alludes as *""lifting-out" of social relations from local contexts of interaction and their restructuring across indefinite spans of time-space"* (Giddens, 1990). The concept of disembedding mechanisms was introduced by Giddens in Consequences of Modernity (CM), whereas opposed to his earlier work, Giddens reflects on the role of technology as a facilitator and contributor to the *"stretching"* of modern social systems (Giddens, 1990, 1991).

There are two types of disembedding mechanisms, symbolic tokens and expert systems, which collectively are termed as abstract systems (Giddens, 1990). Symbolic tokens such as money are defined as *"media of interchange which can be "passed around" without regards to the specific characteristic of individuals or groups that handle them at any particular juncture"* (Giddens, 1990). Expert systems on the other hand are described as *"systems of technical accomplishment or professional expertise that organise large areas of the material and social environments in which we live today"* (Giddens, 1990).

Comparing these definitions, technology can be understood as an expert system that combines *"technical accomplishments"* of equipment such as telehealth and electronic records with *"professional expertise"* of clinicians aimed at organised delivery of care for patients in a modern healthcare system.

It is not clear why the two disembedding mechanisms taken together should be called an abstract system, and therefore this research uses a modified definition where an abstract system is depicted to encompass an expert system organised and structured around care delivery processes. This modified perspective has two inherent advantages. From a theoretical point of view, it acknowledges all the pivotal concepts related to the abstract system without losing its central meaning (no interdependence is ascribed between the expert system and the symbolic token by Giddens, therefore subtracting symbolic tokens from abstract system studied should not affect its functionality). From a practical view, this perspective allows the contextual differences between traditional care delivery practices and technology based practices to be distinguished.

As an expert system, technology based healthcare delivery services create systems where the experts within the system meet lay people (Giddens, 1990). The interactions within such systems are based on trust and expertise.

Within this research, the 'context' may be substituted as the 'social system' which better describes a setting such as healthcare. Furthermore, using the term 'healthcare social system' in place of 'context' in the Triality framework acknowledges that modern organisations are extensions of social systems, and enables various dimensions such as political, economical and social conditions to be incorporated in the definition of context (Pettigrew, 1987; Van de Ven & Garud, 1993).

This approach also provides understanding of the dynamics of technology deployment and organisational discourse as instigator of social change (Avgerou, 2001) and as an interacting entity (Dopson et al., 2008).

2.6.3 Conceptualising user as human agent

The user of technology has been defined within IS deployment research as an actor with habitus, such as in ANT (Schultze & Boland, 2000; Latour, 1999). In this research, the term 'human agent' is considered more appropriate as it acknowledges that humans have an 'agency' that is their inherent capacity to act. This 'agency' is engaged in reflexivity where each action taken is reflected upon and learned from. By doing so, the human agents enact new routines and employ transformative capacity (power) to enhance their knowledgeability (Giddens, 1984; Edmonson et al., 2007). The term "agency" also enables intentionality and choice to be considered. It may also be argued that despite the constrictive nature of context; agents have freedom of choice as "*people are purposive, knowledgeable, adaptive, and inventive agents who engage with technology in a multiplicity of ways to accomplish various and dynamic ends. When the technology does not help them achieve those ends, they abandon it, or work around it, or change it, or think about changing their ends*" (Orlikowski, 2000).

Therefore, if this notion of agency is extended to understand technology deployment at the individual level, it can be argued that human agents will adopt and enact practices that influence the way the technology is used within a given context; and due to the recursive nature of this relationship, these adopted and enacted practices will in turn influence their perceptions of the technology. They weave a complex web of interpretation (which is influenced by previous experience, interaction with colleagues and the work environment) around the technology whilst making sense of it, it's real purpose and usefulness; and in doing so they lay the foundations of their interaction with the IS innovation that ultimately determines whether they will use it and accept its use as part of their daily routine or circumnavigate and resist its use. This notion also suggests that such a process of sense making can be lengthy, as the structures enacted by human agents are temporal in nature, that is, while some structures change quickly, others might change much more slowly (Barley, 1986; Barley & Tolbert, 1997; Black et al., 2004).

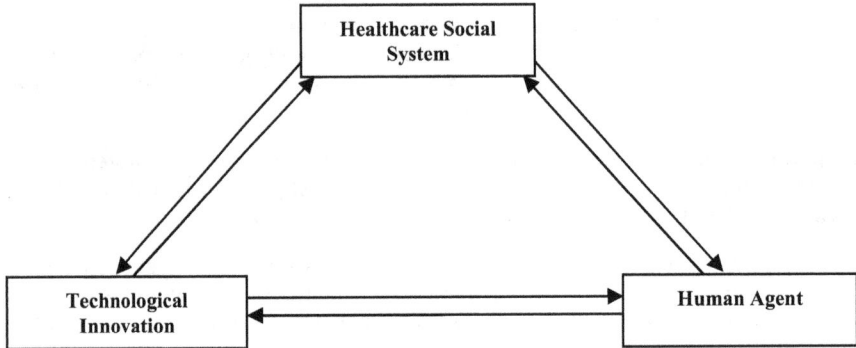

Fig. 1. Triality framework (initial version)

It is noted that when studying the processes within healthcare, in addition to the broader view, understanding dynamics between the three entities is also essential due to the complexities (of healthcare context and processes) involved. Therefore, in the next section, perspectives providing insight into such dynamics are presented. This also enables the perspectives discussed earlier to be aptly summarised and provide a holistic understanding of the deployment and evaluation efforts in the field of healthcare.

3. Theoretical concepts from healthcare and medical informatics literature: Impact of innovation deployment and evaluation

The perspectives discussed in this section contribute to Triality framework in number of ways. These perspectives outline the impact of introducing healthcare intervention such as telehealth and electronic records on routines and work practices of healthcare professionals over time, and also acknowledge the impact of such change on the expertise of healthcare professional, interaction during clinical encounter, and quality of care. It can be argued by drawing on these perspectives that such changes might influence healthcare professionals' decision to accept the IS innovation in daily routines.

In addition, these perspectives provide an understanding of complex processes and dynamics within healthcare systems, which can be attributed to the recursive relationship between the context of social system, the content of IS innovation and the human agent as the user. This enables the Triality framework to be updated.

3.1 Change in work practices and routines: The concept of normalisation, and routinisation

Routines are enacted because they involve guiding, accounting, referring, creation, maintenance, and modification. Individuals continuously adapt their routines with experience and by reflecting on previous outcomes, and routines rather than being stable are subjected to constant change (Feldman, 2000, 2003; Edmonson et al., 2007).

Routines have four attributes that include the repetition, recognisable pattern of action, multiple participants, and interdependent actions. Taking this notion into account, individuals within an organisation can be argued to share understanding of routines and the process of its enactment, and as a result of enacting dependent and interconnected routines lead to organisational routine which encompass ostensive and performative aspect (Feldman & Rafeali, 2002). The Ostensive aspect *"is the ideal or schematic form of a routine. It is the abstract, generalized idea of the routine, or the routine in principle"*. Performative aspect on the other hand, is described to *"consist of specific actions, by specific people, in specific places and times. It is the routine in practice"* (Feldman & Pentland, 2003). As a consequence, whether a change will be accepted and routinised depends upon the *"processes of variation, selection and retention that take place between the ostensive and performative aspect"* (Feldman & Pentland, 2003).

Extending the concept that organisational routines are a result of mutual action that is interdependent, a hospital setting can be considered a as negotiated order, or in other words, the healthcare context can be considered as comprising *"practice(s) in which various actors are expected to be relevant partners in specific work processes"* (Tjora & Scambler, 2009).

A similar approach is advocated by Greenhalgh and colleagues (Greenhalgh, 2008; Robert et al., 2009, 2010), where they considers that routines in medical settings overlap with each other and are impacted by changes at different levels, such as changes in the wider environmental context, organisational context, and individual context.

It is further argued that implementing an IS innovation or complex healthcare intervention such as telehealth and its evaluation strategy such as RCT, changes routine of professionals, and that it takes time for professionals to accept such changes (May & Ellis, 2001; May, 2006a, 2006b). This is called normalisation, and can be understood as *"the embedding of a technique, technology or organisational change as a routine and taken-for-granted element of clinical practice"*, and is argued to consist of a set of endogenous and exogenous processes (May & Finch, 2009; May et al., 2007a, 2007b; Mair et al., 2007a, 2007b). In this context, endogenous processes refer to processes that govern the patient-professional encounter; and exogenous processes refer to organisational structure, culture and division of labour.

The concept of normalisation has now been developed into normalisation process theory (NPT) that provides *"a robust and replicable ecological framework for analysing the dynamic collective work and relationships involved in the implementation and social shaping of practice"*. The main postulate of theory is that the work in enactment of practice is governed by four generative mechanisms of human agency. These are: Coherence which is *"work that defines and organizes the objects of a practice"*, Cognitive participation which is *"work that defines and organizes the enrolment of participants in a practice"*, Collective action which is *"work that defines and organizes the enacting of a practice"*, Reflexive monitoring *"work that defines and organizes the knowledge upon which appraisal of a practice is founded"*.

Having identified how routines and work practices are interrelated, and why they are important, we now move to discussing how work practices change due to IS innovation deployment and evaluation.

3.2 Virtualisation and invisibility

It is argued that deploying IS innovations such as telehealth and electronic records make medicine a virtual practice and predispose these work practices to become invisible (Mort et al., 2003; Mort & Smith, 2009; Sandelowski, 2001).

There are different forms of invisible work, and in this research disembedding background work is considered as one such form. Disembedding background work is the work where although the individual is visible to others, the extent and severity of their work practices is rather invisible (Star & Strauss, 1999; Oudshoorn, 2008; Nicolini, 2006). One example of such work is that of nurses.

Disembedding background work due to IS innovation deployment and evaluation originates due to virtualisation of clinical encounter, particularly in case of telehealth. This virtualisation apart from having an impact on interaction between the patient and nurse, also influences trust as telehealth involves *"faceless commitments"* that are based on trustworthiness vested in technology. Nursing practice without telehealth however, involves trustworthiness established between different individuals based on *"facework commitments"*, due to physical presence of those involved. In other words, introduction of disembedding work such as virtualisation of clinical encounter calls for investing trust *"not in individuals but in abstract capacities"* (Giddens, 1990; Kouroubali, 2002; Nicolini, 2007; Nicolini, 2010; Bhattacherjee & Hikmet, 2007; Berg, 2003).

Disembedding work can also be perceived as threat to expertise. This stems from changes to skill set and predisposing one's experience due IS innovation deployment and evaluation. In addition, individual's resistance to disembedding work could also be enacted due to concerns pertaining to the impact of change on standards of care delivery (Nicolini, 2006, 2007; Wears & Berg, 2005; Waterworth et al., 1999). This notion of change having an impact on the quality of care is well documented by Donabedian in his work, and it is discussed next.

3.3 Structure, process and outcome framework

According to this framework, the impact of IS innovation on quality of care can be evaluated through seven attributes (Donabedian, 2003):

- *Efficacy*
  The ability of science and technology of healthcare to bring about improvements in health when used under the most favourable circumstances
- *Effectiveness*
  The degree to which attainable improvements in health are, in fact, attained
- *Efficiency*
  The ability to lower the cost of care without diminishing attainable improvements in health
- *Optimality*
  The balancing of improvements in health against the costs of such improvements
- *Acceptance*
  Conformity to the wishes, desires, and expectations of patients and their families
- *Legitimacy*
  Conformity to social preferences as expressed in ethical principles, values, norms, mores, laws, and regulations

- *Equity*
  Conformity to a principle that determines what is just and fair in the distribution of health care and its benefits among members of the population

Many researchers have used this *"triad"* to explain the process of IS innovation deployment and evaluation such as electronic prescribing and administration system, in healthcare setting (Runciman, 2010; Cornford et al., 1994; Barber et al., 2007).

This concept allows to establish a link between deployment of healthcare innovations such as telehealth, and electronic records, and perception of healthcare professionals pertaining to impact on quality of care. It therefore, enables defining attributes of recursive relationship between healthcare social system and the human agent, and between the healthcare social system and IS innovation as presented in the next section.

4. Triality framework

By drawing on the all the theoretical concepts discussed in this chapter, the conceptual framework can now be presented in its entirety. Where according to the previous version, the Triality framework acknowledged that the technological innovation, healthcare social system, and human agent are interrelated to each other through recursive relationship, the new version identifies set of attributes attached to each of the recursive relationship. The framework further provides an understanding on how and why these attributes are articulated, and thus facilitates gaining a deeper insight into the deployment and evaluation processes.

In addition, the framework encompasses the notion that each attribute influences work routines which are described here as social practices that are enacted by interactions of individuals whose actions are interdependent with the actions of other individuals within the complex context of healthcare.

The Triality framework has many advantages. It ensures multilevel analysis that is longitudinal in nature; responds to the calls for furthering the use of theoretical perspectives in the field of healthcare (Davidson & Chiasson, 2004; Cho, 2007; Greenhalgh et al., 2009); encompasses the 10-e's of e-health described by Eysenbach (2001), and the notion of complex healthcare intervention described by authors such as May and colleagues (2007a, 2007b), and Campbell et al., (2000) within IS innovation.

4.1 Attributes of relationship between IS innovation and healthcare social system

The three attributes at this level attributes are drawn from SCOT and Donabedian's work on quality of care, and illustrate how the design of IS innovation is influenced by, and influences the healthcare social system; and how the use of IS innovation is assessed on the basis of efficiency and effectiveness, and optimality and equity.

- *Design*
  The balance of IS innovation design against the needs and requirements of the end-user, and its ability to fulfil and meet the outlined needs and requirements through its design (and design features), without being cost intensive.

Innovation design is one of biggest factors that determine whether it will be used by users and assimilated in their work practices (Orlikowski & Icano, 2001; Greenhalgh et al., 2009).

Attributes:
1. Design
2. Effectiveness & Efficiency
3. Optimality & Equity

Attributes:
1. Legitimacy
2. Acceptability
3. Demand & Efficacy

Healthcare Social System

Routines/ Work practices

Technological innovation

Human Agent

Attributes:
1. Expertise
2. Interaction
3. Trust

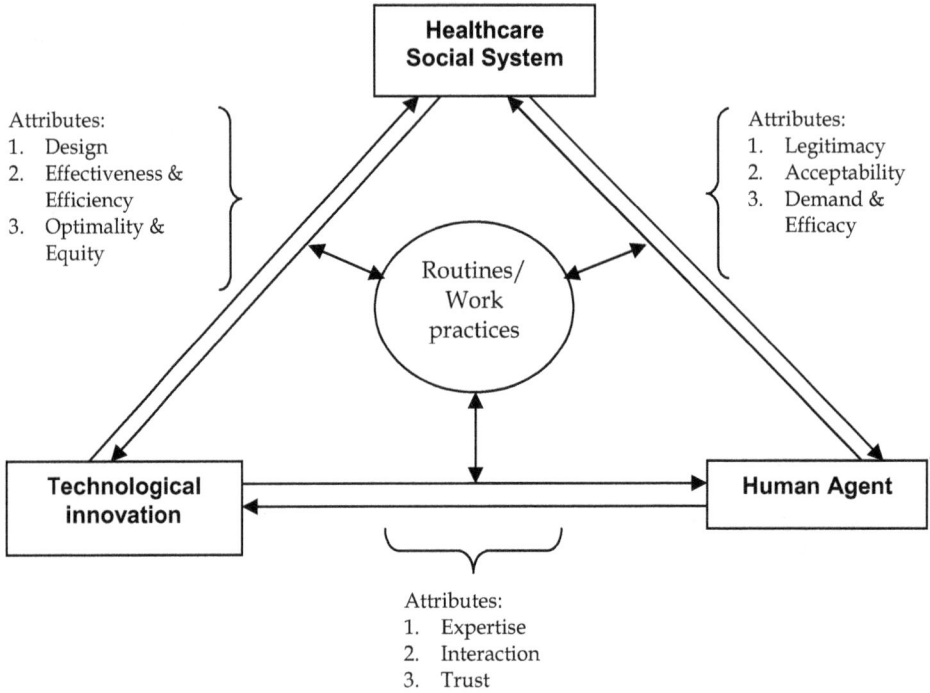

Fig. 2. The Triality framework (final version)

The tension between the design as needed and as provided is created due to many factors such as contextual issues, regulatory issues, financial margins etc; and in many cases it is the end user who decides how successful the design is, as according to Chau and Hu's (2001, 2002) study, healthcare professionals use a technology that they perceive to be designed appropriately and fit well with their work practices.

- *Efficiency*
  The ability of IS innovation to lower the time and allow someone to complete/carry out various different tasks thereby increasing productivity "and using time efficiently" at individual level
  The ability of IS innovation to lower the cost without diminishing attainable improvements at system level, &
- *Effectiveness*
  The degree to which attainable improvements are attained at system level and individual level (adopted from Donabedian, 2003)

It is argued that healthcare professionals assess the efficiency and effectiveness by evaluating how the IS innovation might help them to deliver care and what impact it has on the patients. Often, efficiency is measured in terms of time saved during a routine task and how well it was performed.

Effectiveness on the other hand is measured in terms of quality of improvement in care delivery processes and the outcome (pertaining to patient health) (Bannister & Remenyi, 2003; Mair et al., 2007a, 2007b). Effectiveness in the field of healthcare is measured by applying various evaluating methodologies to an intervention and RCT is one such method (MRC 2000). Applicability of RCT to evaluate technological intervention has been questioned due many reasons. These include gaining ethical approval, defining selection criteria and recruiting patients, training staff and gaining their support (Rosen & Mays, 1998).

- *Optimality*
  The balancing of improvements in health against the cost of such an intervention (adopted from Donabedian, 2003), &
- *Equity*
  The just and fair distribution of health care delivery among the patient population (adopted from Donabedian, 2003).

This attribute outlines that the healthcare professionals often question service procurement decisions due to doubts over cost spend and, benefit promised and realised (McDonald et al., 2006; Mair et al., 2007a, 2007b).

In addition, it is argued that healthcare professionals favour equitable distribution of IS innovation despite the innovation being at evaluation stage as some strategies such as RCT are not promoted well enough and in doing so the managers and teams involved in evaluation fail to adequately address the need for RCT and healthcare professionals role in it (Greenhalgh & Russell, 2010; Mair et al., 2007a, 2007b;, Berghout & Remenyi, 2005).

4.2 Attributes of relationship between healthcare social system and the human agent

The three attributes at this level are deduced from ST and Donabedian's work on quality of care, and they highlight the structures enacted by the human agents as users of IS innovation within the healthcare system, where, by drawing on rules and resources the agents assess the compatibility and the need of such an innovation.

- *Legitimacy*
  Compatibility of IS innovation and its conformity to the ethical issues, laws, regulations and values.

Healthcare professionals are known to make decisions that are bound by ethical and legal issues.

In particular, ethical issues arise due to evaluation strategies such as RCT as it governs the provision and allocation of equipment to patients (Finch et al., 2003, 2006; Heaven et al., 2006; Lankshear & Mason, 2001; Greenhalgh & Russell, 2010; McDonald et al., 2006; Ammenwerth et al., 2003).

- *Acceptability*
  The degree to which IS innovation is compatible with the current job role and work routines, and conforms to the expectations of its users.

It is argued that both, the processes of IS innovation deployment and evaluation impact the work practices of healthcare professionals. This can be caused due to increased workload and redefining work roles (McDonald et al., 2006).

However, it is also suggested that such instances of increased workload and change in roles are negotiated over time and become part and parcel of routinised work practices (Greenhalgh, 2009; May et al., 2007a, 2007b).

- *Demand*
  The degree to which IS innovation is needed to improve healthcare delivery at individual level; and reduce cost at system level, &
- *Efficacy*
  The ability of IS innovation to bring about improvements in health of the patients (adopted from Donabedian, 2003).

It is argued that the demand of IS innovation is mainly dominated by users experience and perception; and the sense of efficacy mainly originates from the scientific evidence provided to back up the IS innovation deployment and evaluation through strategies as RCT (McGrath et al., 2008; Clark & Goodwin, 2010; Cornford et al., 1994; Hibbert et al., 2004; Finch et al., 2008).

4.3 Attributes of relationship between the human agent and the IS innovation

The attributes at this level are conceived from ST and theory of normalisation, and they facilitate understanding the various dynamics that contribute to the human agent's perception and the practice they enact in response IS innovation deployment and/or its evaluation, and the implications this has on the assimilation of IS innovation in work practices.

- *Expertise*
  Skills and knowledge required to use IS innovation in order to deliver better care to the patients (and enhance clinical encounter

Addition of IS innovation could be regarded as an agent responsible for introducing *"expertise asymmetry"* and can cause hindrance towards its assimilation (Lehoux et al., 2002; Sicotte & Lehoux, 2003; Greenhalgh 2009).

In healthcare, expertise can be understood to be of two types: social expertise required during social interaction with patient and colleagues, and technical expertise employed in using the IS innovation correctly.

Expertise is affected by training and support (Gagnon et al., 2003, 2006; Sallas et al., 2007). In addition, Chau and Hu (2001, 2002) found in their study, that staff support was credited for physicians' acceptance of technology; and it can be challenged due to status quo, power dynamics and autonomy.

- *Interaction*
  Establishing new interaction patterns and ways of communicating between various individuals at different levels

Deployment and evaluation processes pertaining to IS innovation impact the normal interaction patterns of healthcare professionals with their colleagues and patients.

Particularly, changes to interactions during clinical encounter are more evident as May (1992) argues that healthcare professionals such as nurses interact with their patients to develop 'knowing', where it refers to developing an understanding of patient not as biomedical subject but as an individual with social and personal background and needs. This requires capturing the holistic picture which entails subjective data such as how did the patient look, and the objective data such as blood pressure, weight etc. However, such 'knowing' is affected by introducing IS innovations in work practices and changing the clinical encounter.

- *Trust*
  Questioning the degree to which IS innovation is perceived to be reliable and safe

According to Giddens, trust serves to "*reduce or minimise the dangers to which particular type of activity are subject*", where "*danger is understood as a threat to desired outcomes*" (Giddens, 1990). A recent study by Sharma et al., (2010, 2011) contributes to this topic by arguing that trust impacts clinical users' decision on using the IS innovation provided, and on a similar note, Gagnon et al., (2006) argues that it affects users' readiness towards IS innovation such as telehealth.

Trust can be envisaged as technology trust, interpersonal trust and organisational trust (Li et al., 2008; Lippert & Davis, 2006; Lippert & Swiercz, 2005). Lippert and Davis (2006) evaluate the impact of trust on technology internalisation which refers to "*the effective and continued use of technology over time*" (Lippert & Davis, 2006; Lippert & Swiercz, 2005), and propose that greater degree of interpersonal trust and technology trust among the members of an organisation would lead to effectiveness in technology adoption and eventually achieve internalisation (Lippert & Davis, 2006).

5. Conclusion

Triality Framework presents an argument that a 'recursive' relationship representing mutual interaction exists not only between two entities, but between all three interacting entities of context, content and the user. This interaction is enacted through various dynamics termed the attributes. The framework further:

- provides an understanding on how assimilation and routinization of technological innovation and its evaluation strategies changes over time.
- Highlights how users' decisions on using and accepting IS innovation and its evaluation strategies is influenced by disturbance in their work practices. These disturbances are mediated through the nine attributes outlined in the framework.
- provides understanding how a healthcare professional makes sense of the change introduced. In the process contributes to the knowledge of change management.
- By application of the nine attributes and the dynamics of their articulation, informs all those involved in e-health deployment and evaluation how to improve their approach.

5.1 Limitations and future work

Triality framework as presented in this chapter does not ascertain the significance of attributes statistically as is the case in models such as Technology Acceptance Model (TAM). Future research could consider using a method that involves statistical analysis and enable evaluating the strength of relationships between the attributes and express them as dependencies.

In addition, the user as human agents in Triality framework mainly encompasses healthcare professionals such as doctors, nurses and technicians. They are viewed as the primary users of innovation. The framework does not consider patient. Future research could consider how to include the patient, as the role that a patient play in modern healthcare system has evolved to that of an expert patient. Research could also consider whether the involvement of patient as one of the primary users influences outcome, and whether there are any differences in ways different group of users perceive the innovation, its deployment and evaluation processes.

Last but not least, it is vital to note that Triality framework has been mainly presented to understand the challenges that the healthcare sector faces when new technological innovations are deployed and evaluated. Future research could consider how Triality framework can be mapped across other areas and whether the challenges facing other sectors are equally challenging.

6. References

AMMENWERTH, E., GRÄBER, S., HERRMANN, G., BÜRKLE, T. & KÖNIG, J. (2003). Evaluation of health information systems - Problems and challenges. *International journal of medical informatics*, 71, pp. 125-135.

AVGEROU, C. (2002). *Information Systems and Global Diversity*. 1 edn. USA: Oxford University Press Inc.

AVGEROU, C. (2001). The significance of context in information systems and organizational change [Online]. *LSE Research Online*.

AVGEROU, C. & MADON, S. (2004). Framing IS studies: understanding the social context of IS innovation. In: C. AVGEROU, C. CIBORRA and F. LAND, eds, *The Social Study of Information and Communication Technology Innovation, Actors, and Contexts*. 1 edn. New York: Oxford University Press, pp. 162-182.

BANNISTER, F., & REMENYI, D. (2003). The Societal Value of ICT: First Steps Towards an Evaluation Framework. *Electronic Journal of Information Systems Evaluation*, 6, pp. 197-206

BARBER, N., CORNFORD, T., & KLECUN, E. (2007). Qualitative evaluation of an electronic prescribing and administration system. Quality & safety in health care, 16, pp.271-278.

BARLEY, S. R. (1986). Technology as an Occasion for Structuring: Evidence from Observations of CT Scanners and the Social Order of Radiology Departments. *Administrative Science Quarterly*, 31, pp. 78-108.

BARLEY, S.R. & TOLBERT, P.S. (1997). Institutionalization and structuration: Studying the links between action and institution. *Organization Studies*, 18, pp. 93-117.

Barlow,J. Bayer,S. Castleton,B. & Curry, R. (2005) Meeting government objectives for telecare in

BERG, M., (2003). The Search for Synergy: Interrelating Medical Work and Patient Care Information Systems. *Methods of information in medicine*, 42, pp. 337-344.

BERGHOUT, E. & REMENYI, D. (2005). The Eleven Years of the European Conference on IT Evaluation: Retrospectives and Perspectives for Possible Future Research" *The Electronic Journal of Information Systems Evaluation*, 8, pp. 81-98

BHATTACHERJEE, A. & HIKMET, N. (2007). Physicians' resistance toward healthcare information technology: a theoretical model and empirical test. *European Journal of Information Systems,* 16, pp. 725-737.

BIJKER, W.E., (1995). *Of bicycles, bakelites, and bulbs : toward a theory of sociotechnical change.* 1 edn. Cambridge:MA: MIT Press.

BLACK, J. L., CARLILE, R. P., & REPENNING, P. N. (2004). A Dynamic Theory of Expertise and Occupational Boundaries in New Technology Implementation: Building on Barley's Study of CT Scanning. Administrative Science Quarterly, 49, pp. 572-607

Boddy, D., King, G., Clark, S. J., Heaney, D., & Mair, F. (2009). The influence of context and process when implementing e-health. *BMC Medical Informatics and Decision Making,* 9:9, doi:10.1186/1472-6947-9-9

CAMPBELL, M., FITZPATRICK, A., HAINES, A., KINMONTH, L. A., SANDEROCK, P., SPIEGELHALTER, D., & TYRER, P. (2000). Framework for design and evaluation of complex interventions to improve health. *British Medical Journal,* 321, pp. 694-696

CHAU, P.Y.K. & HU, P.J. (2001). Examining a model of information technology acceptance by individual professionals: An exploratory study. *Journal of Management Information Systems,* 18, pp. 191-229.

CHAU, P.Y.K. & HU, P.J. (2002). Investigating healthcare professionals' decisions to accept telemedicine technology: An empirical test of competing theories. *Information and Management,* 39(4), pp. 297-311.

CHO, S. (2007). *A contextualist approach to telehealth innovations.* PhD thesis. The Robinson College of Business of Georgia State University.

CHIASSON, M.W. & DAVIDSON, E. (2004). Pushing the contextual envelope: Developing and diffusing IS theory for health information systems research. *Information and Organization,* 14, pp. 155-188.

CIBORRA, C.U. & LANZARA, G.F. (1994). Formative contexts and information technology: Understanding the dynamics of innovation in organizations. *Accounting, Management and Information Technologies,* 4, pp. 61-86.

CLARK, M. & GOODWIN, N. (2010). *Sustaining innovation in telehealth and telecare.* UK: WSD Action Network.

CONSTANTINIDES, P. & BARRETT, M. (2006a). Large-scale ICT innovation, power, and organizational change: The case of a regional health information network. *Journal of Applied Behavioral Science,* 42, pp. 76-90.

CONSTANTINIDES, P. & BARRETT, M. (2006b). Negotiating ICT development and use: The case of a telemedicine system in the healthcare region of Crete. *Information and Organization,* 16, pp. 27-55.

CORNFORD, T., DOUKIDIS, G. & FORSTER, D. (1994). Experience with a structure, process and outcome framework for evaluating an information system. *Omega,* 22, pp. 491-504.

CORNFORD, T. & KLECUN-DABROWSKA, E. (2003). Images of Health Technology in National and Local Strategies. *Methods of information in medicine,* 42, pp. 353-359.

CRUICHSHANK, J. (2010). *Healthcare without walls: a framework for delivering healthcare at scale.* London. 2020health.

DAVIDSON, E. (2006). A technological frames perspective on information technology and organizational change. *Journal of Applied Behavioral Science,* 42, pp. 23-39.

DAVIDSON, E. (2002). Technology frames and framing: A socio-cognitive investigation of requirements determination. *MIS Quarterly: Management Information Systems, 26,* pp. 329-358.

DAVIDSON, E. (1997). Changing frames or framing change? Social cognitive implications of organizational change during IT adoption. *Hawaii International Conference on System Sciences (HICSS)* Volume 3: Information System Track-Organizational Systems and Technology, pp. 475-484.

DESANCTIS, G. & POOLE, M. (1994). Capturing the complexity in advanced technology use: Adaptative structuration theory. *Organization Science,* 5, pp. 121-146.

DONABEDIAN, A. (2003). *An Introduction to Quality Assurance in Health Care.* 1 edn. Oxford: Oxford University

DOPSON, S., FITZGERALD, L., & FERLIE, E. (2008). Understanding Change and Innovation in Healthcare Settings: Reconceptualizing the Active Role of Context. Journal of change management, 8, pp. 213-231

Edmondson, c. a., Bohmer, M, R., & Pisano, P, G. (2001). Disrupted Routines: Team Learning and New Technology Implementation in Hospitals *Administrative Science Quarterly,* 46, pp. 685-716

EYSENBACH, G. (2001). What is e-health? *Journal of Medical Internet Research,* 3:e20.

FAULKNER, A. (2009). *Medical technology into healthcare and society: a sociology of devices, innovation and governance.* 1 edn. UK: Palgrave Macmillan.

FELDMAN, M.S. (2003). A performative perspective on stability and change in organizational routines. *Industrial and Corporate Change,* 12, pp. 727-752.

FELDMAN, M.S. (2000). Organizational Routines as a Source of Continuous Change. *Organization Science,* 11, pp. 611-629.

FELDMAN, M.S. & PENTLAND, B.T. (2003). Reconceptualizing organizational routines as a source of flexibility and change. *Administrative Science Quarterly,* 48, pp. 94-118.

FELDMAN, M.S. & RAFAELI, A. (2002). Organizational routines as sources of connections and understandings. *Journal of Management Studies,* 39, pp. 309-331.

FINCH, T., MAY, C., MORT, M. & MAIR, F. (2006). Telemedicine, Telecare, and the Future Patient: Innovation, Risk and Governance. In: A. WEBSTER, ed, *New Technologies in Healthcare Challenge, Change and Innovation.* 1st edn. P McMillan, pp. 84-96.

FINCH, T.L., MAY, C.R., MAIR, F.S., MORT, M. & GASK, L. (2003). Integrating service development with evaluation in telehealthcare: An ethnographic study. *British medical journal,* 327, pp. 1205-1208.

FINCH, T.L., MORT, M., MAIR, F.S. & MAY, C.R. (2008). Future patients? Telehealthcare, roles and responsibilities. *Health and Social Care in the Community,* 16, pp. 86-95.

FITZGERALD, L., FERLIE, E., & HAWKINS, C. (2003). Innovation in healthcare: how does credible evidence

GAGNON, M.-., DUPLANTIE, J., FORTIN, J.-. & LANDRY, R. (2006). Implementing telehealth to support medical practice in rural/remote regions: What are the conditions for success? *Implementation Science,* 1, doi:10.1186/1748-5908-1-18

GAGNON, M.-., GODIN, G., GAGNÉ, C., FORTIN, J., LAMOTHE, L., REINHARZ, D. & CLOUTIER, A. (2003). An adaptation of the theory of interpersonal behaviour to the study of telemedicine adoption by physicians. *International journal of medical informatics,* 71, pp. 103-115.

GAMMON, D., JOHANNESSEN, L.K., SØRENSEN, T., WYNN, R. & WHITTEN, P. (2008). An overview and analysis of theories employed in telemedicine studies: A field in search of an identity. *Methods of information in medicine*, 47, pp. 260-269.

GIDDENS, A.. (1991). *Modernity and self-identity : self and society in the late modern age.* Cambridge: Polity.

GIDDENS, A. (1990). *The consequences of modernity.* UK: Polity Press in association with Blackwell.

GIDDENS, A.. (1984). *The constitution of society : outline of the theory of structuration.* Cambridge: Polity.

GIDDENS, A. (1979). *Central problems in social theory: action, structure and contradiction in social analysis.* London (etc.): Macmillan.

GREENHALGH, T. (2008). Role of routines in collaborative work in healthcare organisations. British Medical Journal, 337, pp. 1269-1271

GREENHALGH, T., POTTS, H.W.W., WONG, G., BARK, P. & SWINGLEHURST, D. (2009). Tensions and paradoxes in electronic patient record research: A systematic literature review using the meta-narrative method. *Milbank Quarterly,* 87, pp. 729-788.

GREENHALGH, T., ROBERT, G., MACFARLANE, F., BATE, P., & KYRIAKIDOU, O. (2004). Diffusion of Innovations in Service Organizations: Systematic Review and Recommendations. The Milbank Quarterly, 82, pp.581-629.

GREENHALGH, T. & RUSSELL, J. (2010). Why do evaluations of eHealth programs fail? An alternative set of guiding principles. *PLoS Medicine,* 7, doi:10.1371/journal.pmed.1000360.

GREENHALGH, T. & STONES, R.. (2010). Theorising big IT programmes in healthcare: Strong structuration theory meets actor-network theory. *Social Science and Medicine,* 70, pp. 1285-1294.

GREENHALGH, T., STRAMER, K., BRATAN, T., BYRNE, E., MOHAMMAD, Y. & RUSSELL, J. (2008). Introduction of shared electronic records: Multi-site case study using diffusion of innovation theory. *British Medical Journal,* 337, pp. 1040-1044.

HARDCASTLE, M.R., USHER, K.J. & HOLMES, C.A. (2005). An overview of structuration theory and its usefulness for nursing research. *Nursing Philosophy,* 6, pp. 223-234.

HARRISON, I.M., KOPPEL, R. & BAR-LEV, S. (2007). Unintended Consequences of Information Technologies in Health Care – An Interactive Sociotechnical Analysis. *Journal of the American Medical Informatics Association,* 42, pp. 542-549.

HEAVEN, B., MURTAGH, M., RAPLEY, T., MAY, C., GRAHAM, R., KANER, E. & THOMSON, R. (2006). Patients or research subjects? A qualitative study of participation in a randomised controlled trial of a complex intervention. *Patient education and counseling,* 62, pp. 260-270.

HIBBERT, D., MAIR, F.S., MAY, C.R., BOLAND, A., O'CONNOR, J., CAPEWELL, S. & ANGUS, R.M.. (2004). Health professionals' responses to the introduction of a home telehealth service. *Journal of telemedicine and telecare,* 10, pp. 226-230.

JACKSON, H.M., POOLE, M.S. and KUHN, T. (2002). THE SOCIAL CONSTRUCTION OF TECHNOLOGY IN STUDIES OF THE WORKPLACE. *The handbook of new media: social shaping and consequences of ICTs.* Lievrouw, A. L.; Livingstone, M. S. edn. SAGE Publications Ltd., pp. 236-253.

JONES, M., ORLIKOWSKI, W. & MUNIR, K. (2004). Structuration theory and information systems: A critical reappraisal. In: J. MINGERS and L. WILLCOCKS, eds, *Social theory and philosophy for information systems.* 1 edn. UK: John Wiley & Sons, Ltd., pp. 297-328.

JONES, M.R. & KARSTEN, H. (2008). Giddens's Structuration Theory and Information Systems Research. *MIS Quarterly,* 32, pp. 127-157.

KLEIN, H.K. & KLIENMAN, L.D. (2002). The Social Construction of Technology: Structural Considerations. *Science, Technology, & Human Values,* 27, pp. 28-52.

KOUROUBALI, A. (2002). Structuration Theory and Conception-Reality Gaps: Addressing Cause and Effect of Implementation Outcomes in Health Care Information Systems. *Proceedings of the 35th Hawaii International Conference on System Sciences.*

LANKSHEAR, G. & MASON, D. (2001). Technology and ethical dilemmas in a medical setting: Privacy, professional autonomy, life and death. *Ethics and Information Technology,* 3, pp. 225-235.

LATOUR, B. (1999). On recalling ANT. In: J. LAW and J. HASSARD, eds, *Actor Network Theory and after.* 1 edn. UK: Blackwell Publishing, pp. 15-25.

LEE, T. (2004). Nurses' adoption of technology: Application of Rogers' innovation-diffusion model. *Applied Nursing Research,* 17, pp. 231-238.

LEHOUX, P., SICOTTE, C., DENIS, J., BERG, M. & LACROIX, A. (2002). The theory of use behind telemedicine: : how compatible with physicians' clinical routines? *Social science & medicine,* 54, pp. 889-904.

LI, X., HESS, J. T., & VALACICH, S. J. (2008) Why do we trust new technology? A study of initial trust formation with organizational information systems. Journal of Strategic Information Systems 17, pp 39–71

LIPPERT, S.K. & DAVIS, M. (2006). A conceptual model integrating trust into planned change activities to enhance technology adoption behaviour. *Journal of Information Science,* 32, pp. 434-448.

LIPPERT, S.K. & SWIERCZ, M. P. (2005). Human resource information systems (HRIS) and technology trust. *Journal of Information Science,* 31, pp. 340-353.

MAIR, F., FINCH, T., MAY, C., HISCOCK, J., BEATON, S., GOLDSTEIN, P. & MCQUILLAN, S. (2007a). Perceptions of risk as a barrier to the use of telemedicine. *Journal of telemedicine and telecare,* 13, pp. 38-39.

MAIR, F.S., MAY, C., FINCH, T., MURRAY, E., ANDERSON, G., SULLIVAN, F., O'DONNELL, C., WALLACE, P. & EPSTEIN, O. (2007b). Understanding the implementation and integration of e-health services. *Journal of telemedicine and telecare,* 13, pp. 36-37.

MAY, C. (2006a). Mobilising modern facts: Health technology assessment and the politics of evidence. *Sociology of Health and Illness,* 28, pp. 513-532.

MAY, C. (2006b). A rational model for assessing and evaluating complex interventions in health care. *BMC Health Services Research,* 6.

MAY, C. (1993). Subjectivity and culpability in the constitution of nurse-patient relationships. *International journal of nursing studies,* 30, pp. 181-192.

MAY, C. & FINCH, T. (2009). Implementing, embedding, and integrating practices: An outline of normalization process theory. *Sociology,* 43(3), pp. 535-554.

MAY, C., FINCH, T., MAIR, F., BALLINI, L., DOWRICK, C., ECCLES, M., GASK, L., MACFARLANE, A., MURRAY, E., RAPLEY, T., ROGERS, A., TREWEEK, S.,

WALLACE, P., ANDERSON, G., BURNS, J. and HEAVEN, B., 2007a. Understanding the implementation of complex interventions in health care: The normalization process model. *BMC Health Services Research,* 7.

MAY, C., FINCH, T., MAIR, F. & MORT, M. (2005). Towards a wireless patient: Chronic illness, scarce care and technological innovation in the United Kingdom. *Social science & medicine,* 61, pp. 1485-1494.

MAY, C. & ELLIS, N.T. (2001). When protocols fail: technical evaluation, biomedical knowledge, and the social production of 'facts' about a telemedicine clinic. *Social science & medicine,* 53, pp. 989-1002.

MAY, C., GASK, L., ATKINSON, T., ELLIS, N., MAIR, F. & ESMAIL, A. (2001). Resisting and promoting new technologies in clinical practice: the case of telepsychiatry. *Social science & medicine,* 52, pp. 1889-1901.

MAY, C., HARRISON, R., FINCH, T., MACFARLANE, A., MAIR, F., WALLACE, P. & FOR THE TELEMEDICINE ADOPTION STUDY GROUP. (2003a). Understanding the normalization of telemedicine services through qualitative evaluation. *Journal of the American Medical Informatics Association,* 10, pp. 596-604.

MAY, C., HARRISON, R., MACFARLANE, A., WILLIAMS, T., MAIR, F. & WALLACE, P. (2003b). Why do telemedicine systems fail to normalize as stable models of service delivery? *Journal of Telemedivine and Telecare,* 9, pp. S1:25-S1:26.

MAY, C.R., MAIR, F.S., DOWRICK, C.F. & FINCH, T.L. (2007b). Process evaluation for complex interventions in primary care: Understanding trials using the normalization process model. *BMC Family Practice,* 8.

MCDONALD, A.M., KNIGHT, R.C., CAMPBELL, M.K., ENTWISTLE, V.A., GRANT, A.M., COOK, J.A., ELBOURNE, D.R., FRANCIS, D., GARCIA, J., ROBERTS, I. & SNOWDON, C. (2006). What influences recruitment to randomised controlled trials? A review of trials funded by two UK funding agencies. *Trials,* 7:9 doi:10.1186/1745-6215-7-9.

MCGRATH, K., HENDY, J., KLECUN, E., WILLCOCKS, L. & YOUNG, T. (2008). *Evaluating "connecting for Health": Policy implications of a UK mega-programme.*

MEDICAL RESEARCH COUNCIL. (2000). *A framework for development and evaluation of RCTs for complex intervention to improve health.* UK.

MORT, M., MAY, C.R. & WILLIAMS, T. (2003). Remote doctors and absent patients: Acting at a distance in telemedicine? *Science Technology and Human Values,* 28, pp. 274-295.

MORT, M. & SMITH, A. (2009). Beyond information: Intimate relations in sociotechnical practice. *Sociology,* 43, pp. 215-231.

NICOLINI, D. (2010). Medical innovation as a process of translation: A case from the field of telemedicine. *British Journal of Management,* 21, pp. 1011-1026.

NICOLINI, D. (2007). Stretching out and expanding work practices in time and space: The case of telemedicine. *Human Relations,* 60, pp. 889-920.

NICOLINI, D. (2006). The work to make telemedicine work: A social and articulative view. *Social Science and Medicine,* 62, pp. 2754-2767.

ORLIKOWSKI, W.J. & SCOTT, V.S. (2008). "Sociomateriality: challenging the separation of technology, work and organisation". Working paper series. LSE

ORLIKOWSKI, W.J. (2007). Sociomaterial practices: Exploring technology at work. *Organization Studies,* 28, pp. 1435-1448.

ORLIKOWSKI, W.J. (2002). Knowing in practice: Enacting a collective capability in distributed organizing. *Organization Science,* 13, pp. 249-273.

ORLIKOWSKI, W.J. (2000). Using Technology and Constituting Structures: A Practice Lens for Studying Technology in Organizations. *Organization Science,* 11, pp. 404-428.

ORLIKOWSKI, W.J. (1992). The Duality of Technology: Rethinking the Concept of Technology in Organizations. *Organization Science,* 3, pp. 398-427.

ORLIKOWSKI, W.J. & BAROUDI, J.J. (1991). Studying Information Technology in Organizations: Research Approaches and Assumptions. *Information Systems Research,* 2, pp. 1-28.

ORLIKOWSKI, W.J. & GASH, C.D. (1994). Technological Frames: Making Sense of Information Technology in Organizations. *ACM Transactions on Information Systems,* 12, pp. 174-207.

ORLIKOWSKI, W.J. & IACONO, C.S. (2001). Research Commentary: Desperately Seeking the "IT" in IT Research - A Call to Theorizing the IT Artifact. *Information Systems Research,* 12, pp. 121-134.

ORLIKOWSKI, J.W., WALSHAM, G., JONES, R.M. & DEGROSS, I.J. eds. (1996). *Information technology and changes in organisational work.* 1 edn. UK: Chapman & Hall.

OUDSHOORN, N. (2008). Diagnosis at a distance: The invisible work of patients and healthcare professionals in cardiac telemonitoring technology. *Sociology of Health and Illness,* 30, pp. 272-288.

PEDDLE, K. (2007). Telehealth in Context: Socio-technical Barriers to Telehealth use in Labrador, Canada. *Computer Supported Cooperative Work,* 16, pp. 595-614.

PETTIGREW, A.M. (1997). What is processual analysis. Scandinavian Journal of Management, 13, pp 337-348.

PETTIGREW, A.M. (1987). CONTEXT AND ACTION IN THE TRANSFORMATION OF THE FIRM. *Journal of Management Studies,* 24, pp. 649-670.

PETTIGREW, A.M. (1985). Contextualist research: a natural way to link theory and practice. In: E.E.E.A. LAWLER, ed, *Doing research that is useful for theory and practice.* San Francisco: Jossey-Bass, pp. 222-259.

PETTIGREW, A.M., WOODMAN, R.W. & CAMERON, K.S. (2001). Studying organizational change and development: Challenges for future research. *Academy of Management Journal,* 44, pp. 697-713.

PICKERING, A. (1993). The Mangle of Practice: Agency and Emergence in the Sociology of Science. *The American Journal of Sociology,* 99, pp. 559-589

PINCH, J.T. & BIJKER E. W. (1984). The Social Construction of Facts and Artefacts: or How the ociology of Science and the Sociology of Technology might Benefit Each Other. *Social Studies of Science,* 14, pp. 399-441.

POZZEBON, M. & PINSONNEAULT, A. (2005). Challenges in Conducting Empirical Work Using Structuration Theory: Learning from IT Research. *Organization Studies,* 26, pp. 1353-1376.

Robert, G., Greenhalgh, T., MacFarlane, F., & Peacock, R. (2009). Organisational factors influencing technology adoption and assimilation in the NHS: a systematic literature review. Report for the National Institute for Health Research Service Delivery and Organisation Programme.UK.

Robert, G., Greenhalgh, T., MacFarlane, F., & Peacock, R. (2010). Adopting and assimilating new non-pharmaceutical technologies into healthcare: a systematic review. *Journal of Health Services Research & Policy*, 15, 243–250

ROGERS, M.E. (2003). *Diffusion of innovations*. 5 edn. Free Press.

ROSEN, R., & MAYS, N. (1998). The impact of the UK NHS purchaser-provider split on the rational introduction of new medical technologies. Health policy, 43, pp 103-123

RUNCIMAN, B. (2010). *Health Informatics*. UK: British Computer Society.

SALLAS, B., LANE, S., MATHEWS, R., WATKINS, T. & WILEY-PATTON, S. (2007). An iterative assessment approach to improve technology adoption and implementation decisions by healthcare managers. *Information Systems Management*, 24, pp. 43-57.

SANDELOWSKI, M. (2001). Visible humans, vanishing bodies, and virtual nursing: Complications of life, presence, place, and identity. *Advances in Nursing Science*, 24, pp. 58-70.

SCHULTZE, U. & BOLAND, R. J. (2000). Knowledge management technology and the reproduction of knowledge work practices. *Journal of Strategic Information Systems*, 9, pp. 193-212.

SHARMA, U., BARNETT, J. & CLARKE, M. (2011). Clinical users' perspective on telemonitoring of patients with long term conditions: Understood through concepts of Giddens's structuration theory & consequence of modernity, *Information Systems Theory: Explaining and Predicting Our Digital Society*, Vol2.

SICOTTE, C. & LEHOUX, P. (2003). Teleconsultation: Rejected and Emerging Uses. *Methods of information in medicine*, 42, pp. 451-457.

STAR, S.L. & STRAUSS, A. (1999). Layers of Silence, Arenas of Voice: The Ecology of Visible and Invisible Work. *Computer Supported Cooperative Work: CSCW: An International Journal*, 8, pp. 9-30.

Symons, J.V. (1991). A review of information systems evaluation: content, context and process. *European Journal of Information Systems*, 1, pp. 205–212.

THOMPSON, A. (2004). Confessions of an IS consultant or the limitations of structuration theory. *Cambridge Research Papers in Management Studies*, No. 1.

TJORA, A.H. & SCAMBLER, G. (2009). Square pegs in round holes: Information systems, hospitals and the significance of contextual awareness. *Social Science and Medicine*, 68, pp. 519-525.

VAAST, E. & WALSHAM, G. (2005). Representations and actions: The transformation of work practices with IT use. *Information and Organization*, 15, pp. 65-89.

WALSHAM, G. (1997). IT and Changing Professional Identity: Micro-Studies and Macro-Theory. *Journal of the American Society for Information Sceince*, 49, pp. 1081-1089.

WALSHAM, G. (1993). *Interpreting information systems in organisations*. 1 edn. US: John Wiley & Sons Inc.

WALSHAM, G. & HAN, K. (1991). Structuration Theory and Information Systems Research. *Journal of Applied System Analysis*, 17, pp. 77-85.

WALSHAM, G. & WAEMA, T. (1994). Information systems strategy and implementation: a case study of a building society. *ACM Transactions on Information Systems*, 12, pp. 150-173.

WATERWORTH, S., MAY, C. & LUKER, K. (1999). Clinical 'effectiveness' and 'interrupted' work. *Clinical Effectiveness in Nursing*, 3, pp. 163-169.

WEARS, R.L. & BERG, M.. (2005). Computer technology and clinical work: Still waiting for godot. *Journal of the American Medical Association*, 293, pp. 1261-1263.

WHITTEN, P. & ADAMS, I. (2003). Success and failure: A case study of two rural telemedicine projects. *Journal of telemedicine and telecare*, 9, pp. 125-129.

WHITTEN, P., JOHANNESSEN, L.K., SOERENSEN, T., GAMMON, D. & MACKERT, M. (2007). A systematic review of research methodology in telemedicine studies. *Journal of telemedicine and telecare*, 13, pp. 230-235.

WHITTEN, P. and MICKUS, M., 2007. Home telecare for COPD/CHF patients: Outcomes and perceptions. *Journal of telemedicine and telecare*, 13, pp. 69-73.

WHITTEN, P.S. & MACKERT, M.S. (2005). Addressing telehealth's foremost barrier: Provider as initial gatekeeper. *International Journal of Technology Assessment in Health Care*, 21, pp. 517-521.

WILLIAMS, R. & EDGE, D. (1996). The social shaping of technology. *Research Policy*, 25, pp. 865-899.

WOOTTON, R., DIMMICK, L.S. & KVEDAR, C.J. eds. (2006). *Home telehealth: connecting care within the community*. 1 edn. UK: Royal Society of Medicine Press Ltd.

ZUBOFF, S. (1988). *In the Age of the Smart Machine: The Future of Work and Power*. 1 edn. US.

9

Technological Spillovers from Multinational Companies to Small and Medium Food Companies in Nigeria

Isaac O. Abereijo and Matthew O. Ilori
Obafemi Awolowo University, Ile-Ife
Nigeria

1. Introduction

The economic future of developing countries like Nigeria depends to a greater extent on whether and how domestic small and medium enterprises (SMES) benefit from the present liberalisation and globalisation. This is because unlike the previous decades, the most important determinants for survival from the 1990s are now quality, speed and flexibility (Economic Commission for Africa, 2001). Empirical evidences however show that majority of SMEs in developing countries are not well prepared both for these new conditions and for the increased competition of the global markets (UNCTAD, 2005). While the trade liberalisation is increasing the ability of well-established foreign firms to penetrate remote and underdeveloped markets, the SMEs in developing countries are finding it difficult to survive or, at least, maintain their business position in the local market (UNCTAD, 2005). This is because majority of SMEs lack the adequate resources to conduct research and development (R&D) which is traditionally considered as the main source of technological innovation for competitiveness.

The central finding in the literature on innovation indicated that, in most cases, innovation activities in SMEs depend heavily on external sources for them to remain competitive (Fagerberg, 2005; Abereijo and Ilori, 2010). Equally important is the international knowledge flows through foreign direct investment (FDI), trade, licensing and international technological collaborations, which can serve as important determinants of the development and the diffusion of innovations to SMEs (Damija, Jaklič and Rojec, 2005). Infact the international lessons on SME development show that external factors such as inter-firm co-operation, institutional support, and learning from various external sources of knowledge are playing a key role in helping SMEs to build up technological capabilities that will enable them compete in regional and global markets.

FDI, as one of the external sources, is theoretically assumed to play an important role in assisting local firms to experience production externalities from technological spillovers both within industry and across industries. The empirical relevance of the spillover argument is conceptually related to the transfer of non-conventional factors of production, including technology, management skills, and motivation between foreign and domestically owned firms (Vera-Cruz and Dutrenit, 2005).

Among the channels through which technological spillover can occur from the presence of multinational companies (MNCs) are linkage and human capital. The linkage between MNCs and SMEs can help integrate SMEs into international chains of production at various stages of added value. It can also serve as one of the fastest and most effective ways of upgrading the technological and managerial capabilities of local SMEs for innovations (United Nations Conference on Trade and Development, 2006). The spillover through human capital is associated with the continuous training of employees by MNCs and the mobility of these employees toward domestic SMEs. Therefore, apart from contributing to the development of technological and managerial capabilities of the local firms, human capital spillover can also increase their absorptive capacity for technological innovations.

As a result of the potential role of MNCs in accelerating growth and economic transformation, many developing countries in general and Africa in particular, seek this type of investment to accelerate their developmental efforts. This in turn has led many African countries to put in place various measures that they hope will attract MNCs to their economies, including improving their investment environment. Specifically in Nigeria, government legislated two major laws which are meant to guarantee investments against nationalisation by any tie of government, and to ensure the free transfer and repatriation of funds from Nigeria. These laws are the Nigerian Investment Promotion Commission (NIPC) Act 16 and the Foreign Exchange (Monitoring and Miscellaneous Provision) Act 17, both of which were enacted in 1995. The NIPC was established to address the problems of multiplicity of government agencies which foreign investors confront when they come to Nigeria. All these efforts are meant to encourage, promote and coordinate foreign investment and enhance capacity utilisation in the productive sector of the economy (NIPC, 2006). This also provides an opportunity for foreign participation in Nigerian enterprises up to 100 percent ownership.

However, efforts at justifying the incentives being offered by the governments of developing countries at attracting the MNCs have made researchers to conduct various studies to establish its benefits (spillover effects) to the developing economies. Though there were positive technological spillovers from MNCs in some developing countries, the results of empirical researches are far from been conclusive (Hausmann, 2000; Kapstein, 2002; Narula and Marin, 2003). One of the reasons adduced for the inconclusiveness was that impact of MNCs depends on a multitude of factors, such as the levels of technology used in domestic production, education of the workforce, financial sector, and institutional development in the host country (Krogstrup and Matar, 2005). These factors determine whether the host country can absorb and hence benefit from MNCs.

Aside from the above factors, two main issues have been identified recently in the literature, which are also responsible for the inconclusiveness of the results (Gachino, 2007). This concerns the conceptualisation of the spillover occurrence. The first issue is the methodological approach employed in the studies; and the second is the inherent weaknesses in spillover analysis. On the issue of methodology, the occurrence and impact of MNCs spillovers on local enterprises cannot be appropriately explained using simple linear aggregate analysis in the case of non-pecuniary (technological) spillovers. This is because non-pecuniary spillovers are exceptionally difficult to deduce from aggregate macro economic data. Such spillovers include knowledge flows that are invisible, imperfectly understood, determined by multiple factors, and difficult to track, hence difficult to investigate (Gachino, 2007).

The inherent weaknesses, according to Gachino (2007), are, first, the tendency towards 'single factor exponentiation', which makes the presence of MNCs to be the only major factor in determining occurrence of spillovers in a host country. Second, there was also a weakness due to 'automaticity or exogeneity problem', where spillovers and their effects were thought to occur automatically, thereby making the process of spillover occurrence quasi-inevitable. The third problem relates to the 'narrow conceptualisation of spillovers' phenomenon, where MNCs were the only firms taken into consideration while analysing spillovers, thereby disregarding the role and effort of local firms and other supportive factors within the national innovation system of host countries.

On the basis of the above issues, an alternative approach was suggested which will enable an appropriate assessment of the influence of interactions, learning and capability development in the spillover occurrence process (Gachino, 2007). Following on this re-conceptualisation, this chapter presents the empirical result of the assessment of the various forms of technological spillovers from MNCs to small and medium food companies (SMFCs) in Nigeria, and examined the factors that influenced the occurrence of these spillovers. The spillover occurrence was based on the presence of foreign firms (MNCs), as well as, the actual effect of the spillover channels on the production capability of the domestic SMFCs. Therefore, the productivity of these domestic SMFCs depended on their accumulated technological capabilities as a result of continuous learning due to the influence of the spillover channels.

2. Theoretical framework

2.1 Technological spillovers

Technological spillover is defined as transfer of knowledge and skills (technical and organisational) from MNCs that result in an improvement in the performance of MNCs partners, suppliers or competitor firms, as well as of the other agents that interact with them (Vera-Cruz and Dutrénit, 2005). The technological spillover thus generates productivity or efficiency benefits for the host country's local non-affiliated firms. The availability of new foreign knowledge through MNCs may benefit domestic firms as they can learn the technology from them, which allow them to upgrade their own production process, and as a result, improve their productivity.

The theoretical and empirical literature identifies two major concepts of technological spillovers, which are rent-spillovers and knowledge-spillovers (Griliches, 1992). Rent-spillovers occur when new goods are purchased at prices below those that would fully reflect the value of technological improvements from R&D investments. They can be considered as a pecuniary externality from upstream industries, whose competitive market structure may not allow firms to fully transform higher quality into higher prices. Knowledge-spillovers occur when innovation by one firm is adopted by "adjacent" firms, thus enhancing their productive and innovative capabilities. Knowledge spillovers arise exclusively as an intangible transmission of ideas; in principle, they are not embodied in traded goods, and thus they do not necessarily require economic transactions.

2.2 Forms of spillover effects

The spillover effects of MNCs to the local industries can be divided into two, namely, inter- and intra-industry spillover effects. Inter-industry (vertical) spillovers occur through foreign companies' impact on the local suppliers in different industries. Through creation of linkages between the foreign company and domestic firms, spillovers may occur when the local suppliers have to meet the demand from the foreign firm in the form of higher quality, price and delivery standards (Smarzynska, 2003). Another implication of inter-industry spillover effects is the increased demand by the MNCs for local intermediate inputs, thus increasing production possibilities in the host economy (Barrios, 2000).

The intra-industry (horizontal) spillovers result from the presence of MNCs in a particular sector and its influence on the host industry's competitors (local companies in the same sector). There are five transmission channels through which intra-industry spillover effects may occur. These are competition, demonstration and imitation effects, transfer of technology and R&D, human capital, and labour turnover (Blomström, Globerman, and Kokko, 1999).

Since the presence of MNCs in any country usually results to an increased competition in the host economy, the less efficient domestic firms might be forced to improve on their production techniques. That is, the superior technology of the MNCs may stimulate efforts of domestic companies to compete, which may lead to new innovations. Such effort could be investment in human and physical capital; and the efficient use of their existing resources. This can raise the productivity of the local firms and thereby assist them to compete with MNCs. It should be noted however that the increased competition could 'crowd out' the domestic firms, especially if the market is populated with inefficient domestic firms (Taymaz and Lenger, 2004).

Demonstration and imitation effects can occur when domestic firms observe and imitate the superior proprietary technology, management and marketing skills possess by the MNCs. This channel of spillover represents "learning by watching effect" (Blomström, Globerman, and Kokko, 1999). Technological spillover effects can then occur through imitation, reverse engineering and copying of foreign companies' products or production processes.

Transfer of technology and R&D can also bring about spillover effect when the local companies, in the same industry, are aware of the existence of a particular technology or result of MNCs' R&D activities. This might enable local firms to increase productivity and build competitiveness in new areas (Mansfield & Romeo, 1980). Also, the existence of technology and productivity gaps between the foreign and local firm may stimulate spillover effects when the domestic firms are making efforts to catch up through imitation of the technology of foreign leaders. However, the risk of this channel is that if the MNCs' advanced technology is beyond the local firm's absorptive capacity, this could lead to adverse consequences for the domestic firms' market position (UN-ECE, 2001).

MNCs often invested in their employees through various trainings that cannot be easily replicated in domestic firms. The knowledge and skill gained by the local employees through these trainings can lead to technological spillover when labour turnover occurs, through "brain-drain in reverse" to the local economy (Dunning, 1970). That is, domestic employees that were trained by the MNCs can start their own business or be employed by domestic companies. This human capital development can play a crucial role not only in the dissemination of technological knowledge from MNCs to the domestic companies, but also in the dissemination of best practices and other organisational innovations which are more difficult to disseminate in other ways.

2.3 Determinants of spillover occurrence

Based on the analytical framework developed by Gachino (2007), the occurrence of spillover in a technically underdeveloped country does not only depend on the presence of MNCs, but also on absorptive capacity, presence of support structures and institutions, and presence of interactions and trade orientation. Others include firm size and age, ownership structure, performance, labour market conditions, firm strategy and industry structure.

The level of absorptive capacity of the local firms will assist them to exploit new knowledge and technology from the MNCs. Hence, only the local firms who have accumulated technological knowledge in human resources as a result of strong R&D base can benefit from the technological spillovers from the MNCs. Moreover, beyond the internal efforts of the local firms, interactions among economic agents within the host country, as well as the infrastructural and institutional supports structures are important determinants to spillover occurrence. The interactions among the economic agents can serve as channel to technological innovations or serve as stimuli for learning and innovation. Also, the support structures such as productivity centres, technology transfer bodies, training programmes and investment promotion councils can play important role towards facilitation of innovation based on knowledge acquired in the spillover process.

Another important determinant of spillover occurrence is the strong network cohesion which supports generation and diffusion of knowledge (Freeman, 1991; Lundvall, 1992). This is important because spillover is an interactive and dynamic process, hence systemic interactions among firms, institutions, and business associations can stimulate the process of spillover occurrence. Closely related to the interactions is clustering, which can promote new product development and make diffusion of new technologies possible through information exchange and joint problem solving between firms in the same industry or different industries (Saxeniaan, 1991; Mytelka and Farinelli, 2000).

The importance of firm's size on its ability to compete and for the occurrence of spillover occurrence is also established in the literature. That is, attainment of a certain minimum efficiency scale by firms is required for competitiveness (Scherer, 1980 quoted in Gachino, 2007). While the attainment of this scale is possible in large firms because of their ability to mobilise productive resources and other services that are either external or internal to them; majority of the small firms have inadequate resources to improve their technological capabilities. Closely related to the firm's size is the influence of age on the spillover occurrence. The accumulated stock of knowledge and experience over time can increase the absorptive capacity necessary to recognise external knowledge, absorb it and utilise it for productive purposes (Gachino, 2007). However, as noted by this author, the level of experience of a firm is more important than how old the firm is. This is because it is the experience that will position the firm to enjoy greater experiential and tacit knowledge, which in turn determines the likelihood of spillovers occurrence.

Another factor that has strong bearing on spillover occurrence, learning and technological capability building is industrial specificity. This is because industries are different with significant differences in technological capabilities and capacities to undertake technological learning and absorption. While some industries are becoming high technology intensive, others have become knowledge intensive.

The firm performance is characterised by high capacity utilisation and high output performance in terms of sales and profits (Gachino, 2007). This enables such firm to undertake dynamic strategies, perform basic R&D, recruit well-trained professionals like scientists and engineers, and undertake human resource development and other enrichment programmes. Hence, a firm with high performance offers more room for learning, acquisition of tacit and experiential knowledge, all of which enhance firm's absorptive capacity. Moreover, a firm with a demonstrated strong path dependence leading towards accumulation of absorptive capacity will likely benefit from spillover occurrence. This is reflected in the firm strategy like ability to modernise its operations, diversify its products, and capture new market. Other strategies could be ability to lower overhead cost, improve quality, and broaden its knowledge base.

Moreover, participation in the export market is also noted to stimulate a dynamic learning process which can assist the local firms to benefit from spillover from MNCs. Hence, the trade orientation towards export in international markets can make the local firms to pay attention to the global tastes and preferences. It can also force them to increase their technological effort in order to learn continuously and master techniques required in maintaining international competitiveness at the world market.

2.4 Empirical studies on spillovers

Empirical studies on technological spillovers have been made with different techniques and methodologies, covering both developed and developing countries that have and have not received substantial FDI inflows. The studies also covered different time periods and used different endogenous as well as exogenous variables. The first group of empirical studies used cross-sectional data in a single year and found positive spillovers (Caves, 1974; Globerman, 1979; Blomström and Person, 1983; Blomström and Wolff, 1994; Nadiri, 1991; Blomström and Sjöoholm, 1998). But the set of second group of studies, which used panel data, found negative spillovers (Aitken and Harrison, 1999; Djankov and Hoeckman, 2000; Kathuria, 2000; Konings, 2000). The findings pointed out that many of the earlier studies, that found positive spillovers, did not introduce control variables of sectoral nature. Moreover, the third group of empirical studies considered the technological and/or productivity gap between local frims and MNCs to discriminate the existence or non-existence of spillovers (Kokko, Tansini, and Zejan, 1996; Castellani and Zanfei, 2001; Girma, Greenaway and Wakelin, 2000; Haskel, Pereira and Slaughter, 2002).

Furthermore, all of the above studies focus on intra-industry spillovers. The studies in inter-industry spillovers, through backward linkages found positive spillover effects, and negative for forward linkages (Schoors and Van der Tol, 2002; Smarzynska, 2003; Kugler, 2000). This is because spillover effect is dependent on the local absorptive capability and the level of sectoral openness (Smarzynska, 2003).

In summary, the pioneer studies showed evidence of positive spillover effects because they were based on cross-sectional data. But the more recent studies, which are based on panel data techniques, tend to show a more heterogeneous reality. While some found negative spillovers, others showed that spillovers may exist but are contingent on different factors, mostly related with technological and innovation variables.

Within the Nigerian environment, the results of the survey conducted by Narula (1997) could not give a precise result as to whether or not technology is being transferred from the foreign firms to the domestic firms. Infact there is historical suspicion that MNCs possess skills and tangible technology that are available locally and are therefore making above normal profits based on better access to capital, and because of their sustained and growing presence (Biersteker 1987, Onimode 1982). Akinlo (2004) examined the effect of FDI on growth in Nigeria, using data from 1970 until 2001. The result of this study pointed out that it cannot unambiguously be said that FDI is growth enhancing. This is because FDI environment in Nigeria is characterised by its focus on oil industry, which is an extractive industry. The results further show that FDI in Nigeria only has a positive impact on growth after a considerable lag. Hence, FDI in the Nigerian case does not have the same effect as it has had in Asia and Latin America. Akinlo (2004) therefore speculates that this was due to the nature of the extractive oil industry, which has very little linkages with other sectors because; as with most natural resource industries there is rarely a requirement for substantial inputs and intermediate materials, procured from local suppliers.

Furthermore, a study by Ayanwale and Bamire (2004) examined the impact of FDI on productivity at the firm-level in the agro/agro-allied sector of the Nigerian economy. Data were obtained from those companies listed in the first tier market (comprising firms with some foreign components), and the second tier foreign exchange markets (involving domestically owned firms) as contained in the publications of the Nigerian Stock Exchange Commission and Central Bank of Nigeria. The result of this study showed that there was positive and significant spillover effect at the firm level, but with little or no spillover effect on labour productivity.

3. Conceptual framework

Contrary to the traditional technique where spillovers were conceptualised in terms of productivity gains, spillovers is re-conceptualised in terms of learning and capability building within the firm (Gachino, 2007). This is because firm's productivity depends on the accumulated technological capabilities over time where constant and continuous learning leads to a dynamic process of technological accumulation. Based on this re-conceptualisation, spillover from foreign firms is accepted to bring about learning in domestic firms by providing raw materials, resources or specific stimuli that triggers various forms of technological changes in the domestic firms.

Firm level technological capabilities which can be improved through spillover effects are investment, linkage, production, and complimentary capabilities (Lall, 1992; Rasiah, 2005). Investment capability includes skills and knowledge used in the project identification, feasibility studies and preparation, design, setting up and commissioning of a new industrial project or the expansion and/or modernisation of existing ones. Linkage capability refers to skills, knowledge and organisational competence needed to transfer information, skills and technology to, and receive them from, component or raw material suppliers, subcontractors, consultants, service firms, and technology institutions. Such linkages affect both the productive efficiency of the enterprise and the diffusion of technology through the economy, and also deepen the industrial structure.

Production capabilities include basic skills such as quality control, operation, and maintenance. It also includes advanced skills such as adaptation, improvement or equipment stretching; and the most demanding activities such as research, design, and innovation. All these skills determine how well technologies are operated and improved; and how in-house efforts are utilised to absorb technologies bought or imitated from other firms. The complementary capabilities include organisation and marketing capabilities. The former consists of skills that are required to relate and co-ordinate the necessary functions so as to utilise effectively various existing capacities both in the firm and outside the firms. Marketing capabilities includes the knowledge and skills required for collecting market intelligence, the development of new markets, the establishment of distribution channels and the provision of customer services.

Based on the above categorisation of technological capabilities, occurrence of spillovers is likely to place domestic enterprises on a learning path, which will then increase their potential to learn, and to accumulate experiential tacit knowledge. In this study spillover was conceptualised in terms of learning and production capabilities building only. The four channels of spillover occurrence identified from the spillover literature were considered, which include linkage, labour mobility, competition, and demonstration effects (training). For each of these spillover occurrence channels, five types of technological changes associated with production capability for spillover occurrence were considered. These are product changes (product innovation); process changes (process innovation); industrial engineering; new marketing strategies; and management or organisation changes.

3.1 Model specification and measurement of variables

The magnitude and nature of FDI spillovers have been identified by employing various direct and indirect approaches. The direct approaches relate productivity measures of host country firms or industries to, among other things, the extent of foreign ownership in the host country. Indirect approaches examine different aspects of the interaction between MNCs and domestic firms that are reasonably related to FDI spillovers. These include technology licenses, vertical linkages, copying of technology introduced by foreign investors, impact of FDI on host country market structure, especially competitiveness, labour training, and performance of R&D by MNC affiliates in the host country. While the direct approach has been investigated through statistical studies, the indirect approach is investigated through more structurally oriented studies (Kathuria, 1998).

However, given the weaknesses in the analysis of spillovers, as discussed before, it is clear that occurrence and impact of MNCs spillovers on local enterprises cannot be appropriately explained using simple linear aggregate analysis. It is only firm level analysis that is capable of offering a well-grounded understanding of relationship among firms, including MNCs' influences on local firms (Gachino, 2007).

3.2 Study variables and measurements

This study used the following categories of variables to identify the various forms of technological spillovers that were from the MNCs to local small and medium scale food companies (SMFCs) in Nigeria. These were knowledge spillover, SMEs' absorptive capacities, and spillover index (SI).

3.2.1 Knowledge spillovers

Knowledge spillovers to local firms happen when local firms get the benefits from higher knowledge related to product, process, or market technologies from MNCs (Blömstron and Kokko, 2003). Direct indicators were used, and they were related to some of the spillovers channels that have been identified in the literature, which are competition, linkage, labour mobility and demonstration effects. Qualitative analysis of knowledge spillover is based on the work of De Fuentes and Dutrénit (2006), and was carried out through the following factors: entrepreneur's previous experience and training in multinational companies; employee's experience and training in multinational companies; formalisation of linkages with multinational companies; and kind of linkages established with clients.

The first two factors are related to the spillovers mechanism of human capital mobility and training effect, while the last two are related to the backward linkages mechanism. Table 1 presents the variables that were used to build these four factors.

The variables used for spillovers through human capital mobility include:

i.   Whether or not owner and/or employees had worked with MNCs.
ii. Years of experience in MNCs by the owner and/or employees (in years).
iii. Specific management levels where the employees had worked. That is, whether policy, management (middle), or operational (low) level.
iv. Types of experience in the MNCs. That is, whether in production, product development, quality improvement/assurance, or management section.
v.   Whether or not they had undergone training courses while in MNCs. This was measured in terms of number and focus of the training, whether it was production- or management-related.

Linkage with MNCs by the SMEs can occur through backward or forward link, which could affect the local firms positively in terms of efficiency and quality of outputs (De Fuentes and Dutrenit, 2006). Therefore, the variables measured were:

i.   Whether or not there was any linkage with MNCs by the domestic SMFCs.
ii. Average number of such relationship.
iii. Type of linkage, whether it was a contractual or an informal relationship.

The kind of linkage established is measured in terms of whether the domestic SMFCs:

i.   shared the MNCs production and laboratory facilities.
ii. received technical advice/assistance from MNCs, such as product quality analysis.
iii. received assistance from MNCs for quality improvement of their products.
iv. received assistance in procuring processing equipment.
v.   had joint projects with MNCs.

vi.   received training for their employees by MNCs.

vii. received assistance in entering new markets.

3.2.2 SMEs' absorptive capacities

Absorptive capacities are the ability of firms to recognise the value of new information, assimilate it and apply it to commercial ends, which is critical to their innovative capabilities (Cohen and Levinthal (1990), cited by De Fuentes and Dutrenit, 2006). Direct indicators were employed for the analysis of absorptive capacities using four indicators. These are human resources, embodied technology, characteristics of the firm's owner, and other organisational features of the firm. These indicators were entrepreneur (owner) and employees' experience and studies; embedded technology in equipment; and organisational capabilities. Others are learning and innovation activities; and linkages established with other local agents.

The variables for owners and employees experience and training include the educational qualifications of the owners and that of employees in charge of production, marketing, and administration. They also included various trainings that the owners and the employees had undergone, especially in the area of quality improvement, innovation, and marketing strategies. The variables measured included type (tertiary or non-tertiary) and highest educational qualifications acquired by the owner and the employees. Others were their areas of specialisation, number of on-the-job and off-the-job training programmes undergone by the owners and employees within the last five years. Tertiary education included highest qualification from Polytechnic or University. Area of specialisation was indicated as science/engineering/technology related or management/finance/art related. The trainings undergone by the owners and employees were specified as product development, quality improvement, and quality assurance. Full time employees including scientists/engineers and others were specified in terms of numbers.

The variables under embodied technology in equipment were level of automation and age of processing equipment. The level of automation was measured in terms of number of computerised equipment, and average age of the processing equipment was measured in years (De Fuentes and Dutrénit, 2006).

The variables under organisational capabilities were age of business, relationship with MNCs, existence of R&D activities, and types of training programmes attended by owners/employees, existence of interactions with MNCs and receipt of quality award.

Learning takes place through internal and external sources. Internal learning takes place through learning from experience in the process of production, commercialisation and use; and in the search of new technical solutions through R&D. External learning occurs through interaction with suppliers, competitors, customers, consultants, associates, universities, research institutes offering technological services, agencies and governmental laboratories, business development centres among others. Therefore, the variables that were considered included:

i.   Types of training programme attended by the owner and employees. These could be products and process, and new marketing programmes, human resource management, product development, quality maintenance, strategic planning, and marketing.

Factors	Variables	Measurement
a. Entrepreneurs' mobility	• Work experience with MNCs	Either Yes or No
	• Years of experience with MNCs	Number of years
	• Specific management level where the entrepreneur worked.	Either at policy level, management level, or operational level
	• Types of experience in the MNCs.	Either experience in R&D, or production, or management.
	• Training courses attended while working with MNCs	Number of training attended
	• Focus of the training.	Whether in product/process development, market development, strategic planning or general business management.
b. Formalisation of linkages with MNCs	• Existence of relationship with MNCs	Yes or No
	• Years of relationship with MNCs	Number of years of relationship.
	• Types of relationship.	Whether contractual or informal.
c. Kind of linkages established with MNCs	Existence of: • Backward linkages • Forward linkages • Access to MNCs facilities • Technical support or advice • Development of joint projects • Sharing knowledge of export	• Whether or not the local SMFCs: - Supplied production input(s) to MNCs - Purchased production input(s) from MNCs - Shared the MNCs production and laboratory facilities - Received technical advice/ assistance from MNCs - Received assistance in quality improvement of the product - Received assistance in entering new markets - Were assisted by MNCs in procuring processing equipment - Staff were trained by MNCs

Table 1. Study Variables and their Measurement

ii. Interactions with external agents. This was indicated by whether or not they had interactions with universities, research institutes, and business associations.

The indicators for this variable were number and type of relationship with suppliers, competitors, research institutions, industrial associations. All the variables used to build the absorptive capacity indicator and their measurements are shown in Table 2.

Factor	Variable(s)	Measurement
Owner's and employees't experience and training	Educational background of Owner and the employees. That is, i. Non-tertiary education (Primary or Secondary or Technical or NCE*) ii. Tertiary education (Polytechnic or University or Postgraduate	Non-tertiary = 1 Tertiary = 2
	Area of specialisation of owner and the employees. Whether: i. Non-science or engineering or technology (that is, management or finance or art or social science), or ii. Science or engineering or technology	Non-science = 1 Science = 2
	Previous training undergone by owner and the employees within last 5 years. That is, whether in: i. General business management, or ii. Product development or Quality assurance/improvement	General business management = 1 Product development or Quality improvement or assurance = 2
	Skills of workforce. i. Number of technicians ii. Number of engineers iii. Number of scientists	Percentage of technician (technologist)/scientists/engineers of total workforce
Embedded technology in the processing equipment	i. Level of automation, that is: (a) Manual (b) Semi-automation (c) Full automation	Manual = 1 Semi-automation = 2 Full automation = 3

Factor	Variable(s)	Measurement
	ii. Average age of the processing equipment	More than 5 years = 1 Between 3 and 5 years = 2 Less than 3 years = 3
	Number of years in business (age of company)	Age of business in year
Organisational capabilities	Relationship with MNCs i. No relationship	No relationship = 1 Existence of Relationship = 2
	ii. Description of relationship	Informal = 1 Formal = 2
Learning and innovation activities	Internal sources through: i. Research and development (R&D) activities	• No R&D activities = 1 • Existence of R&D activities = 2
	ii. Attendance of training programme by owner and employees in the areas of new products or process or marketing development, quality assurance or maintenance, and strategic planning.	• Training related to new product or process or marketing development, quality assurance or maintenance, and strategic planning = 2 • Training not related to the above = 1
	External sources through: i. Interactions with universities, research institutes, and business associations	Level of importance of each interaction to the acquisition of knowledge in the company. • Not important = 1 • Important = 2 • Very important = 3

† Employees here are production or marketing or administrative managers
* NCE means National Certificate of Education (middle level teacher)

Table 2. List of Variables Used to build Absorptive Capacity Indicator of SMFCs

3.2.3 Effects of the spillover mechanisms on the learning and technological changes

In order to establish the effect that spillover channels had on the learning and technological changes of domestic SMFCs, the spillover index was calculated. This was based on the work of Gachino (2007). Each of the four channels of spillover occurrence, which are competition, linkage, labour mobility and demonstration effects, was conceptualised in terms of learning and dynamic technological changes that had taken place in the production capacity of each of the SMFCs surveyed.

For each of the spillover occurrence channel considered, five types of technological changes associated with production capability were used as proxies for spillover occurrence. These were product changes, process changes, industrial engineering, new marketing strategies, management, and organisation changes. The degree to which each change took place was also determined subjectively in the firms on a continuous gradual ordinal scale ranging from a minimum score of 1 representing "nothing happened" to a maximum score of 7 representing "very much had happened". An index was then computed and used in the quantitative determination of spillover occurrence as well as spillover determinants.

3.3 Computation of the spillover Index

In the questionnaire, each of the SMFCs was asked to evaluate the effect of each of the four channels of spillover occurrence on the five types of technological changes associated with production capability. For example, the firms were asked to rank the effect of linkage with MNCs on the product changes; process changes; industrial engineering changes; new marketing strategies; and management and organisation changes in their companies. This was premised on the assumption that due to linkage with MNCs, each of these domestic SMFCs would react by undertaking changes ranging from production to organisational changes. For each factor, each firm was asked to indicate subjectively the degree of perceived change due to linkage on the basis of scale provided (Table 3).

The spillover index (SPO Index) was then developed from the ranking indicated by each SMFCs. The use of index to evaluate firm level processes and activities is used when dealing with complex technological capability issues in developing countries (Gachino, 2007). The average spillover indices, C, L, M, and D computed for competition, linkage, labour mobility, and demonstration effect respectively were calculated (Table 4). The arithmetic average of all the four channels of spillover occurrence was taken as the composite spillover index. That is, SPO Index equals Composite average of C, L, M, and D.

The calculated spillover index (SI) of each SMFC ranged from 1 to 7. That is, SI value of 1 indicated that the combined influence of all the four channels of spillover had no effect on the production capacity of such SMFC. While SI values of 2 and 3 indicated an insignificant effect and little effect on the production capability respectively. Also, SI values of 4, 5, 6, and 7 were indication that the channels of spillover had moderate, considerable, much, and very much effect respectively on the SMFCs' production capability.

Type of Production Capability	Effects on each Production Capability	Ranking by importance No effect ----> Much effect 1 2 3 4 5 6 7
Production Changes	• Developing new products. • Improving our products.	

Type of Production Capability	Effects on each Production Capability	Ranking by importance No effect ----> Much effect 1 2 3 4 5 6 7
	• Copy or imitate competitor's products.	
Process Changes	• Improving processing techniques.	
	• Improving raw material and quality control.	
	• Upgrading our technology and equipment to raise productivity.	
Industrial engineering changes	• Replacement of our processing equipment.	
	• Upgrading our processing equipment.	
	• Repair and maintenance of our processing equipment	
New marketing strategies	• Improve our marketing department with new ideas, skills, and knowledge in domestic or foreign marketing.	
	• Diversify our products.	
Management and organisation changes	• Undertake organisational changes for better management and implementation of production and other routine activities that enhance the firm's efficiency.	
	• Introduction of information technology for quick and better decision making.	

Source: Gachino (2006), but modified.

Table 3. Reaction to Competition Pressure Ranked by Order of Importance

Spillover conceptualization	Channels of Spillover Occurrence				
	Competition (C)	Linkage (L)	Labour Mobility (M)	Demonstration (D)	Average Score
Product Changes (Pd)	Pd_c	Pd_l	Pd_m	Pd_d	PD
Process Changes (Pr)	Pr_c	Pr_l	Pr_m	Pr_d	PR
Repair & Maintenance (Rm)	Rm_c	Rm_l	Rm_m	Rm_d	RM
Marketing Strategy (Ms)	Ms_c	Ms_l	Ms_m	Ms_d	MS
Management & Organisation (Mo)	Mo_c	Mo_l	Mo_m	Mo_d	MO
Average Score	C	L	M	D	SPO Index

Source: Gachino (2006)

Table 4. Composition of Spillover Index (SPO Index)

3.4 Evaluation of factors responsible for the occurrence of technological spillovers

The spillover index is taken as a proxy for spillover occurrence in the literature (Gachino, 2007). Also, spillover occurrence is a function of individual firm's resource endowment and their interactions with socio-economic agents, which is also determined by a number of factors relating to the absorptive capacity of domestic firms. Therefore, spillover index (SI) was taken to be a function of SMFCs' absorptive capacity. The variables considered were age of company; percentage of Nigerian ownership; and percentage of technicians, scientists, and engineers. Others were highest qualification of entrepreneurs (owner), production manager, marketing manager, and administrative manager; area of specialisation of entrepreneurs (owner), production manager, marketing manager, and administrative manager. Other variables used were previous work experience of entrepreneurs (owners), production managers, marketing manager, and administrative manager with MNC. The average age of the main processing equipment, level of automation of the processing equipment, and number of year in relationship with MNC were also used.

The influence of these variables on spillover index of SMFCs was then estimated using categorical regression model. Correlation technique was also employed to determine the relationship between the dependent variable (SI) and independent variables.

4. Methodology

The study was carried out in Southwestern Nigeria, which comprises of Lagos, Oyo, Osun, Ogun, Ondo and Ekiti States. However, the study was limited to Lagos, Ogun, and Oyo States because the activities of MNCs are most prominent in this part of the country. The sample population for this study consisted of domestic small and medium scale food manufacturing companies operating in this part of Nigeria. The samples were drawn from the database and directories of National Association of Small Scale Industries (NASSI), National Association of Small and Medium Scale Enterprises (NASME), and Manufacturing Association of Nigeria (MAN), specifically from the directory of the Association of Food, Beverage and Tobacco Employers (AFBTE). However, only small and medium companies with more than 10 full time employees were surveyed (CBN, 2004).

Considering the sub-sectors where the SMFCs are most prominent, the methodology for sampling was stratified random sampling with the stratification based on 7 sub-sectors, which are Roots and Tubers products; Fruit juices and Drinks; Bakery products; Beverage; Fat and oil; Wines and Spirit; and Dairy products. Based on the Baseline Economic Survey of SMEs in Nigeria by the Central Bank of Nigeria (CBN) in 2004, a total population of 455 companies was identified within the Food, Beverages and Tobacco sectoral group in the study area (CBN, 2004). Out of this population, 200 were randomly selected from the directories.

The primary data were collected through interview and structured questionnaire, directed at the Managing Director and/or the Production, Marketing and Personnel Managers. The questionnaire was designed to elicit information on the educational background, experience and training of the business owners and key employees, especially those in charge of production, management, and marketing. Data were collected on the technology embedded in the processing machinery and equipment, and organisational capabilities. The questionnaire also elicited information on learning and innovation activities, new market programme, product and process innovation; linkages established with MNCs and other local agents, and types and format of these linkages. Others data collected were owner's and employees' job mobility, in terms of experience in MNCs and position and the various job-related training undergone; and reaction of the each firm to the spillover occurrence channels in terms of technological changes effected in their production capability. The completed questionnaires were analysed using the descriptive and inferential statistics using Statistical Package for Social Scientists (SPSS) version 15.

5. Results and discussion

Out of the 200 questionnaires administered, 150 were retrieved, with only 112 usable, representing 56% of the whole questionnaire administered. Within the usable ones there were 4 food companies (3.6%) from Roots and Tubers sector, 22 (19.6%) from Fruits juices and Drinks sector, 51 (45.5%) from Bakery products sector, 4 (3.6%) from Beverage sector, 15 (13.4%) from Wines and Spirit sector, and 16 (14.3%) from Dairy products sector. The main products of these companies included white and yellow garri, plantain chips, apple juice, blackcurrant, flavoured milk, orange and pineapple juice, and milky juice. Other products were baking powder, biscuit (coaster and sweet cream), bread (buttered, chocolate and sliced), and sausage. The remaining products included chocolate drinks, sweet, gin, wine, rum, yoghurt, ice-cream, milk (liquid and powdered), and strawberry.

5.1 Channels of technological spillovers from the MNCs to SMFCs in the study area

Technological spillovers from MNCs occurred in the food companies through two main channels, which were linkage and labour mobility.

i.   Linkages

About 45% of the SMFCs had one or more forms of linkage with the MNCs in the studied area (Table 5). Among the MNCs that SMFCs had linkage with, 68% of the MNCs operated within the Food, Tobacco, and Beverage (FTB), while 8% and 4% operated within the Chemical and Pharmaceutical, and Electrical and Electronics industries respectively. Also, 12% of the SMFCs had linkages with only one MNC, while 60%, 24%, and 4% of them had linkages with 2, 3, and 5 MNCs respectively.

The various types of linkages included purchasing of inputs (raw materials) from the MNCs (36.6%), being subsidiary of MNC (3.6%), and supply of inputs (raw materials) to MNCs (1.8%). Other forms of linkage indicated were outsourcing whereby some parts of the production of MNCs were done by some SMFCs; and provision of assistance to SMFCs by MNCs in the purchase of processing equipment (0.8%). Some SMFCs also had access to one or more facilities of the MNCs. About 25% of the SMFCs indicated that the MNCs provided training for their staff, and 6.3% received technical assistance from them. About 6.0% also had access to the MNCs' laboratory facilities to conduct quality control, physical and chemical analyses of their products. One of the owners of the SMFCs indicated that a MNC assisted in pushing the product of his company into the international market.

These types of relationship between the SMFCs and MNCs were expected to serve as stimulus for learning and innovation in the local food companies. Earlier empirical results from some developing countries, including Nigeria, also confirmed this assertion (Blalock and Gertler, 2008; Javorcik, 2004; Ajayi, 2001, 2007). Spillovers occurred through vertical relationship (through backward linkage) rather than horizontal relationship. Specifically in Nigeria, there was adoption of production sub-contracting among the food, beverages and tobacco, chemicals and pharmaceuticals and textiles, wearing apparel and leather industry groups after the introduction of structural adjustment programme in 1986. This sub-contracting of production among firms was perceived as very important in reducing the costs of production (Ajayi, 2007).

	No of SMFCs	Percent (%)
Any Linkage with MNCs?		
Yes	50	44.6
No	62	55.4
No of MNCs that SMFCs had Linkage with		
None	62	55.4
1	6	5.4
2	30	26.8
3	12	10.7
4	0	0.0
5	2	1.7
Sector where the MNCs belong		
No response	72	64.3
Food, Tobacco and Beverages	34	30.4
Chemical & Pharmaceutical	4	3.6
Electrical and Electronic	2	1.7
Type of relationship between SMFCs and MNCs		
No response	62	55.4
Supplier of inputs (raw materials) to the MNCs	2	1.8
Purchase inputs (raw materials) from the MNCs	41	36.6
Subsidiary of a MNC	4	3.6
MNC outsourced part of the production from our company	2	1.8
Purchase equipment through MNCs	1	0.8
Facilities having access to in MNCs		
No response	67	59.8
Product certification	-	0.0
Quality control and analysis	5	4.5
Sharing of laboratory	2	1.8
Assisting in procuring processing equipment	1	0.9
Assisting in entering foreign market	1	0.9
Receiving technical assistance	7	6.3
Development of joint projects	-	0.0
Providing training for our staff	29	25.8

Table 5. Types of Linkage between the SMFCs MNCs

ii.   Labour turnover and human capital development

About 38% and 8.9% of the owner managers and production managers of the SMFCs had working experience from MNCs (Table 6). Within the owner manager who had worked in the MNC, 50.0% worked in R&D department, and 38.1% in production or operation department. The remaining 9.5% and 2.4% worked in quality control and administration departments of the MNC respectively. Majority (69.0%) of these owner managers worked at operational level, while 28.6% and 2.4% had worked at management and policy levels of the MNCs respectively. Furthermore, among the production managers that had previous working experience with MNCs, 60% and 40% worked in production/operation and R&D departments respectively; and majority (80%) worked at operational level. This result also agreed with an earlier study which reported that small and medium scale firm owners within the same study area had diverse backgrounds, which included previous experiences with MNCs (Oyelaran-Oyeyinka, 2004).

Working Experience of SMFCs with Food MNCs & Level of Management	Owner's Manager		Production Manager		Marketing Manager		Administrative Manager	
	No of SMFCs	Percent (%)	No of SMFCs	Percent (%)	No of SMFCs	Percent (%)	No of SMFCs	Percent (%)
Working Experience with Food MNCs	42	37.5	10	8.9	-	0	-	0
Yes	70	62.5	102	91.1	112	100	112	100
No								
Department where worked	-	-	-	-	-	-	-	-
Administration	1	2.4	-	-	-	-	-	-
Marketing	16	38.1	6	60	-	-	-	-
Production or Operation	21	50.0	4	40	-	-	-	-
Research & Development	4	9.5	-	-	-	-	-	-
Quality control								
Level of Management worked	29	69.0	8	80	-	-	-	-
Operational Level	12	28.6	2	20	-	-	-	-
Management Level	1	2.4	-	-	-	-	-	-
Policy making Level								
Training Courses/Seminar Attended	12	28.6	-	-	-	-	-	-
Strategic Planning	4	9.5	2	20.0	-	-	-	-
Product Development	-	-	-	-	-	-	-	-
Marketing	16	38.1	1	10.0	-	-	-	-
Human Resource Management	10	23.8	7	70.0	-	-	-	-
Quality Maintenance								

Table 6. Working Experience of the key Personnel of SMFCs with Food MNCs at various Level of Management

About 38.1%, 28.6%, 23.8%, and 9.5% of these owner managers attended human resource management, strategic planning, quality maintenance, and product development training courses/workshops respectively. Also 70%, 20% and 10% of the production managers attended quality maintenance, product development, and human resource management respectively while working with the MNCs. It has been established that labour turnover brings about spillover when owner managers in local firms started their careers in foreign companies and/or there was brain-drain in reverse to the local economy (Dunning, 1970;

Ikiara, 2003). Hence, with the physical movement of workers from MNCs, the knowledge embodied in these workers could be transferred to the local economy. Based on the result above, the owner managers and production managers would have acquired knowledge and skill as a result of the training courses/seminars attended while working with the MNCs.

iii. Changes effected by SMFCs in their Production Capabilities due to the Channels of Spillover

Influence of Competition from MNCs on production capabilities

About 43.8% of the SMFCs modified their products to reduce the production cost as a result of competition from the MNCs (Table 7). Also, 52.7% changed the design of the product packaging, 64.7% introduced new equipment to improve production efficiency, and 31.2% introduced automation in certain areas of production. Other changes undertaken were upgrading of processing equipment (42.6%) and regular repair and maintenance of the processing equipment (40.5%). Small percentage (1.2%) of the food companies embarked on new product development in order to sustain their market share and remain competitive in the market place.

Influence of linkages with MNCs on production capabilities

About 44.8% of the local SMFCs modified their products as a result of their linkages with MNCs (Table 7). Also, 52.0% changed the design of the product packaging. The linkage channel also brought about introduction of new equipment to improve production efficiency (54.6%). It also led 40.2% of the SMFCs to introduce automation in certain areas of their production processes. Other changes which resulted from linkages with MNCs included upgrading (56.3%) and constant repair and maintenance (32.1%) of processing equipment.

Influence of previous working experience of owner/staff of SMFCs with MNCs on production capabilities

The previous working experience of owner/staff with MNCs assisted 40.2% of the SMFCs to modify their products so as to reduce production cost. Also, 54.1% of small and medium food companies changed their product packaging design. Other changes that resulted from the influence of previous working experience with MNCs were introduction of new equipment to improve production efficiency (65.5%), upgrading of processing equipment (42.2%), and repair and maintenance of equipment' (46.2%).

In summary, the above information provided evidences that there were efforts by the local food companies at effecting changes in their production technology as a result of the spillover channels. The spillover channels that brought about new product formulation in very few SMFCs were staff experience with MNCs and training received from MNCs. All the four spillover channels however resulted in improvement of production capabilities in majority of the SMFCs. The SMFCs indicated that their modification of product packaging was to make their products attractive and appealing to the consumers as those of imported substitutes. Some even sourced their packaging materials from overseas. This is consistent with some empirical studies that local firms are forced to learn and introduce appropriate changes to achieve allocative and/or technical efficiency, especially in response to competition from MNCs (Wang and Blomstrong, 1992; Gachino, 2006). Therefore, important observation from this result is that some of the domestic SMFCs were being placed on learning function thereby increased their potential to learn. This is a form of spillovers occurrence.

Changes to Product, Process and Industrial Engineering	Percentage of SMFCs based on changes introduced as a result of the Spillover Channel of:			
	Competition from MNCs	Linkage with MNCs	Staff Experience with MNCs	Training Received from MNCs
(a) Types of Changes in Product				
Product modification through enrichment	1.2	0.2	0.2	3.2
New product formulation	-	-	1.0	1.5
Quality improvement of the product	2.3	3.0	4.5	2.3
Modifying the product to reduce the production cost	43.8	44.8	40.2	40.4
Changing the design of the product packaging	52.7	52.0	54.1	52.6
(b) Types of Changes in Production Technique				
Improvement of traditional methods of processing	-	-	-	-
Introduction of automated machines throughout the production line	0.6	0.8	2.4	2.6
Introduction of automation only at a certain area of production	31.2	40.2	28.6	28.1
Introduction of new equipment to improve production efficiency	64.7	54.6	65.5	66.2
Laying out the machines on the factory floor in a better order	3.5	4.4	3.5	3.1
(c) Types of Changes in Industrial Engineering				
Replacement of Processing Equipment	16.9	11.6	11.6	14.7
Upgrading of Processing Equipment	42.6	56.3	42.2	42.3
Repair and Maintenance of Equipment	40.5	32.1	46.2	44.0

Table 7. Various Changes to Product, Process and Industrial Engineering due to different Channels of Spillover from Multinational Companies

5.2 Factors responsible for the occurrence of technology spillovers

Organisational capabilities and working experience

More than half of the SMFCs (69.6%) indicated that their owner managers had post graduate qualification (Table 8). Also, 93.8%, 99.1%, and 96.3% of these food companies indicated that their production, marketing, and administrative managers respectively had tertiary education. This level of education of the management team was an indication that majority of the food companies in the study area had some basic requirements for building absorptive capability for spillover. Oyelaran-Oyeyinka (2004) also found the same result within the same study area when he reported that about 63.2% of firms' owners had bachelor degree. In addition, more than half (63.4%) of the owner managers of the SMFCs specialised in science/engineering while 35.7% specialised in management related discipline (Table 8). Only one of them (0.9%) had both science/engineering and management related

	Owner Manager		Production Manager		Marketing Manager		Administrative Manager	
	No of SMFCs	(%)	No of SMFCs	(%)	No of SMFCs	(%)	No of SMFCs	(%)
Highest Qualification								
Secondary	0	0	0	0	0	0	1	1.0
Technical	0	0	3	2.7	0	0	0	0
Tertiary (polytechnic/university)	34	30.4	105	93.8	108	99.1	105	96.3
Post graduate	78	69.6	4	3.6	1	0.9	3	2.7
Area of Specialisation								
Science or Engineering	71	63.4	94	83.9	1	0.9	1	0.9
Management or Finance related	40	35.7	18	16.1	107	98.2	108	99.1
Science & Management	1	0.9	0	0	0	0	0	0
Others	0	0	0	0	1	0.9	0	0
Previous Work Experience								
SMEs Company	33	29.6	73	65.8	58	53.7	67	61.6
Large corporation/MNCs	41	36.5	31	27.8	30	28.0	10	9.0
University/Research Institutes	15	13.5	5	4.5	4	3.7	10	9.2
Government Ministry/Parastatals	16	14.5	2	1.9	16	14.6	22	20.2
Small and Large	7	5.9	0	0	0	0	0	0
Training Courses/Workshops attended by the Management Team								
Human resources management	48	42.9	-	-	1	0.9	16	15.0
Product Development	1	0.9	97	87.4	2	1.9	-	-
Quality Maintenance	-	-	2	1.8	1	0.9	1	0.9
Strategic planning	2	1.8	-	-	-	-	84	78.5
Marketing	-	-	1	0.9	99	91.7	-	-
Combination of above	61	54.4	11	9.9	5	4.6	6	5.6
Total	**112**	**100**	**111**	**100**	**108**	**100**	**107**	**100**

Table 8. Highest Qualifications, Areas of Specialisation, Previous Work Experiences and Training Courses attended by the Members of the Management Team of the Small and Medium Food Companies (SMFCs)

discipline. Majority (83.9%) of their production managers had science and engineering background. However, the remaining 16.9% of the SMFCs indicated that their production managers specialised in management/finance related disciplines. Similarly, majority of the firms also indicated that their marketing (98.1%) and administrative (99.1%) managers had relevant areas of specialisation.

About 37% of owner managers had previous working experience with MNCs, while 29.6% had worked with small and medium companies, 13.5% with university/research institutes, and 14.5 per cent with Government ministries/agencies (Table 8). Majority of other management teams worked with small and medium companies. That is, 65.8%, 53.7%, and 61.6% of production managers, marketing managers, and administrative managers had previous working experience with small and medium companies respectively. The SMFCs indicated that 27.8% and 28.0% of production manager and marketing managers respectively had previous working experience with MNCs, while 20.2% reported that their administrative managers worked with Government ministry. This result indicated that there was sizeable number of labour turnover from MNCs, especially among the owner managers. By this it can be assumed that these owner managers would have acquired some knowledge and skills from these MNCs.

Status of Embedded Technology in Equipment	No of SMFCs	Percent (%)
Sources of Processing Equipment (N=112)		
Locally Fabricated	37	33.0
Imported	70	62.5
Combination of Locally Fabricated and Imported	5	4.5
Average Age of Processing Equipment (N=112)		
Less than 3 years	21	18.8
Between 3 and 5 years	23	20.5
More than 5 years	68	60.7
Level of Automation (N=112)		
Fully Automatic	13	11.6
Semi Automatic	94	83.9
Manually Operated	2	1.8
Fully Automatic and Semi Automatic	2	1.8
Fully Automatic, Semi Automatic, Manually Operated	1	0.9
Whether the Processing Equipment are the best in the Market (N=111)		
Yes	12	10.8
No	99	89.2
Reasons for not Acquiring Recent Equipment (N=111)		
High cost of acquisition	2	1.8
Insufficient Capital to Acquire them	98	88.3
Not Applicable	11	9.9

Table 9. Status of Embedded Technology in Processing Equipment of SMFCs

All the MNCs reported attendance of training courses by their management team. These courses included human resources, production development, quality maintenance, strategic planning, marketing, and combination of these. Within the owner managers of these food companies, 42.9% attended human resources management programme and 54.4% attended combination of these programmes. Also, 87.4% of production managers attended product development related programme, 91.7% of marketing managers attended marketing-related programme, and 78.5% and 15.0% of administrative managers attended strategic planning programme.

ii. Embedded technology in processing equipment and linkages with external agents

More than half (62.5%) of the food companies reported (Table 9) that their processing equipment were imported and had been in operation for more than five years (60.7%). They further indicated that majority of the processing equipment were semi-automatic and were not the best in the market. The high cost (1.8%), and insufficient capital (88.3%) were indicated as limiting factors for the acquisition of best equipment. This probably limits the technological capabilities of these food companies to produce for export.

About 75.0% of the SMFCs had relationship with other SMFCs, 35.7% had relationship with SME business associations (such as NASSI, NASME), and 98.2% had relationship with financial institutions (Table 10). These types of relationship included support in setting up plant (92.0%), provision of technical consultancy (38.4%) and sharing of exporting knowledge (30.4%).

Organisational Capabilities	No of SMFCs	Percent (%)
Formal Establishment with Agent (N=112)		
Small and Medium Companies (SMEs)	85	75.0
Multinational food companies (MNCs)	107	95.5
Business development service (BDS) Providers	2	1.8
Universities/Research Institutes	1	0.9
Business Associations (e.g. NASSI, NASME, etc)	40	35.7
Financial Institutions	110	98.2
Types of Relationship (N=109)		
Sharing of production equipment	1	0.9
Sharing of laboratory	1	0.9
Joint development of product and processes	4	3.6
Support in setting up our plant	103	92.0
Provision of technical consultancy	43	38.4
Share knowledge of export	34	30.4

Table 10. Organisational Capabilities of SMFCs

5.3 Relationship between the factors determining absorptive capacity and spillover index

The calculated spillover indexes, the frequency and percentage of each value are shown in Table 11. Majority (61.5%) of the SMFCs had composite spillover index of 4. This showed that the influence of the spillover channels on majority of the SMFCs' production capability was moderate. Other effects were little effect, considerable effect, and insignificant effect on the production capabilities of 30.4%, 4.5%, and 3.6% of the SMFCs respectively.

Composite Spillover Index	No of SMFCs	Percent of SMFCs (%)
1 (No effect)	0	0
2 (Insignificant effect)	4	3.6
3 (Little effect)	34	30.4
4 (Moderate effect)	69	61.5
5 (Considerable effect)	5	4.5
6 (Much effect)	0	0
7 (Very much effect)	0	0

Table 11. Percentage Distribution of Composite Spillover Index among the SMFCs

i. Relationship between absorptive capacity determinant and spillover index (SI) of SMFCs

The correlation between the dependent variable (SI) and each of the independent variables showed that there were weak but significant relationships ($p \leq 0.05$) between SI and age of company (r=0.295), percentage of Nigerian ownership (r=-0.527), area of specialisation of owner (r=0.301), work experience of owner (r=0.249), work experience of marketing manager (r=0.272), level of automation of the main processing equipment (r=-0.320), or number of years in relationship with multinational companies (r=0.329).

The positive relationship between age of the small and medium food companies and SI indicates that as the number of years in business increases the SMFCs were accumulating technological knowledge which resulted in increased absorptive capacity. The results also showed that the area of specialisation and previous work experience of the owners of SMFCs in MNCs were important for spillover occurrence. Most of the time the small and medium entrepreneurs (owners) exercise controlling influence on all the activities of their businesses. Hence, the level of scientific and technological knowledge possessed by these owners as a result of their specialisation and experience had influence on the spillover occurrence.

In addition, the positive correlation between SI and work experience of marketing manager indicates the important contribution of relevant knowledge in other human resources to the absorptive capacity of these SMFCs. However, the low (weak) value of correlation coefficients for previous working experience of owners in MNCs indicates that it is not enough to have experience in MNCs, ability to absorb tacit knowledge, codify it and apply it at their own firms is equally important. Also, the positive influence of the number of years in relationship with MNCs is a demonstration that the interactions arising from the linkage had served as a stimulus for learning and innovation among the small and medium food companies, and hence had influence on the spillover occurrence.

On the other hand, the strong, negative and significant relationship between SI and percentage of Nigerian ownership (r=-0.527) indicates that as the percentage of Nigerian ownership increases there was a decrease in the absorptive capacity of the SMFCs. The reason could be attributed to low level of technological competence of the Nigerian owners, which is not adequate to recognise valuable new knowledge from MNCs. This is because the firm's level of absorptive capacity depends upon its level of technological competence as well as its learning and investment efforts undertaken to be able to use new knowledge from MNCs productively (Hamida, 2007). Moreover, minority foreign ownership could also serve as a disincentive for the MNCs' parent firms to transfer more advanced technology to its affiliate due to its reduced control over the management (Javorcik and Spatareanu, 2003).

Also, negative relationship between SI and the level of automation of the main processing equipment (r=-0.320) indicates that investment in new equipment for product/process innovation is an important determining factor for the absorptive capacity. As noted from the empirical results from the literature, relatively high technological firms benefit from spillovers through demonstration and/or competition effects (Mody, 1989). This is because such firms are not far behind the technological frontier of the industry. This could have assisted them to imitate and/or to improve their production efficiency needed for competition with MNCs' products. Using investment in new equipment as a proxy for the absorptive capacity of domestic firms, Narula and Marin (2003) also observed positive spillovers for domestic firms in Argentina which had high investment in new equipment oriented to product/process innovation.

Moreover, the regression results obtained showed that the percentage of scientists over the total workforce, area of specialisation of owner, works experience of owner, and no of year in relationship with MNCs showed statistically significant relationship. The coefficient values (β) were -0.287, 0.434, 0.432, and 0.315 for the percentage of scientists, area of specialisation of owner, works experience of owner, and no of year in relationship with MNCs respectively. The negative coefficient value for the percentage of scientists could be that the scientists employed by these SMFCs were not skilled enough to make any significant contribution to the spillover occurrence in the companies. They could also be performing routine work without any opportunity to be involved in R&D activities. Since the absorptive capacity required by firms in developing countries for spillover occurrence depends on the complementary role of the level of technological knowledge in human resources and physical capital investment (Gachino, 2007). Hence, the insignificant contributions of the technical and engineering personnel from this study could be due to inadequate physical investment in most of the SMFCs.

The positive values of correlation coefficient and significant relationship for the specialisation of the owner in science/technology/engineering and their previous work experience indicated the important of these factors. De Fuentes and Dutrénit (2006) also discovered that SMEs with high level of absorptive capacities had most of the owners with professional degree in engineering. Also, the knowledge and skill acquired, as a result of previous experience of the owners of SMFCs in MNCs, were important for knowledge spillovers. Earlier theory had established that technological or knowledge spillover could occur in the domestic firms when there is movement of employee from the MNCs to local firms.

Moreover, the interactions between small enterprises and MNCs is also an important mean through which interactive learning, information and technology can be exchanged or jointly exploited for the purpose of productive activities, which can stimulate the process of spillover occurrence.

The insignificant relationship of age of the SMFCs indicated that age had not contributed to the spillover process in these firms. The result is consistent with the assertion that accumulation of knowledge might not be taken as a simple function of firm age in developing countries, because most firms in developing countries might not be in position to accumulate knowledge over time due to lack of resources (Gachino, 2007). Also, the technical personnel did not have relevant previous work experience that could bring about spillover occurrence in these SMFCs. Therefore, having experience in MNCs is not enough for gaining the required knowledge that can be diffused through technological spillovers. This result could be explained from the type of training received by workers at MNCs. That is, as noted by Fosfuri, Motta, and Ronde (2001), if they received training in a more firm-specific technology, local firms might have less advantage in obtaining that technology as it might be costly to adapt to their own production process. The level of automation of equipment of SMFCs was not high enough to contribute to spillover occurrence. Spillover was observed to occur in Argentina domestic firms where there was high investment in absorptive capacities inform of training activities or new equipment (Narula and Marin, 2005).

6. Conclusion

This study re-conceptualised technological spillover from MNCs by linking the spillover occurrence to the technological changes associated with the production capability building in the SMFCs in southwestern Nigeria. It was observed that there was only a moderate building process of production capability among these companies. This further confirms that the occurrence of spillovers does not depend just on the presence of MNCs alone but also on absorptive capacity of the local firms.

The factors with the highest influence on the absorptive capacities among the SMFCs were area of specialisation and previous work experience in the MNCs of the owners, and year of relationship with MNCs. These results showed that, though MNCs had played an important role in stimulating learning and capability building in the local food companies in Nigeria, but promoted a minimal innovation in these companies. Hence, in order to make FDI have greater impact on future opportunities for catching up technologically there is need to re-assess and strengthen the national linkage promotion programmes and institutions in Nigeria. This will assist in smoothen the linkage relationship between SMEs with MNCs, and also with universities and research institutes. This is because a strong network which supports generation and diffusion of knowledge can only stimulate the process of spillover occurrence to SMEs. Equally important is the policy that will encourage regular update of the list of MNCs' local suppliers, and encourage joints venture partnership between the MNCs and local food companies whenever the former are embarking on expansion and upgrade of their production activities.

The study also revealed that many of the SMFCs were not able to acquire better processing equipment even when they were aware that their processing equipment were not the best in the market. The implication of this is that even when the workforce had sufficient innovation and learning capacities, the level of automation in the processing equipment could still affect the production efficiency. Therefore, there is need for policy measure to encourage the financial sector to assist the SMFCs to invest in the upgrade of their processing equipment which could assist them to imitate the MNCs' production technologies.

The various technological capability building that can occur in small firms as a result of technological spillover occurrence includes investment, production, linkage, and complimentary capabilities (such as innovation, organisation and marketing capabilities). However, due to the magnitude and scope of work, this study focused only on the production capability of the SMFCs in the southwestern Nigeria. It is therefore suggested that further studies should focus on any or combination of these firm level capabilities.

7. References

Abereijo, I.O. and Ilori, M.O. (2010). *New Technology Acquisition by SMEs in Nigeria: Ability and Competencies to Innovate*, LAP Lambert Academic Publishing, Germany, ISBN: 978-3-8383-9135-9.

Aitken, B. and A. Harrison (1999). Do Domestic Firms Benefit from Direct Foreign Investment? Evidence from Venezuela. *American Economic Review*, Vol. 89, pp. 605 – 618.

Ajayi, D.D. (2001). Industrial Subcontracting Linkages in the Lagos Region, Nigeria. *The Nigerian Journal of Economic and Social Studies (NJESS)*, Vol. 42, No. 1, pp. 95-111.

Ajayi, D.D. (2007). Recent Trends and Patterns in Nigeria's Industrial Development. *African Development*, Vol. XXXII, No. 2, pp. 139-155. Publication of Council for the Development of Social Science Research in Africa (ISSN 0850-3907). Available on

Akinlo, A. (2004). Foreign Direct Investment and Growth in Nigeria: An Empirical Investigation. *Journal of Policy Modeling*, Vol. 26, pp. 627-639.

Ayanwale, A. B. and Bamire, S. (2004). Direct Foreign Investment and Firm-level Productivity in the Nigerian Agro/Agro-allied Sector. *Journal of Social Sciences, India*, Vol. 9 No. 1, pp. 29-36.

Barrios, S. (2000). Foreign Direct Investments and Productivity Spillovers: Evidence from the Spanish Experience. Fundacion de Estudios de Economia Aplicada (FEDEA), *Working paper 2000-19*

Benhabib, J. and Spiegel, M.M. (1994). The Role of Human Capital in Economic Development: Evidence from Aggregate Cross-Country Data. *Journal of Monetary Economics*, Vol. 34, pp. 143-173.

Biersteker T. (1987). *Multinationals, The State, and Control of the Nigerian Economy*, Princeton University Press, Princeton.

Blalock, G. and Gertler, P.J. (2008). Welfare Gains from Foreign Direct Investment through Technology Transfer to Local Suppliers. *Journal of International Economics*, Vol. 74, Issue 2, March, pp. 402-421.

Blomström, M. and Person, H. (1983). Foreign Investment and Spillover Efficiency in an Underdeveloped Economy: Evidence from the Mexican Manufacturing Industry. *World Development*, Vol. 11, pp. 493-501.

Blomströom, M. and Sjööholm, F. (1998). Technology Transfer and Spillovers: Does Local Participation with Multinationals Matter? *Working Paper Series in Economics and Finance*, No. 268, Stockholm School of Economics.

Blomström, M. and E. Wolf (1994). Multinational Corporations and Productive Convergence in Mexico. In: *Convergence of Productivity: Cross-National Studies and Historical*

Evidence, W. Baumol; R. Nelson; and E. Wolff (Eds.), Oxford University Press, Oxford.

Blomström, Magnus; Globerman, Steven; and Kokko, Ari (1999). The Determinants of Host Country Spillovers from Foreign Direct Investments: Review and Synthesis of the Literature. The European Institute of Japanese Studies, Stockholm School of Economics, *Working paper No. 76.*

Blomström, M. and Kokko, Ari (2003). The Economics of Foreign Direct Investment Incentives. *Working paper 168*, Bundesbank-Conference, Stockholm, Sweden

Castellani, D. and A. Zanfei (2001). Productivity Gaps, Inward Investments and Productivity of European Firms, *Mimeo*, University of Urbino, Italy.

Caves, R. (1974). Multinational Firms, Competition, and Productivity in Host-Country Markets, *Economica*, Vol. 41, pp. 176-193.

Central Bank of Nigeria (2005). Small and Medium Industries Nigeria Information System (SMINIS). Available from http://smi-nigeria.org/cbnsmi/mainform.aspx

Cohen, W. M. and Levinthal, D.A. (1990). Absorptive Capacity: A new Perspective on Learning and Innovation. *Administrative Science Quarterly*, Vol. 35, No. 1, pp. 128-152.

Damijan, J.P., Jaklič, A. and Rojec, M. (2005). Do External Knowledge Spillovers Induce Firms' Innovations? Evidence from Slovenia. Centre for International Relations (CIR) *Working Paper Series No. 3.* Centre for International Relations, Faculty of Social Sciences, University of Ljubljana, Slovenia. Available from http://www.mednarodni-odnosi.si/cmo/WP/CIRWP3_PDamijan_Jaklic_Rojec.pdf

De Fuentes, Claudia and Dutrénit, Gabriella (2006). SMEs' Absorptive Capacities and Large Firms' Knowledge Spillovers in a Mexican Locality. *Proceedings of the paper presented at Innovation Systems for Competitiveness and Shared Prosperity in Developing Countries* (GLOBELICS INDIA 2006), October 4-7, at Trivandrum, Kerala, India.

Djankov, S. and B. Hoekman (2000). Foreign Investment and Productivity Growth in Czech Enterprises, *World Bank Economic Review*, Vol. 14, pp. 49-64.

Dunning, John H. (1970). *Studies in International Investments*, (The Pitman Press, Bath, George Allen and Unwin Ltd., USA)

Economic Commission for Africa (2001). Enhancing the Competitiveness of Small and Medium Enterprises in Africa: A Strategic Framework for Institutional Support, *Working paper* ECA/DMD/PSD/TP/00/04

Fagerberg, J. (2005). Innovation: A Guide to Literature. In: *The Oxford Handbook of Innovation*, J. Fagerberg; D.C. Mowery; and R.R. Nelson (Eds.), Oxford University Press, Oxford.

Fosfuri, A. and Motta, M. and Ronde, T. (2001). Foreign Direct Investment and Spillovers through Workers' Mobility, *Journal of International Economics*, Elsevier, Vol. 53, No. 1, pp. 205-222.

Freeman, C. (1991). Networks of Innovators: A Synthesis of Research Issues, *Research Policy*, Vol. 20, pp. 499-514.

Gachino, Geoffrey (2007). Technological Spillovers from Multinational Presence: Towards a Conceptual Framework, *Working Paper*, United Nations University – Maastricht Economic and Social Research and Training Centre on Innovation and Technology (UNU-MERIT), The Netherlands. Available from http://www.merit.unu.edu

Girma, S.; Greenway, D.; and Wakelin, K. (2000). Who Benefits from Foreign Direct Investment in the UK? *Paper presented at the Royal Economic Society Annual Conference, St. Andrews,* July.

Globerman, S. (1979). Foreign Direct Investment and Spillover Efficiency Benefits in Canadian Manufacturing Industries, *Canadian Journal of Economics,* Vol. 12, pp. 42-56.

Griliches, Z. (1992). The Search for R&D Spillovers. *Scandinavian Journal of Economics,* Vol. 94 (supplement), pp. 29-47.

Hamida, B.L. (2007). Inwards Foreign Direct Investment and Intra-Industry Spillovers: The Swiss Case. *Unpublished Ph.D. Dissertation,* University of Fribourg.

Haskel, J. E.; Pereira. S.C.; and Slaughter, M.J. (2002). Does Inward Foreign Direct Investment Boost the Productivity of Domestic Firms?, *NBER Working Paper Series,* Working Paper 8724.

Hausmann, R. (2000). Foreign Direct Investment: Good cholesterol? *Inter-American Development Bank Working Paper.* Washington, DC.

Ikiara, Moses M. (2003). Foreign Direct Investment (FDI), Technology Transfer, and Poverty Alleviation: Africa's Hopes and Dilemma. *African Technology Policy Studies Network (ATPS) special paper series No. 16.* Published by ATPS Communications Department, Kenya. (Newtec Concepts, Kenya)

Javorcik, B. and Spatareanu, M. (2003). To Share or Not To Share: Does Local Participation Matter for Spillovers from Foreign Direct Investment? *World Bank Policy Research Working Paper No. 3118.*

Javorcik, B. S. (2004). Does Foreign Direct Investment Increase the Productivity of Domestic Firms? In Search of Spillovers Through Backward Linkages, *American Economic Review,* Vol. 94, No. 3, pp. 605–627.

Kapstein, E.B. (2002). Virtuous Circles? Human Capital Formation, Economic Development and the Multinational Firms, *OECD Development Centre Working Paper,* OECD, Paris.

Kathuria, Vinish (1998). Foreign Firms and Technology Transfer Knowledge Spillovers to Indian Manufacturing Firms. The United Nations University (UNU), Institute for New Technologies (INTECH) *Discussion Paper Series No. 9804.* Available from http://www.intech.unu.edu/publications/discussion-papers/9804.pdf

Kathuria, Vinish (2000). Productivity Spillovers from Technology Transfer to Indian Manufacturing Firms, *Journal of International Development,* Vol. 12, pp. 343-369.

Kokko, A., R. Tansini and M. Zejan (1996). Local Technological Capability and Productivity Spillovers from FDI in the Uruguayan Manufacturing Sector, *Journal of Development Studies,* Vol. 32, pp. 602-611.

Konings, J. (2000). The Effects of Foreign Direct Investment on Domestic Firms: Evidence from Firm Level Panel Data in Emerging Economies, *Economics of Transition,* Vol. 9, pp. 619-633.

Krogstrup, S. and Matar, L. (2005). Foreign Direct Investment, Absorptive Capacity and Growth in the Arab World. *Working Paper No. 02,* Graduate Institute for International Studies, Geneva.

Kugler, M. (2000). The Diffusion of Externalities from Foreign Direct Investment: Theory Ahead of Measurement, *Mimeo,* Department of Economics, University of Southampton.

Lall, S. (1992). Technological Capabilities and Industrialization, *World Development*, Vol. 20, No. 2, pp. 165 – 186.

Lundvall, B. A. (1992). *National Systems of Innovation: Towards a Theory of Innovation and Interactive Learning*, Francis Pinter, London.

Mody, A. (1989). Strategies for Developing Information Industries. In: *Technology and Development in the Third Industrial Revolution*, Cooper, C. and Kaplinsky, R. (Eds.), Frank Cass, London.

Mytelka, L. K. and Farinelli, F (2000). Local Clusters, Innovation Systems and Sustained Competitiveness. Paper presented at a Conference in Rio de Janeiro.

Nadiri, M. (1991). Innovations and Technological Spillovers, *Mimeo*, New York University.

Narula, Rajneesh (1997). The Role of Developing Country Multinationals in the Acquisition of Industrial Technology in Nigeria: A Pilot Study. Available from http://edata.ub.unimaas.nl/www-edocs/loader/file.asp?id=68

Narula, Rajneesh & Marin, Anabel (2003). FDI Spillovers, Absorptive Capacities and Human Capital Development: Evidence from Argentina. *MERIT-Infonomics Research Memorandum Series 2003-16*. Amsterdam.

Narula, Rajneesh and Marin, Anabel (2005). Exploring the Relationship between Direct and Indirect Spillovers from FDI in Argentina. *Research Memoranda 024*, MERIT, Maastricht.

NIPC (2006). Invest for Success in Nigeria. Available on http://www.nipc-nigeria.org/page002.html

Onimode, B. (1982). *Imperialism and Underdevelopment In Nigeria*, Zed Books, London.

Oyelaran-Oyeyinka, Banji (2004). Networking Technical Change and Industrialisation: The Case of Small and Medium Firms in Nigeria, *African Technology Policy Studies Network (ATPS) Special Paper Series* No. 20.

Ramachandran, V. (1993). Technology Transfer, Firm Ownership, and Investment in Human Capital, *Review of Economics and Statistics*, vol. 75, No. 4, pp. 664-670.

Rasiah, R. (2005). *Foreign Firms, Technological Capabilities And Economic Performance: Evidence From Africa, Asia and Latin America*. Edward Elgar Publishing, USA.

Saxeniaan, A. L. (1991). The Origins and Dynamics of Production Networks in Silicon Valley, *Research Policy*, Vol. 20.

Scherer, F. M. (1980). *Industrial Market Structure and Economic Performance*, Rand McNally, Chicago.

Schoors, K. and B. Van der Tol (2002). FDI Spillovers Within and Between Sectors: Evidence from Hungarian Data, Ghent University *Working Paper 2002/157*

Smarzynska, B.K. (2002). Does Foreign Direct Investment Increase the Productivity of Domestic Firms? In Search of Spillovers through Backward Linkages. *Policy Research Working Paper No. 2923*, The World Bank Dev. Research Group. Available from http://www-wds.worldbank.org/servlet/WDSContentServer/WDSP/IB/2002/ 11/ 22/000094946_02111304010628/Rendered/PDF/multi0page.pdf (Accessed Oct 2011)

Taymaz, Erol and Lenger, Aykut (2004). Multinational Corporations as a Vehicle for Productivity Spillovers in Turkey, Danish Research Unit for Industrial Dynamics (DRUID) *Working paper No. 04-09*.

UNCTAD (2005). Linkages, Value Chains and Outward Investment: Internationalisation Patterns of Developing Countries' SMEs, *Working paper* TD/B/COM.3/69.

United Nations Economic Commission for Europe (UN-ECE) (2001). The Environment for FDI spillovers in the Transition Economies, *Working paper*, Geneva.

United Nations Conference on Trade and Development (UNCTAD) (2006). Promoting TNC-SME Linkages to Enhance the Productive Capacity of Developing Countries' Firms: A Policy Perspective. *Issue Note* by the UNCTAD (TD/B/COM.3/75), Tenth session of Commission on Enterprise, Business Facilitation and Dev. at Geneva, Feb 21 – 24.

UNCTAD (2006). World Investment Report – Overview, *Publication of United Nations Conference on Trade and Development (UNCTAD)*, United Nations, New York and Geneva. Available from http://www.unctad.org/en/docs/wir2006overview_en.pdf

Vera-Cruz, A.O. and Dutrenit, G. (2005). Spillovers from MNCs through Worker Mobility and Technological and Managerial Capabilities of SMEs in Mexico. *Innovation: Management, Policy and Practice.* Vol. 7, Issue 2-3, April, pp. 274-297.

Wang and Blomstrong, (1992). Foreign Investment and Technology Transfer: A simple Model. *Economic Review*, Vol. 36, pp. 137-155.

Part 3

Enablers of Technological Innovation

10

Open Innovation in the Automotive Industry: A Multiple Case-Study

Alfredo De Massis[1], Valentina Lazzarotti[2],
Emanuele Pizzurno[2,*] and Enrico Salzillo[3]
[1]Università di Bergamo, Center for Young & Family Enterprise (CYFE)
[2]Università Carlo Cattaneo – LIUC
[3]Business Integration Partners (BIP)
Italy

1. Introduction

Our chapter aims at exploring the concept of Open Innovation (OI) and evaluating whether, why and how it is adopted in the automotive field. Moreover, the intent is identifying which kinds of potential advantages and risks automotive companies should face when choosing Open Innovation strategies. As regards its basic goal, the study attempts to enrich the existing empirical evidence because few studies about the topic were conducted.

Our research is carried out with a combination of literature analysis (i.e. bibliographic research of journal articles, books, and official companies' press) and face-to-face interviews, conducted with a semi-structured protocol and addressed to three well-known companies operating in the automotive industry (i.e. an Italian automotive company[1], Pininfarina, Bosch).

Figure 1 represents our main research questions:

1. whether and why open approaches can be adopted;
2. how openness can be set-up;
3. what are the obstacles firms face and the advantages companies achieve while adopting an OI approach.

The underlying idea is that firm-specific strategic goals as well as external factors (i.e. environmental/industry features) can affect (i.e. explaining whether and why) the adoption of OI approaches and that openness can lead to some advantages despite some obstacles and risks. Both literature and case studies helped to define the framework and to set it in the automotive industry. Case studies in particular allowed us to grasp in practice these pressures toward OI and to describe in detail how OI is set up (this is possible in many different conceptions as it will be explained later in this chapter: e.g. in terms of prevalent partners; phases of the innovation funnel on which the collaboration is in place; adopted

---

* Corresponding Author
[1] Named "Company A" because confidentiality reasons force the authors to hide the real name of the company.

organizational modes, only to give some examples of questions); the obstacles and risks on one side as well as the potential advantages on the other side.

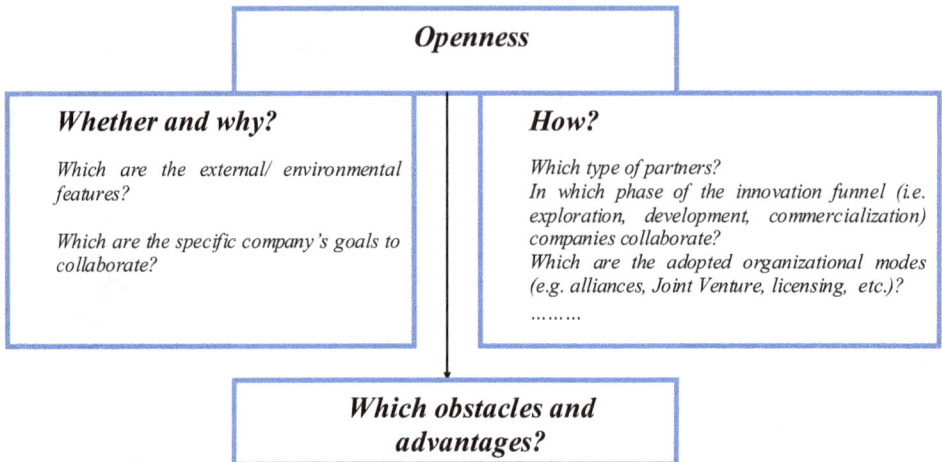

```
┌─────────────────────────────────────────┐
│              Openness                    │
└─────────────────────────────────────────┘

┌──────────────────────────┐   ┌──────────────────────────────┐
│  Whether and why?        │   │  How?                        │
│                          │   │                              │
│  Which are the external/ │   │  Which type of partners?     │
│  environmental features? │   │  In which phase of the       │
│                          │   │  innovation funnel (i.e.     │
│  Which are the specific  │   │  exploration, development,   │
│  company's goals to      │   │  commercialization)          │
│  collaborate?            │   │  companies collaborate?      │
│                          │   │  Which are the adopted       │
│                          │   │  organizational modes        │
│                          │   │  (e.g. alliances, Joint      │
│                          │   │  Venture, licensing, etc.)?  │
│                          │   │  .........                   │
└──────────────────────────┘   └──────────────────────────────┘

        ┌──────────────────────────────┐
        │   Which obstacles and        │
        │   advantages?                │
        └──────────────────────────────┘
```

Fig. 1. Research questions

The chapter is divided in four main sections. It begins with the literature analysis in order to identify suggestions by previous studies about our research questions. Methodology section follows, with the description of the investigation protocol. Then, chapter goes on by presenting the results of the three case studies developed through semi-structured interviews. Finally, the chapter ends with conclusions and limitations of our study.

2. Literature review

Literature analysis is aimed at capturing suggestions from previous studies about three main aspects consistent with our research questions:

1. the main general competitive factors leading nowadays to the adoption of the OI approach in the automotive industry (i.e. why);
2. the firm-specific factors or strategic automotive companies' goals leading to the adoption of an OI approach (i.e. why); studies belonging to this stream of contributions usually try to find whether the expected advantages have been achieved and which types of obstacles companies eventually faced;
3. the manner in which openness can be set up (i.e. how), this interpreted in many different conceptions (e.g. prevalent typology of partners; organizational modes to be adopted for cooperation – Joint Venture, R&D contracts, etc., only to give some examples).

Open innovation is a phenomenon that has become important for both practice and theory over the last few years in many industries. One of its most often used definitions is: "the use of purposive inflows and outflows of knowledge to accelerate internal innovation and to expand the markets for external use of innovation" (Chesbrough and Crowther., 2006). In 2003, Henry Chesbrough coined the term "open innovation" in order to create a contrast with closed innovation strategy, supposed its predecessor, where companies generate their own innovation ideas and then develop, build, market, distribute, service, finance and support them on their own (Chesbrough, 2003a). Open innovation deals instead with relying on a firm's capability to carry out internally and externally technology management tasks along the innovation process. In this way, the company interacts and "collaborates" with its environment from different points of view; this leads to external technology acquisition and exploitation, on the one hand, and to share its core competencies with other companies, on the other. Chesbrough (2003b; 2006) goes on with recognizing that an open innovation process involves variegated internal and external technology sources and commercialization channels. By adopting open innovation approaches, the boundaries between a company and its environment, including customers, suppliers and competitors, becomes porous; thus, technological knowledge should be considered as an economic good itself (Chesbrough 2003b). In the recent years, many industries such as computers, semiconductors, telecommunications equipment and biotechnology, are more and more abandoning the closed innovation approach in favour of the open one, as well documented by several studies (Chesbrough et al., 2006; Diaz-Diaz et al., 2006). Conversely, in the literature there are still not many empirical studies regarding the adoption of Open Innovation in the automotive industry. Thus, the question whether an Open Innovation approach could be more adequate in the attempt to achieve a better R&D performance for automotive companies than a closed innovation model is not completely answered. As mentioned above, the first stream of literature contribution tries to study whether the OI model is appropriate also for the automotive industry and which are the main factors leading to this approach. Ili et al. (2010) depict the actual scenario in the automotive industry concerning the way of generating innovations. They consider the automotive industry as trapped by cost and innovation pressure and forecast a revolutionary discontinuity in generating innovations and a change in the way of creating and profiting from innovations themselves. Two of the most important factors leading the change are customers and globalization; customers demand more and more cars for the same old price. As a consequence, their demands and expectations are mirrored on the Original Equipment Manufacturers (OEM) who increase more and more their challenge to innovate. This aspect is also influenced by marginal growth provided by mature markets in industrialized countries, as they are almost saturated. Besides, strict environmental protection guidelines and safety conditions are affecting more and more the way of approaching innovation. Moreover, the pressure for OEMs is increased by the need to build and support the most important brands with innovations that are considered by customers worth of those names. However, adopting strategic approaches aimed at supporting innovations implies significant costs for companies. In 2006, the R&D expense for one innovation added up €70-80 million at Porsche, BMW and VW, and Daimler spent more than €150 million (Bratzel and Tellermann 2007). In the meanwhile, the cost is dramatically reduced by price erosion and shorter product life cycles. Moreover, the technology intensity and fusion (different technologies need to be combined to give the final product, i.e. the car) are crucial factors in the automotive industry. Following Gassmann (2006), Ili et al. (2010) conclude that, because of increasing innovation, cost pressure, globalization, technology intensity and fusion, the automotive industry needs to look outside their own boundaries and OI should be considered a good opportunity even though some obstacles/barriers are still remaining. The issue of barriers and obstacles to OI adoption is well elucidated by the second line of investigation mentioned above, that primarily aims to identify the automotive companies' goals to collaborate and then tries to evaluate the achievement of such goals (i.e. the expected advantages from collaboration). A relevant contribution is from Dilk et al. (2008) that, through a series of semi-structured interviews with managers from European automobile firms, find the most important goals include "flexible access to technologies", "intensified contact with clients and markets", "long term bonding of suppliers and clients", "access to other competencies (besides technology)", "improving quality of R&D", "reducing R&D costs", "reducing R&D time". The investigated companies report a fairly good performance by declaring that the strategic goals are almost achieved even though several deficits are identified. In fact, collaborations are successful only when the goals, responsibilities and tasks are clearly set among the partners and monitored as it is common practice in project management procedures. Soft and cultural aspects are of crucial importance: authors find that the human resource management of the automobile companies has not yet adapted to the specific challenges that employees are facing during collaborations. In most times, employees do not get adequate training preparing them for working together with external people and incentives to motivate employees to this purpose are also used rarely.

Bartl et al. (2010) investigate the opening of the innovation process when companies' goal is increasing knowledge, creativity and skills by co-creation with users and other stakeholders in new product development. Similarly to Dilk et al. (2008), Bartl et al. (2010), highlight the importance for success of collaboration of adapted innovation processes, organizational routines and cultures in order to change a company's attitude from "not invented here" syndrome to an enthusiasm for ideas and innovations which were found elsewhere. In summary, and also expanding the investigation beyond the automotive sector, literature shares the view that the implementation of Open Innovation is in any case quite difficult, whatever is the prevailing objective. As a matter of fact, recent studies stress the importance of the "right conditions" (in terms of company's strategy, capabilities, organizational factors, managerial tools, etc.) to make any open approach successfully carried out (Chesbrough & Crowther, 2006; Dahlander and Gann, 2010; Pisano and Verganti, 2008; Raasch et al., 2008; Bilgram et al., 2008).

And we finally get to the third line of investigation, which is indeed how Open Innovation is practically setting up. About this, the innovation & technology literature in general is really vast, while that specifically devoted to automotive industry is confirmed still poor. Just to give an idea of this broad set of conceptions, without any claim of completeness, we mention that the way the innovation process can be opened has been studied in innovation & technology literature according to the following main perspectives:

1. the kind or direction of openness (Chesbrough & Crowther, 2006; Gassmann and Henkel, 2004; Lichtenthaler, 2008): inbound (i.e. technological acquisition, where new ideas flow into an organization); outbound (i.e. technological commercialization, where unused technologies can be acquired by external organizations with business models that are better suited to commercialize a given technology); coupled as a combination of the previous two, with the innovation and ideas exchanges at the same time in both directions by establishing cooperation with complementary partners e.g. to co-develop projects;

2. the organisational form of acquisition or commercialization and consequent level of integration and time horizon (i.e. contractual agreements, licensing, alliances, joint ventures, etc: Chiesa and Manzini, 1998; van de Vrande et al., 2006);
3. the number and typologies of partners (von Hippel, 1988; Laursen and Salter, 2004; Pisano and Verganti, 2008; Enkel et al., 2009; Keupp and Gassman 2009);
4. the phases of the innovation process (i.e. exploration, development, commercialization) actually open (Gassmann and Henkel, 2004);
5. the kind of governance of the innovation networks: e.g. hierarchical, in which anyone can offer ideas but only one company defines the problem and chooses the solution; or flat model, in which anyone can generate ideas, and no one has the authority to decide what is or is not a valid innovation (Pisano and Verganti, 2008).

As specifically dedicated to automotive industry, it is important to quote again the contribution of Ili et al. (2010). The authors identify the typology of partners currently being followed by the automotive industry in order to improve idea generation and innovation: customers are considered as the most important sources for innovation; competitors and suppliers follow. Moreover, the study finds that there is high innovation potential from examples of other industries. While the tendency to look outside own boundaries for external sources to increase innovativeness (i.e. inbound openness) is confirmed, authors find that the external paths to outside the current business with own intellectual property is still hard and rare (i.e. outbound openness). In fact, there are many unused patents and companies are not even aware of their potential of external exploitation. As organizational modes of collaboration, reciprocal license agreements, alliances and joint ventures are the most common, while at individual organizational level the human profile defined as gatekeeper has become increasingly important. Learning journeys and trend scouting are promising methods to grasp innovation from different sources. This is the reason why companies such as BMW, Daimler and VW gave birth to the "trend-scouts" in variegated technological areas like Palo Alto, North America and Tokyo. In addition to this, passive web-based methods are adopted; for instance, BMW adopts the so called "Virtual Innovation Agency" and VW uses an online interface allowing engineers from different locations all over the world to provide their contribution with new ideas and innovations.

Another relevant work about the role of partners in the automotive industry is that by Heneric et al., (2005) which depict a clear scenario about the automotive industry evolution and trends. They claim that R&D performance is very important in the automotive industry, as it is an important factor in order to measure technological performance and gain competitive advantage. For this reason, many European companies invested more in research and innovation, carried out both internal and through to external collaborations. The cited study outlines the growing importance of collaboration with suppliers during the product development stages, and not only for the production. Due to the fact that in the last years customers have expected more and more additional improvements from manufacturers, but were not willing to pay higher prices, companies concentrated on core activities and increased efficiency by delegating other activities to partner companies. In this context, the vehicles manufacturers have the special ability to manage the complexity of the production process, which requires to co-ordinate up to 2500 suppliers for the most advanced models (Womack and Jones 1991). The increasing collaboration with suppliers let emerge a pyramid of manufacturers, with first tier suppliers directly involved in production process of vehicle manufacturers, and second and third tier suppliers with no direct contact with automotive companies (Terporten 1999). Future trends identified by McKinsey & Company (2003) estimate an average decrease of 10% of the value proportion added by the vehicle manufacturers. This decline is mainly explained from the potential spin-offs of tasks for the chassis and engine technology areas to suppliers. Moreover, it is forecasted an increase of strategic alliances in the form of cross-borders developments and components sharing. Roland Berger & Partners (2000) conducted a study on technological innovations for different components of the car (see Figure 2).

It is possible to notice that Information Technology (IT) will take a higher and higher margin in the automotive innovation; it is forecasted that 90% of all future innovations in the automobiles will be driven by IT. McKinsey & Company (2003) claims that vehicle manufacturers are trying to gain important positions in the electronic engine control, but with low success; the growth in this and other IT areas will be so occupied by specialized suppliers, that will gain a higher and higher importance in adding value to the product.

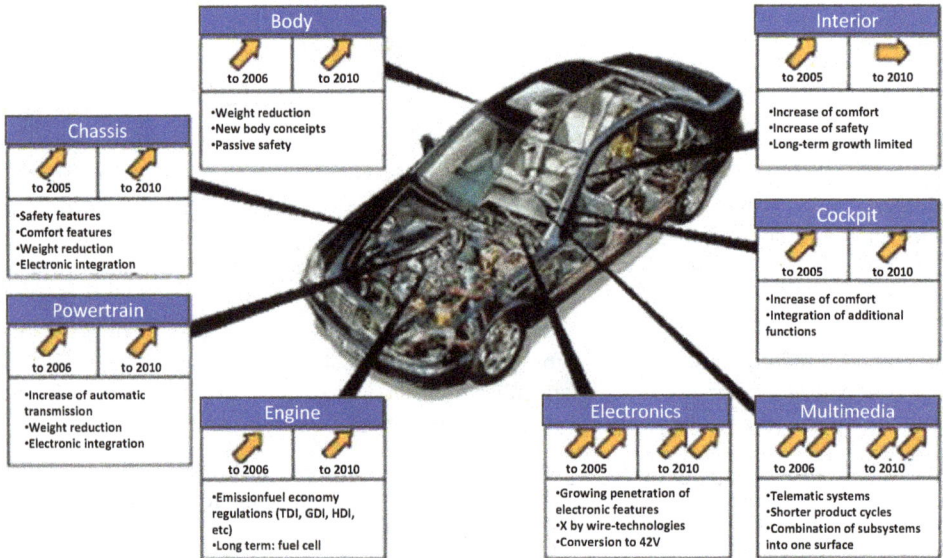

Fig. 2. Technological innovations

In any case, literature is unanimous in believing that a series of factors, i.e. the shorter innovation cycle time, the increasing product complexity, the downward cost pressure combined with increasing demands for performance and quality, deeply increased the number of challenges related to innovation that firms have to face today. In this context the access to new technologies has become crucial. Very interestingly, Gassmann et al., 2004 find that a particular type of partner, i.e. firms operating in other industries, can provide a valuable contribution to this goal. Thus, the choice of external sources is amended by adopting the so called "cross industry innovation (CII)", that is the deliberate combination of the potentials of companies operating in different industries (Gassmann et al., 2004).

Combining the different potentials of young and mature firms, the possibility of introducing innovative products using new and/or disruptive technologies is enabled. In 2004, the authors conducted a study over 12 automotive companies and found out that for certain projects companies combined complementary potentials of young and small with large and mature firms of different industries. These collaborations allowed to realize a radical product innovation in a deeply reduced time to market. A relevant example of this opportunity is that of *iDrive* developed and commercialized by BMW. Herrler (2001) depicts the *iDrive* as an innovative control device allowing drivers to access many different functions of their cars in an intuitive and interactive way, with a single hand and using a computer-like screen (*Control Display*). *iDrive* is aimed at simplifying and improving the interaction between the driver and the car, as the number of functionalities in the cockpit of modern cars has increased significantly over the last few years. Besides gear shift, lights, basic air conditioning and radio, CD music, telephone, navigation and advanced climate control systems are now present. The number and complexity of functionalities is in constant increase, as telematics and online applications are already available in several cars. The *iDrive* represents a radical innovation in the man-machine interfaces within the automotive industry, as it radically reduces the complexity and quantity of the control elements in the car's cockpit. This innovation was possible thanks to the integration of a new technology in the automotive environment. BMW involved many companies for the development of the device; a not specifically related to car projects research was initiated and different concepts for systems and concepts for in-car controls were investigated. The aim of the project was the development of a control system inspired by the personal computer's user interface of mouse and screen, and launches it in the new BMW 7 series in the autumn 2001. The BMW Group's technology scouting office in Palo Alto, California, identified *Immersion* company as a potential partner for the development of the project. *Immersion* had never operated in the automotive industry, but developed the proprietary *TouchSense* technology that was mainly applied in joysticks, flights sticks and steering wheels for video games, providing force feedback to the players. *Immersion* started the development of first prototypes very soon, and presented the first of them to BMW, that launched a feasibility study. *Immersion*'s engineers adapted the technology for the automotive industry very quickly, and soon they created a prototype integrated into an experimental car. To demonstrate that the implementation of the collaboration is never easy, it is important to note that the final agreement for collaboration was reached only in the early 2000. Moreover, as the collaboration became closer, several problems became more visible. On the one hand, *Immersion*'s management had difficulties in dealing with a large number of people in BMW. Because of the many BMW's departments involved, they were not able to identify a clear responsibility; on the other hand, BMW's engineers were still sceptical in adopting that technology, as it had never been adopted in the automotive. Once the BMW requirements were fully satisfied, the serial production could be started. However, *Immersion* was not able to become a supplier of complete systems or components, but only a technology supplier; for this reason, other partners' know-how was needed. As a consequence, BMW decided to delegate the *iController*'s serial production to the Japanese electronic group ALPS and purchased the rights for the *TouchSense* application in cars, with the exclusive for a certain period of time. Today, the *iController* can be found in all BMW cars, and Immersion receives royalties for each device built into a BMW's car.

Finally, a last topic studied by literature in the automotive industry concerns the kind of governance of the innovation networks. Dilk et al. (2008) provides interesting evidence: in their sample of European automobile firms, flat configurations, mainly based on trust, seem to dominate. The trust network is based on a collaboration contract, pre-defined assignments or, even if in few cases, oral agreements only. The partners keep their organizations independent among themselves. To work together, they usually set up teams with employees from the involved companies. The network progress is monitored by a committee and budgeting is the instrument that is commonly applied. For conflict resolution, contractual rules build the main basis even though mutual trust between partners is a more important parameter for the network success.

In summary, literature review surely shows a tendency to adopt an open approach, and this appears useful for increasing the firm innovativeness. However, except for the case BMW, examined in detail by Gassmann et al., 2004, other studies investigate groups of companies in order to highlight a set of issues and trends, definitely interesting, but not so in depth. Thus, we believe that the detailed study of three case studies can usefully enrich the empirical evidence on a subject that seems very relevant nowadays in the automotive industry. In particular, "understanding why, how and with what benefits and barriers" open innovation has been adopted in real case studies can provide useful insight for managers operating in automotive industry.

3. Methodology

To carry out our case study investigation, we adopted the triangulation method that is defined as the adoption of different data collection techniques within the same study in order to ensure the truthfulness of the gathered information (Saunders, Lewis and Thornhill 2007). For this reason, we combined both information provided from secondary sources and face-to-face interviews. The need of direct interviews emerged as a consequence of the gap between the literature and the aims of the research, as there were not found any similar studies, and the gap needed to be filled with the direct experience of people working in important automotive companies. In particular, we opted for the adoption of semi-structured interviews. The aim is to combine the advantages of both structured and unstructured interviews. Moreover, the possibility of conducting telephone or mail interviews was excluded; we believed that a direct contact could make the information sharing process easier, as the interviewed people could directly provide potential useful material of the company. Besides, the respondents could lead the discussion to areas that were not previously considered but are significant for understanding. Furthermore, they could feel more confident in providing information in a face-to-face context as they can receive personal assurance about the way the information will be used (Saunders, Lewis and Thornhill 2007).

We developed three case studies respectively about an important Italian automotive company, Pininfarina design and Bosch. The intent was to obtain diverse points of view about the theme from actors operating in different parts of the supply chain. After the initial contact with the human resource offices, we interviewed responsible specialists working in the R&D departments and management of the company.

3.1 Interview structure

The interview is going to be divided in three parts (see Table 1), consistently with our research questions reported in previous Figure 1: whether and why; how; obstacles/advantages.

Interview section	Research questions
Understanding whether and why company searches for collaboration	• Identify whether the company considers the possibility of collaborating with external actors • Identify the goals for technological collaborations with external partners
Understanding how	• Identify the typologies of partners • Identify the phases of the innovation process in which the company collaborated with external partners • Identify the kind of collaboration (i.e. inbound and/or outbound direction) • Identify the organizational modes adopted by the company for the technological collaborations • Identify the kind of governance
Understanding obstacles and advantages	• Identify the main reasons determining the failure of technological collaborations • Evaluate the main benefits achieved with technological collaborations

Table 1. Interview's structure and objectives

4. Findings

After a brief overview of its context and innovation strategy, each case study will be described in detail by trying to answer to the research questions reported in Table 1.

4.1 Case study 1: Automotive company A

For many years, the company's strategy was aimed at being a follower; however, after the change in the management, things evolved in better, and Company A managed to develop and to patent successful technologies that were also licensed to competitors. At the moment, the company retains a leadership position in diesel, LPG and natural gas engines; the lower success of petrol engines is attributed to an investment reduction in their research, as a consequence of the fuel price increase and the customers' tendency to opt for alternative engines such as diesel, LPG and natural gas. Moreover, Company A has been pursuing for many years an international policy aimed at accessing new emerging markets, by locating there the production and introducing products according to their needs. However, in those countries the innovation is not as important as in the European countries and, thus, it is just an adaptation of the existent models (or also past models) to specific needs. The core activity of innovation is in Europe, where Company A feels more and more the pressure for innovating, as the demand for new product increases and the time to market is dramatically reduced. Nowadays, the economies of scale are a critical success factor for the company. However, their pursuit implies significant investments for every new product. Anyway, the high flexibility levels obtained thanks to the versatility and standardization allow the pursuit of economies of scale and the production of different models on the same production line. Partnership with other automotive companies supports this aim. The expansion strategy focused at reaching economies of scale is often mirrored in the collaboration agreement with local companies. This approach is followed for two main reasons: on the one hand, it is possible to reduce designing, production and investment costs, as local companies already own the plants. On the other, it is possible to overcome some government barriers as many states prohibit foreign companies to produce in their territory unless they do not set a joint venture with a local company.

The innovation process is different according to the kind of project the company is working at. Efforts are paid to both radical innovations (breakthroughs) and incremental innovations. The company perceives innovation as a new or existent product, process, methodology or service applied at a lower cost, and not necessarily as a new technology application. The innovation might also be a recombination of existing processes and technologies.

The external sources of innovation are various and the company has a specific department to identify them. However, Company A tends to directly and secretly control the so called "perceived distinctive functionalities", considered as belonging to the mosaic that provides value added to the brand. These variables directly affect the customer in the choice of the product[2]. The company has internal and external technology intelligence departments that conduct every day opportunity/risks analysis, as well as technology scouting. Moreover, a market intelligence area explores existing and new market segments in which a new technology might have potential success. Economic incentives often support these analyses, in order to increase productivity and motivate employees. Especially with regard to short-medium projects (3-4 years), directed at improving technologies or carrying on products' restyling, the voice of customer (VOC) has a significant impact. Thanks to the support of appropriate market analyses, the potential customers express their needs providing an external point of view; their ideas are intersected with the technicians and engineers' ability, which are responsible for evaluate whether ideas are realizable in practice and carry on the development. The designing cycle is usually performed in co-design with the suppliers. The objectives are enlarging company's competence base, stimulating creativity and capability of generating new ideas, reducing and sharing risks related to innovation activities and costs of innovation process. In this case, the company decides whether to set a task force dedicated to a project, composed of members of both the company and the supplier, or to name supervisors operating at the suppliers' premises. One of the latest company's innovation is the "Virtual Car", that is a software able to perfectly simulate the dynamic behaviour of

---

[2] For instance, Company A retains very successful patents in the field of diesel medium-powered engines and therefore keeps a strong control over all the components related to the engine monitoring. Besides, the design is also a strongly controlled variable as it allows to differentiate the components that are visible to the customer.

the car in different driving conditions. It gives the opportunity to perform all the electronic setups preventing potential faults. Until the 2008, Company A was the only one adopting this kind of electronic systems. The importance of the suppliers in providing value added to the product is witnessed by the slogan "global purchasing to the global partnership". Company A adopted a structured collaboration plan with suppliers, offering them three years contracts in order to plan more realistically their growth and development strategies. At the same time, the company developed a co-sourcing strategy, allowing Company A and its suppliers to join each other in the materials purchase, in order to obtain prices that are more competitive. Moreover, Company A's managers carry on training to the suppliers' employees in order to improve the quality of the service.

In addition to this, Company A relies on other external actors such as companies operating in other industries (i.e. from aeronautic sector), universities and research centres. It considers their collaboration very useful, as they are able to provide a significant contribution in the user functions' development. In particular, universities' collaboration concerns the idea generation and testing phases.

As concerns the kind of collaboration (i.e. inbound and/or outbound), Company A declares the adoption of both the approaches by showing its willingness in exchanging innovations and ideas on both directions. Ranking them in order of their use (i.e. less employed to most), contractual agreements, reciprocal license agreements, licensing, patent sales, alliances, and Joint Ventures are the organizational modes employed for opening up innovation process in both directions. The lesser adoption of alliances and Joint Ventures is justified by the fact that they imply a higher resource employment and therefore are chosen for high-relevant long-term projects only. As proof that outbound open innovation is adopted, Company A reports that the number of ideas "on the shelf" is definitely reduced as well as the number of un-used patents.

With regards to the kind of governance, Company A has in general the main function of coordinating all the actors' work in order to respect timescales, by adopting a hierarchical approach. However, the governance approach depends on the importance of the project and the partners' capabilities. In case of relevant projects, the decision making process is shared and the work is carried out in co-development. Moreover, when the company A considers the supplier as able to safely carry out the project on his own, it cooperates with him only in the phase of requirement specifications and final testing in order to verify whether the product is effectively interfaced with the vehicle. In this case, the supplier takes the decisions for the other phases on his own and carries on the development.

Finally, based on its expressed goals, Company A was asked to evaluate its degree of satisfaction in reaching these goals. Company A declares that the initial expectations have been partially achieved even thought some problems emerged. For example, disagreement on prices or misunderstandings related to supplying timescales and payments are very frequent especially with low-importance partners. Furthermore, the non-compliance of the specifics by suppliers is also a much discussed theme. In these cases, it results very difficult identifying the guilty, as responsibilities are not often well distributed. Besides, the high and increasing complexity of the product does not help in solving this issue. However, this phenomenon is not reported for important suppliers, as it is their primary interest to carry on a successful collaboration in order to have high sales: in fact, if Company A reports low sales, the failure is reflected on the supplier too. Effective times longer than what planned and cultural differences between partners are also reported.

4.2 Case study 2: Pininfarina

Pininfarina is one of the most famous designing companies operating in the automotive industry as well as in other industries. Differently from the other automotive constructors and designers, Pininfarina operates in a diverse reality. Its aim is not directed at involving mass production products, but niche markets that are not attractive from an industrial point of view. Pininfarina always retained a leadership position in its activities. In most of the cases, it tried to behave as first mover, by introducing innovations in terms of design and technology before competitors. Since the seventies, Pininfarina gained a leadership position, as it was one of the first companies adopting personal computers and calculators. In the same period, it built a wind tunnel that is still working and implies a yearly 2 million € investment for research and innovation. The tunnel is on a 1:1 scale, and allows performing virtual simulations, avoiding testing prototypes on the street, as there are many unpredictable hidden risks. Moreover, the company tried to influence the structure and rules of the sector by pursuing the mission of eco-compatibility. This pattern has been followed for about thirty years; among the most important innovations, the Ethos and m³ in collaboration with the FIAT research centre are worth to be remembered. The radical innovations introduced in the eco-compatibility field were so successful that now it constitutes the company's core business. The company is actually working on the launch in the market of "Nido", one of the first electric cars. Since its origins, Pininfarina dealt with innovation, as it constitutes the core element of the company's mission. Until the 2007, the whole innovation process was sponsored by auto-financing. However, collaboration modalities were set with strategic partners, with whom Pininfarina started co-development projects aimed at building prototypes based on an emerging technology. Furthermore, public financing was also an important sponsor.

The idea generation does not present a fixed pattern. A new idea may come from specific market needs in a certain time or an internal intuition that is proposed to the customers. In general, it may bear internally, as well as be commissioned by the customer. It is very frequent that customers ranging from automotive companies to private entrepreneurs delegate Pininfarina to design a coach according to their particular needs. However, a new idea can be developed also for particular occasions.[3]

Before the creation of industrial prototypes needed to test the serial production, the company creates prototypes for show purposes, that are used for exhibitions and marketing initiatives. In this way, the customer has the possibility to directly evaluate the product according to the previously set requirements and suggest changes or new ideas. This strategy is successful; in fact, the company worked for the most important worldwide automotive companies: customers' collaboration can be witnessed in almost all the phases of the innovation process (Figure 3; Pininfarina 2010a and Pininfarina 2010b).

---

[3] For instance, in order to celebrate the 80 years of Alfa Romeo, Pininfarina designed an unique prototype.

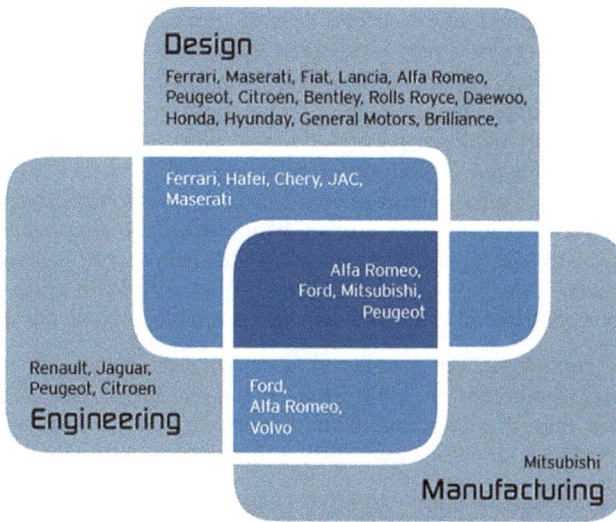

Fig. 3. Pininfarina' s customers for design, engineering and manufacturing

The time needed to complete a project, from the idea generation to the serial production is more and more shortening in the last years. However, the effective time is influenced by the presence of an existing chassis; in this case, the cycle usually lasts 18 months with an average investment of 100 million €. Moreover, the interviewed people report that a completely new project may take up to 36 months to be completed and investments up to 500 million €.

Due to the fact that Pininfarina has always worked as a full service provider to automotive companies (i.e. customers), these are the most common partners. They have the possibility to ask for services ranging from the program management to the full design of a car. In the latter case, the project is often carried on in co-development and the most used organizational mode is the "equal sharing alliance". Key figures are the program manager, that keeps the contacts with customers and deals with timescales and costs, and the engineering manager, who deals with the engineering. To favour co-development, in 2003 Pininfarina built a structure in which it is possible to give hospitality to the partners' employee in order to work together and facilitate the data exchange. The structure is very flexible, in order to best adapt itself to the different projects. However, the opening of innovation process is not always easy, as the company has to learn how to work with the partners' software and vice-versa. Anyway, Pininfarina developed IT interfaces where both the company and the customer can share projects and ideas in a definitely secure way.

The project cycle is usually carried on with external supplier partners; each of them is specialized in a different area and thus provides the necessary technology competencies to the projects. Among the most important partners, it is worth to remember: OSRAM, for the vehicle external and internal lighting; NUVERA that provided fuel cells with a reformer able to produce hydrogen; BI Technology that contributed to the vehicle energetic efficiency improvement; REICOM that developed vehicle to vehicle interfaces technologies. Their cooperation deeply contributed to the success of many important projects. Obviously, there are also more occasional external suppliers, called upon to satisfy peaks of demand. When Pininfarina is confident that the supplier is able to carry on the work on his own, job order contracts are set. In the opposite case, there is integration on the production platform, in order to ensure the requirement respect and the error reduction. Pininfarina' s purchasing office does not adopt a formal procedure for selecting suppliers, but it relies on the one presenting the best quality/price balance chosen from the supplier checklist developed after direct visits to supplier facilities. On the other hand, when the supplier is considered of strategic importance, stronger organizational modes of collaboration, such as joint ventures, are set up. An important example of joint venture is that with WEBASTO, which saw the creation of a new society (50%-50%), OASIS, for the retractable top realization.

In addition to this, commercial agreements are used to be set. It is worth to name the one with Ansaldo Breda, who bought the right for avoiding that Pininfarina would work for its competitors. Besides, there is a tacit agreement with Ferrari that has been lasting for seventy years about "not working" for competitors such as Porsche or Lamborghini.

The strategy of continuously dealing with external actors has always been pursued with the basic goal of enlarging the company's competence base. The choice of partners is usually influenced by the possibility of acquiring complementary competencies. Besides, CII is an important source of innovation as Pininfarina adapted railway and naval technologies to the automotive.

As concerns the kind of collaboration (i.e. inbound and/or outbound), the co-development is a clear example of coupled type of Open Innovation: through its collaborations, Pininfarina acquires ideas and knowledge, but also it transfers knowledge by making available its patents to benefit cooperation. However, this does not mean that Pininfarina is an active seller of patents to exploit those that are not suitable to its current portfolio. In other words, there is no mindset to actively exploit intellectual property.

With regard to the kind of governance, it depends on the type of partner, i.e. supplier or automotive customer, with which Pininfarina is collaborating. As concerns suppliers, Pininfarina usually holds the ranks of the projects, even if the granted delegation is strong when it comes to highly competent actors. In the case of automotive customer partner, as it could be for example Company A, Pininfarina has to submit to customer' s governance conception although, in turn, its distinctive competencies favour the adoption of participative approaches.

When we asked about the benefits of collaborations, Pininfarina never reported significant failures in agreements, as it tries to build partner loyalty by establishing long collaborations[4] and following a common pattern.

However, an unsuccessful cooperation with GM (General Motors) can be cited as example. GM commissioned to Pininfarina the development and production of a car to be partially assembled in the Pininfarina Italian plants and completed in Detroit with the engine assembling. The shipping costs had a too high impact to GM, which did not manage to forecast the costs well, and implied the failure.

---

[4] 70 years with Ferrari, 30 with Ansaldo Breda, 80 with Peugeot.

Moreover, Autoblog (2010) reports the failure in the collaboration between Pininfarina and Mitsubishi for the production of the Colt CZC. The latter complained with Pininfarina about delays in production and quality issues, demanding a 43,4 million € refund. Pininfarina replied with a 100 million € damages refund for the investments performed. The international Paris law-court accepted both damages requests, establishing a 19,2 million € refund in favour of the Italian company. Finally, the interviewed people explain that in the past the company invested in collaborations aimed at producing models that were not successful. However, they also claim that the failure was exclusively linked to a Pininfarina's mistake, rather than a partner fault.

4.3 Case study 3: Robert Bosch S. p. A.

Robert Bosch S.p.A. is a company operating worldwide that supplies the most important automotive companies. It deals with many industries, and the automotive field constitutes one of its main businesses. Bosch supports automotive companies by researching and developing 360° innovative solutions for the cars of the future and supplying standard components in a large scale.

As observed before, the recent financial crisis has led to a significant increase in the cost of technology development, as customers are demanding more and more reliable and sophisticated products but, at the same time, they do not accept any increase in the price. Therefore, companies supplying commodities in big quantity and high precision tend to perform investments in order to reach economies of scale and lower costs for the new released products.

The interviewed people report that, for the automotive division, Bosch aims to be a leader for each product. Its constant engagement in innovation makes that automotive companies rely in greater and greater part to Bosch's components: reported data show that in every car the Bosch's components contribute to the product value creation for at least the 15%. However, while the constant research allows keeping a leadership position for the innovations, there is a higher risk of imitation for the standard components. As the competition increases, especially from eastern countries, Bosch is trying to acquire new markets and penetrate them as first mover.

The innovative ideas do not come from a research "with an end in itself", but they can be originated in different ways. In the most of the cases, they come from a specific need of the customer; moreover, the company has a specific office that analyses market trends and press releases, from which it draws to suggest potential ideas to develop.

The Bosch innovation process presents a series of innovation steps at the end of which the idea development is approved. However, the interviewed preferred not to go in depth to these steps for confidentiality reasons.

At the end of these steps, the company looks for customers with who would carry on the idea development and would make an initial application of it. The high know-how of the company allows always finding a customer; obviously, in case the idea is generated from a customer need, the search for customers is not needed.

Before the designing phase, a business plan is developed. This phase is carried out in cooperation with the customer, as the success of the product depends on how many cars the customer manages to sell. Partners evaluate together ideas, Bosch's products and innovations as well as information about both Bosch and the customer competitors.

The project cycle follows different patterns according to the source of the idea. In case the idea is developed according to a customer need, the project cycle is carried on in cooperation with the customer. The company usually opts for co-development contracts or joint ventures according to the importance of the project. In this case, the idea is specifically developed for the customer and it does not need to be adapted to its product. In case the idea is proposed to the customer, the project cycle deals with adapting it to its vehicles.

The testing cycle presents two different phases. First of all, the component is tested internally in order to verify whether the technical specifications provided by the customer are respected, and the objective quality level has been reached. Bosch has also an internal society (ETAS) that develops and tests simulation systems and the software needed for the electronic components. After that, the component is assembled on the vehicle and technicians verify whether it well interfaces with the other components. It is very important to cooperate with the customer technicians, as they are the only ones who can express better the needs for a certain product. This phase presents a high degree of complexity, especially for components with a relevant percentage of electronic parts. When all the tests provide positive results, it is possible to plan the production cycle. This is internal or performed by an external actor depending on the production is completely carried out by Bosch or outsourced to an external supplier.

The innovation process time-horizon depends on the time to market of the car. When working in co-development with a specific customer, it usually lasts 14-15 months. However, when Bosch establishes joint ventures, there is not always a specified time horizon, as there is the hopefulness to collaborate as long as possible.

Bosch cooperates with a large range of partners for its innovations. In many countries, it constitutes the main supplier for the local automotive companies; for instance, in Italy it is the most important among FIAT suppliers.

As mentioned above, it sets different kinds of contracts and organizational modes for collaboration according to the importance of the project. The company does not adopt a formal procedure for the risk evaluation, but it depends from case to case. If the product specifications are clear, Bosch usually prefers a co-development agreement. Instead, for longer collaborations, the joint venture is the favourite form. In all these collaborations, Bosch keeps the control or at the last the 50% of the share by confirming, similarly to Pininfarina, a more hierarchical style of governance with suppliers (especially with those whose competencies are not so distinctive) and more participative with (automotive) customers. Among the most successful joint ventures, it is worth to remember the one with Samsung, for the development of lithium-ion batteries, lasting ten years, and the one with ZF Lenksysteme (ZFLS), in 1999. The latter provided its knowledge in steering systems and Bosch offered its ability in dealing with the management of electric components. Besides, in 2008, as reported on the website (Bosch Press 2010) the company established a fifty-fifty joint venture with MAHLE GmbH to develop, manufacture and commercialize exhaust gas turbochargers for gasoline and diesel engines.

For all its joint ventures, Bosch sets the objective to gain complementary knowledge; thanks to the acquired knowledge it will have a stronger contractual power in the future. Moreover, it usually adopts the policy of employing people of different nationalities within the same project, in order to stimulate creativity and integrate different kind of knowledge, improving the ability to respond to market requirements.

Finally, also collaboration with universities has been set-up. However, the people interviewed claim that "the general tendency of universities to provide general and sometimes useless knowledge" leads Bosch to set this kind of cooperation only if there is a specific project with detailed requirements.

Again similarly to Pininfarina, the co-development proves the coupled type of open innovation: through its collaborations, Bosch acquires ideas and knowledge, but also it transfers knowledge by making available its knowledge to benefit cooperation. However, Bosch perceives a very high risk of imitation and therefore it has a careful management of its highly innovative intellectual property (IP): detailed agreements for IP protection are subjected to all the partners from the Bosch's headquarter in Germany and any project will not start if the agreement is not signed by the participants. In 2009, Bosch applied for 3872 patents. People interviewed reinforce this concept by saying that "without our IP policy all the investments would have no sense". As next step, however still not done today, a more active exploitation of IP is expected by licensing products and ideas as opportunity for further revenues.

Despite the lack of a formal procedure for evaluating partners, Bosch reports that the highest part of cooperation agreement was successful. The management-training program for dealing with foreign cultures and international teams are the key of success for good cultural exchanges. As negative aspects, sometimes management makes mistakes in costs evaluation and the company does not obtain the forecasted margins. Besides, in spite of accurate specifications of technical requirements, Bosch experiences collaborations with partners who do not provide the expected quality performance. In addition to this, supplier delays cause effective times higher than the planned ones; the interviewed people said that in these cases the risk of image damages is very relevant.

5. Conclusions and limitations

Except for some relevant studies analyzed in the literature review, there is general consensus on the need to produce more empirical evidence about the automotive industry and its propensity to adopt Open Innovation approaches, how they are implemented and with what results (Ili et al., 2010). With the aim of developing this evidence, we studied the experience of three well-known companies operating at different levels in the value chain of the automotive industry: a final car-producer on one hand; two suppliers (i.e. a designer and a supplier of components) on the other.

First paper goal was to identify whether companies operating in the automotive industry consider the possibility to collaborate with external actors and with which main goals. In this regard, respondents confirm literature suggestions: the automotive industry is reported as being trapped by cost and innovation pressure by customers. As a matter of fact, factors such as globalization, technology fusion (i.e. the need of integrating different technologies in the final product - the car) and technology intensity seem to force companies to search for external sources of knowledge. Main declared goals for collaboration by the respondents are the enlargement of company's competence base, the stimulation of creativity and capability of generating new ideas, the reduction and sharing of risks related to innovation activities and costs of innovation process.

The second paper objective was understanding how Open Innovation approaches could be set-up. In this regard, the typology of partners has been firstly investigated. As result, the partner variety is considerable high. Company A probably presents the highest openness degree by collaborating with the highest number of different partners. In particular, the company relies on the collaborations with a lot of suppliers, customers, firms operating in other industries (cross industry innovation, i.e. CII) and universities' research centres.

With regard to the other respondents, they cooperate with a minor number of partners. As suppliers of automotive companies, Pininfarina and Bosch keep the main objective of satisfying the customer that, in this case, is represented by the automotive company itself. For this reason, they establish the highest number of agreements with them. However, Pininfarina deeply relies on CII, as its prototypes are aimed at showing the highest concentration of technology in order to impress automotive companies and induce them in introducing those innovations in their vehicles. In particular, Pininfarina retains that collaborating with partners from other industries is a daily activity, rather than a rare experience. To be noted is the almost total absence of competitors among the possible partners identified by the respondents. This is in contrast with the cited findings of Ili et al. (2010), who show competitors as the second source of ideas (after customers) for automotive companies. Probably the respondents have not yet overcome the strong aversion to collaborate with competitors as they perceive a very high risk of imitation.

However, the management of intellectual property is rapidly getting better and, although an active exploitation of IP (i.e. outbound open innovation) is not yet adopted, it is perhaps only a matter of time. With adequate protection mechanisms of IP, also collaborations with competitors will not seem so risky.

As concerns the openness of the innovation process phases, all the respondents agree with the fact that all the stages can be opened according to the project and the kind of agreement among partners. In general, the phase of idea generation always takes information from external sources, ranging from the analysis of customer needs to the adaptation of solutions from other industries. The other phases can be conducted internally, as well as in cooperation with other partners: for example, Company A specifies that it tends to exclude partners when phase activities are dealing with perceived distinctive functionalities.

Also the adopted organizational modes are various: they range from contractual agreements to more integrated forms as alliances and joint ventures, these chosen only for high-relevant long-term projects because of the high amount of required resources.

Finally, both hierarchical and participative kinds of governance are used by the respondents, with the tendency towards stronger delegation when partner competencies are highly distinctive, not standardized and therefore difficult to be replaced.

Lastly, despite the identification of some obstacles, partnerships are described as successful by proving Open Innovation is an appropriate approach to be adopted in the automotive industry.

Our work obviously has severe limitations because it is focus on a limited set of companies. The fact that the investigated companies have different roles in the same value chain allowed at least to provide a broader view on the industry. However, future researches could deal with validating these findings by interviewing other actors of the same dimension of the respondents. In this way, it would be possible to provide more significant evidence on the industry. Moreover, similar analysis could be conducted on small-medium companies operating in the automotive industry, as well as on emerging automotive companies in the Eastern countries. More extended investigations, i.e. surveys, with the use of quantitative tools of data analysis should be also recommended.

6. References

Autoblog (2010). *Mitsubishi contro Pininfarina: in tribunale per la Colt CZC*. [online] Milan: Autoblog. Available from: http://www.autoblog.it/post/28620/mitsubishi-contro-pininfarina-in-tribunale-per-la-colt-czc.

Bartl, M.; Jawecki, G. & Wiegandt, P. (2010). Co-Creation in New Product Development: Conceptual Framework and Application in the Automotive Industry. *Conference Proceedings R&D*, Manchester, 30 June – 2 July.

Bilgram, V.; A Brem & K Voigt (2008). User-centric innovations in new product development- systematic identification of lead users harnessing interactive and collaborative online-tools. *International Journal of Innovation Management*, Vol. 12, No. 3, pp. 419-458.

Bosch Press (2010). *Bosch and MAHLE plan joint venture to develop and manufacture exhaust gas turbochargers*. Stuggart: Bosch Presse. Available from: http://www.bosch-presse.de/TBWebDB/en-US/PressText.cfm?id=3455.

Bratzel, S. & Tellermann, R. (2007). *The innovations of the global automotive firms*. Bergisch Gladbach: FHDW Center of Automotive.

Chesbrough, H. (2006). *Open Business Models: how to Thrive in the New Innovation Landscape*. Boston: Harvard Business School Press.

Chesbrough, H. (2003a). *Open Innovation: The New Imperative for Creating and Profiting from Technology*. Boston: Harvard Business School Press.

Chesbrough, H. (2003b). The era of Open Innovation. *MIT Sloan Management Review*, Vol. 44, No. 3, pp. 35-41.

Chesbrough, H. & Crowther, A.K. (2006). Beyond high-tech: Early adopters of open innovation in other industries. *R&D Management*, Vol. 36, No. 3, pp. 229-236.

Chiesa, V. & Manzini, R. (1998). Organizing for technological collaborations: A managerial perspective. *R&D Management*, Vol. 28, pp. 199-212.

Dahlander, L. & Gann, D. (2010). How Open is Innovation? *Research Policy*, Vol. 39, No. 6, pp. 699-709.

Diaz-Diaz, N.L.; Aguiar-Diaz, I. & De Saa-Perez, P. (2006). Technological knowledge assets in industrial firms. *R&D Management*, Vol. 36, No.2, pp. 189-203.

Dilk, C.; Gleich, R.; Wald, A. & Motwani, J. (2008). State and development of innovation networks. Evidence from the European vehicle sector. *Management Decision*, Vol. 46, No. 5, pp. 691-701.

Enkel, E., Gassmann, O. and Chesbrough, H. (2009), "Open R&D and Open Innovation: Exploring the Phenomenon", *R&D Management*, Vol. 39, No. 4, pp. 311-316.

Gassmann, O.; Stahl, M. & Wolff, T. (2004). The Cross Industry Innovation Process: Opening up R&D in the Automotive Industry. *Proceedings of the R&D Management Conference (Lissabon)*. 12 July 2004. Oxford: Blackwell.

Gassmann, O. & Henkel, E. (2004). Towards a theory of open innovation: three core process archetypes. *Proceedings of the R&D Management Conference*, Lisbon, Portugal, July 6–9.

Gassmann, O. (2006). Opening up the innovation process: towards an agenda. *R&D Management*, Vol. 36, No. 3, pp. 223-228.

Heneric, O.; Licht, G. & Sofka, W. (2005). *Europe's Automotive Industry on the Move Competitiveness in a Changing world*. Mannheim, Germany: Physica-Verlag Heidelberg.

Keupp, M.M. and Gassmann, O. (2009). Determinants and Archetype Users of Open Innovation. *R&D Management*, Vol. 39, No. 4, pp. 331-341.

Herrler, M. (2001). *User-Interface-Gestaltung: Cockpit der Zukunft – kunftige Richtungen in der Anzeigen- und Bedienelement-Gestaltung im Automobil*. Munchen: BMW AG.

Ili, S., Alberts, A. and Miller, S. (2010). Open Innovation in the automotive industry. *R&D Management*, 40(3), pp. 246-255.

Laursen, K. and Salter, A. (2004), Searching High and Low: What Type of Firms Use Universities as a Source of Innovation? *Research Policy*, Vol. 33, No. 8, pp. 1201-1215.

Lichtenthaler, U. (2008), Open Innovation in Practice: an Analysis of Strategic Approaches to Technology Transactions. *IEEE Transactions on Engineering Management*, Vol. 55, No. 1, pp. 148-157.

McKinsey & Company (2003). *HAWK 2015 – Knowledge-Based Changes in the Automotive Value Chain*. Frankfurt.

Pininfarina (2010a). *I Clienti, Auto*. Cambiano, Italy: Pininfarina. Available from: http://www.pininfarina.it/index/servizi/clienti/clientiAuto .

Pininfarina (2010b). *Magazine*. Cambiano, Italy: Pininfarina. Available from: http://sintesi.pininfarina.com/docs/magazine.pdf.

Pisano, G. P. and Verganti, R. (2008). Which Kind of Collaboration Is Right for You? *Harvard Business Review*, December, pp. 1-9.

Raasch C, C.; Herstatt, C. & Lock, P. (2008). The dynamics of user innovation: Drivers and *impediments* of innovation activities. *International Journal of Innovation Management*, Vol.12, No. 3, pp. 377-398.

Roland Berger & Partners (2000). *Automotives Supplier Trend Study*. Detroit.

Saunders, M.; Lewis, P. & Thornhill, A. (2007). *Research methods for business students*. 4th ed. Harlow: Pearson Education.

Terporten, M. (1999). *Wettbewerb in der Automobilindustrie*. In: Heneric, O.; Licht, G. & Sofka, W. (2005). *Europe's Automotive Industry on the Move Competitiveness in a Changing world*. Mannheim, Germany: Physica-Verlag Heidelberg.

van de Vrande, V., Lemmens, C. & Vanhaverbeke, W. (2006). Choosing Governance Modes for External Technology Sourcing, *R&D Management*, Vol. 36, No. 3, pp. 347-363.

von Hippel, E. (1988). *The Sources of Innovation*, Oxford University Press: New York.

Womack, J. P. & Jones, D. T. (1991). *Die zweite Revolution in der Automobilindustrie*. Konsequenzen aus der weltweiten Studie aus dem Massachusetts Institute of Technology. In: Heneric, O.; Licht, G. & Sofka, W. (2005). *Europe's Automotive Industry on the Move Competitiveness in a Changing world*. Mannheim, Germany: Physica-Verlag Heidelberg.

11

The Impact of ICT on Productivity: The Moderating Role of Worker Quality and Quality Strategy

Ana Gargallo-Castel and Carmen Galve-Górriz
University of Zaragoza
Spain

1. Introduction

Technological innovation has generated considerable interest among academics and practitioners in recent years. In recent decades, Information and Communication Technologies (ICT) such as computer terminals, e-mail and the Internet and their applications have become the major drivers of innovation, growth and social change. Moreover, as the OECD points out (OECD, 2010), in times of crisis there must be a focus on the contribution of ICT to innovation and growth. However, although interest in this subject has grown substantially, research on the importance of the combination of organizational change together with technological innovation has been less common. Some studies suggest that technological innovation is not an isolated source of improvement, but part of a system or cluster of mutually-reinforcing organizational approaches.

Then, the simply availability of new ICT does not necessarily lead to success. On the contrary, it requires that firms accompany innovation with the development of best organizational practices (Brynjolfsson et al., 2002; Huerta et al., 2008), and it is therefore necessary to study the influence of this question in more detail, as we explore in this chapter. Our main goal is to examine whether technological innovation and organizational change are complementary, and whether they are associated with better performance.

In order to reach this objective, survey data from more than a thousand Spanish firms has been used. In particular, we analyze the impact of technological innovation and diverse complementary elements on organizational productivity. The main questions to answer are:

Is firm performance improved through ICT?

Are organizational characteristics complementary to ICT in improving firm performance?

These questions are especially important in today's highly competitive environment. Over the last decade, competition has intensified and companies have found the need to restructure and improve their business practices in order to find new and more efficient ways to obtain competitive advantage as a condition of survival.

We undertake diverse descriptive analyses, which allow us to observe the characteristics of more efficient firms. Following prior studies, we use a standard Cobb-Douglas production function to model the production process, considering ICT as a factor of production (Brynjolfsson et al., 2011; Hitt and Brynjolfsson, 1996; Dewan and Min, 1997). With this analysis, we will asses the impact of these technologies on company efficiency and productivity.

To test organizational complementarities, we use a methodology that allows us to model the joint effects of ICT and organizational and human resources on performance. The results show the importance of organizational human capital in order to increase the benefits of the technological innovation.

Our results contribute to the international body of research and offer a new conceptualization and empirical evidence of the technology innovation performance.

We have structured the work into five parts, including this introduction. In the next section we review the main literature. We then discuss the theoretical relationship between ICT and competitive advantage, stating our hypothesis using the complementarities perspective as the main theoretical framework. Next, we present the research design, sample and definition of variables and we discuss the main results. In the final section, we present our conclusions, discuss certain limitations of our study, and propose several directions in which to continue our research.

2. Literature review

There have been many challenges and variations in the forces for globalization during the last decade. One that has garnered substantial attention over the past few years is concerned with the impact of information and communication technology on economic growth and on firm performance (OECD, 2004). The widespread diffusion of the Internet, the mobile phone and the broadband networks shows how pervasive these technologies have become.

According to much theoretical and empirical evidence, ICT offers benefits for a wide range of business processes and improves information and knowledge management within the firm, leading to better performance. Firms can manage their processes more efficiently and, as a consequence, they increase their operational efficiency. Moreover, ICT reduces the coordination costs of the firm because of lower procurement and inventory costs and closer coordination with suppliers (Tachiki et al., 2004; OECD, 2003, 2004). In addition, communication based on ICT and the Internet can also improve external communication, reducing the inefficiencies resulting from lack of co-ordination between firms, and increasing the speed and reliability of information processing and transfer. In general, ICT reduces transaction and coordination costs, maximizing the value of the transactions (OECD, 2004).

However, according to the literature review on the impact of ICT at the firm level, we can confirm the diversity of theoretical approaches and empirical evidence on the role of ICT in the improvement of the firm performance.

Much of the early literature on ICT, mainly in the 1980s and early 1990s, theoretically justified the advantages of information technologies, but they obtained contradictory empirical evidence, especially weak or with no link between ICT and firm performance (Brynjolfsson, 1993; Davenport, 1994; Kettinger et al., 1994; Loveman, 1994; Roach, 1987;

Strassmann, 1985, 1990; Weill, 1992; Wilson, 1993, 1995). This empirical evidence led to the so-called Productivity Paradox, well summarized by Nobel Prize Robert Solow (1987), who said, "You can see the computer age everywhere but in the productivity statistics."

According to Brynjolfsson (1993) and Brynjolfsson and Yang, (1996) the various explanations that have been proposed for this apparent lack of relationship can be grouped into four categories:

1. Mis-measurement of outputs and inputs
2. Lags due to learning and adjustment
3. Redistribution and dissipation of profits
4. Mismanagement of information and technology

However, moving beyond the productivity paradox, there is growing new evidence that ICT generates large positive returns that are even in excess of the returns from other types of investments (Dewan and Min, 1997). Authors such as Lichtenberg (1995) and Brynjolfsson and Hitt (1995, 1996) offer empirical evidence of the positive impact of ICT on a variety of measures of firm performance.

In any case, the notion that ICT per se does not generate sustainable competitive advantage has received important support in recent research (Kettinger et al., 1994; Powell and Dent-Micallef, 1997), giving rise to what has become known as the "strategic necessity hypothesis". According to the Resource Based View, also known as RBV (Rumelt, 1984; Wernerfelt, 1984; Peteraf, 1993; Barney, 1991) a resource must possess certain characteristics to be qualified as strategic. Specifically, in order to both generate and sustain competitive advantage over a period of time, a resource must be "valuable", "scarce", "unable to imitate" and "complementary with other elements or resources of the firm.

Today, ICT is so widely available that it can hardly be described as being scarce, or difficult to imitate, and thus it does not satisfy the necessary criteria to be considered critical. Conversely, critical value may reside in the complementary or synergic effects of ICT with internal resources and capabilities of the firm.

Recently, the empirical literature has begun to re-assess the association between ICT and a wide variety of complementary factors (Arvanitis and Loukis, 2009; Giuri et al., 2008; Aral and Weill 2007), with a consensus emerging that, in order for ICT to be properly utilised, it must be used in conjunction with complementary resources such as organisational structure, human resources or organisational resources (Walton, 1989; Bélanger and Collins, 1998; Bresnahan et al., 2002; Mata et al., 1995; Ramírez et al., 2001; Peppard and Ward ,2004; Aral et al., 2010).

Focusing on the retail industry in the US, Powell and Dent-Micallef (1997) find that ICT alone has not produced sustainable performance advantages, while some firms have gained competitive advantages by using these technologies with complementary human and organisational resources. Also for the US, Bresnahan et al. (2002) offer empirical evidence about the positive effects of combining ICT and organisational design on increasing firm productivity. The same results are offered by Black and Lynch (2001), who examine the impact of ICT, human resource practices, and firm reorganization on productivity.

Crespi et al. (2007) examine the relationships between productivity growth, ICT investment and organisational change in UK firms, and their results support the idea that gains from IT need re-organisation to produce measured productivity growth.

Gretton et al. (2004) obtain empirical evidence of the positive impact of complementarities between the use of ICT and human resources, innovative business practices, and intensity of organizational change on the productivity growth of Australian companies.

In a comparative study, Arvanitis and Loukis (2009) offer empirical evidence of the positive impact of ICT capital, human capital and new organizational practices on labour productivity in Greece and Switzerland, while they observe that the Swiss firms are more mature and more efficient at combining these new production factors.

In general, all the studies analyzed contain the idea that, to achieve a more competitive position, the firm should complement ICT investments with an appropriate use of these technologies, for which, implicitly, complementary resources are required.

According to Soh and Markus (1995), we can establish three different processes that include, first, the conversion process in which ICT expenditures are converted to ICT assets; second, the ICT use process, where a higher or lower impact of ICT is obtained depending on the appropriate or inappropriate use of these technologies; and finally, the authors highlight the importance of the competitive process, in which any number of factors beyond the firm's control may result in failure to realize improved organizational performance. These three processes, and the integration of the complementarities model, are shown graphically in Figure 1:

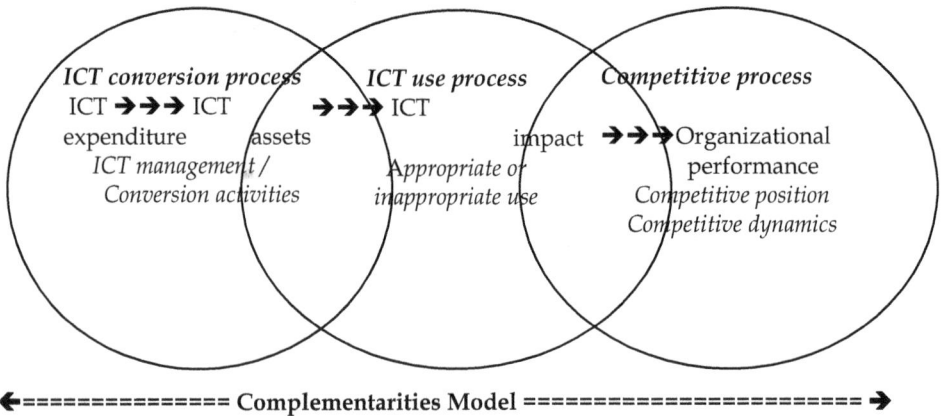

Source: Adapted from Soh and Markus (1995).

Fig. 1. ICT management and competitive position process

In any case, it is generally accepted that ICT in itself is not a panacea and there is still a serious debate about how its adoption improves firm performance. In general, it is supposed that complementary investments in skills, organisational change, and innovation are the key to making these technologies work. In this study, we argue that the simultaneous presence of complementary resources increases the positive effect of ICT investment in performance.

3. Hypothesis development

As we have seen, although some research concludes that ICT has negative or irrelevant effects, since the last decade there has been a growing consensus that there is a strong positive link (Atzeni and Carboni, 2006). ICT can help improve productivity and the efficiency of all stages of the production process by reducing set-up time, run time and inspection time, which will help efficient firms to improve their market position. Accordingly, we propose our first hypothesis:

Hypothesis 1: "The impact of information and communication technologies on productivity will be positive"

However, within the theoretical framework of RBV and the complementarities discussed above, we can argue that the firm-level impacts of ICT may be higher in those companies that have been able to reorganise. In order to implement effectively and to reap benefits from ICT investments, complementary resources and capabilities will be needed.

Investment in human resources has been one of the factors most commonly explored, because of its possibilities for enhancing the effectiveness of organizational practice. Furthermore, human resources management is considered to be not only a determining factor of productivity, but also a complementary element of ICT (OECD, 2001a). According to 'skill-biased' technological change, we can argue that ICT is associated with a greater demand for highly educated workers (Acemoglu, 1998; Black and Lynch, 2001; Bresnahan et al., 2002; Caroli and Van Reenen, 2002). Skilled employees allow firms to integrate ICT more effectively in the planning process of the business (Bharadwaj, 2000). Available empirical evidence suggests that improvements in the composition of the labour factor have directly contributed to labour productivity growth in many countries (OECD, 2005). At the firm level, various authors have shown that ICT use can have positive effects on performance, but those positive effects may be smaller if the necessary complementary investments in skills have not occurred to a sufficient degree (Arvanitis, 2005; Powell and Dent-Micallef, 1997; Pinsonneault and Kraemer, 1997; Francalanci and Galal, 1998 and Bresnahan et al., 2002). As these authors show, ICT has substantial impacts on firm performance when it is combined with a higher level of better-qualified personnel. Therefore, we define the next hypothesis as follows:

Hypothesis 2: "The impact of ICT on results will be greater for organisations that combine ICT with a high level of worker qualification".

Research has also highlighted the importance of strategic alignment, defined in terms of ICT support for the business strategy (Tallon and Kraemer, 2008). Specifically, ICT adoption has been positively associated with those firms that attach greater importance to the quality of their products (Lal, 1999). ICT investments provide significant opportunities to decrease the costs of overseeing production and distribution. Insofar as quality systems are closely related to the availability of flows and information processing, ICT will offer new approaches to the implementation of quality assurance systems. In this context, implementation of ICT is most effective when introduced in conjunction with quality systems (Bruque and Moyano, 2007). Moreover, as Ramírez et al. (2001) stress, the impact of ICT on results is positively influenced by the adoption of TQM and re-engineering. Thus, we propose the following hypothesis:

Hypothesis 3: "The impact of ICT on results will be greater for organisations that combine ICT with an assurance and quality control strategy"

4. Research design and methodology

First, we are interested in estimating the effects of ICT, in conjunction with appropriate organizational and human resources. Following prior studies, the present paper deals with the productivity approach (Gurbaxani et al., 1998; Lichtenberg, 1995; Ramírez et al., 2001). In this sense, most econometric studies that attempt to assess the contribution of ICT to productivity rely on a standard Cobb-Douglas production function, created in 1927 by Paul Douglas and Charles Cobb, as the basic analytical framework (Loveman, 1994; Dewan and Min, 1997). It is the most commonly used form for the production function, and has the advantages of both simplicity and empirical robustness for the calculation of firm performance[1]. Recent studies, such as Badescu and Garcés-Ayerbe (2009) and Brynjolfsson et al. (2011) demonstrate the consistency of this methodology in formulating and estimating the impact of ICT capital on productivity.

In line with the majority of empirical studies, we assume that the production function for manufacturing firms can be described by an extended Cobb-Douglas function, in which we include ICT capital as a factor of production (Brynjolfsson and Hitt, 1996, 2003). In order to incorporate the interaction of ICT with the complementary resources, various dummies are added [$e^{\alpha_i \sum Di}$].

The coefficients of the multiplicative dummy variables show the direction of the joint effect of their components. A negative/positive estimated coefficient implies that the interaction of both variables will negatively/positively affect productivity.

In order to test previous hypotheses, the production function to be estimated is defined according to the following equation (Equation 1) :

$$Q = AK_{ICT}{}^{\beta_1} K_{NICT}{}^{\beta_2} e^{\alpha_i \sum DilCT_{01}} e^{\alpha_{leg} \sum D_{leg}} e^{\alpha_s \sum D_s} L^{\beta_3} e^{\mu} \tag{1}$$

Where β_i represents the output elasticity of ICT capital, non-ICT capital and labour, respectively, A is a parameter measuring the total productivity factor, a_{leg} and a_s are the coefficients of legal structure and industry dummy variables, and α_i refers to the multiplicative dummy variables.

The logarithmic transformation of equation (1) provides a log-linear form which is convenient and commonly used in econometric analyses using linear regression techniques. Performing the necessary transformations, we obtain the basic production model, including the effects of ICT and the importance of multiplicative variables (Equation 2).

$$Ln\ (Q/L) = \alpha + \beta_1 Ln\ (K_{ICT}/L) + \beta_2 Ln\ (K_{NICT}/L) + \beta_3 Ln\ L + \alpha_1 ICT_{QUALIF} + \alpha_2 ICT_{QUALIT} + \sum \alpha_{leg} D_{leg} + \sum \alpha_s D_s + \varepsilon \tag{2}$$

---

[1] See Brynjolfsson and Hitt (1996) for a discussion of this formulation.

4.1 Sample and definition of the variables

The empirical analysis is based on firm-level longitudinal data, from which we have information on the necessary variables. The sample comes from the Survey on Business Strategies (SBS)[2], carried out annually by the Spanish Ministry of Science and Technology [Ministerio de Ciencia y Tecnología], containing information on a wide range of firms Spanish manufacturing firms with 10 or more employees. The SBS has the advantage of a statistically representative sample of manufacturing firms in Spain (Fariñas and Jaumandreu, 1999). The sample covers those firms with 200 or more employees and firms with less than 200 employees are selected by a random sampling method.

The total number of manufacturing firms in the SBS for the year 2002 was 1,724. Excluding those observations for which the database does not provide information for all necessary variables, we are left with a final sample of 1,269 firms.

We have defined a performance measure that, in accordance with the specialised literature, could have a significant impact on the analysis of firm performance. Specifically, to test the validity of the propositions, the following variables have been defined:

Output Variable

Productivity is one of the performance measures most frequently used in the literature (Brynjolfsson and Hitt, 2003), as evidenced by its use in different samples, sectors and methodologies (Barua et al., 1995; Krueger, 1993; Mahmood and Mann, 2000). In line with previous research, in this study we define the output variable according to labour productivity, measured using the value added of the firm divided by the total number of employees *(Q/L)*.

Information and Communication Technologies Capital

We use the ICT capital intensity variable measured by ICT stock divided by number of employees of the firm (K_{ICT} / L). On the one hand, we define K_{ICT} as the amount of ICT capital per unit of labour input, where ICT capital is computed using the perpetual inventory method. This estimation allows for cumulative investments carried out over time in computer equipment and data processing equipment by the firm. The annual depreciation rate is assumed to be 0.20[3]. On the other hand, labour input is defined as the total number of workers at the firm at the end of the year.

Rest of Capital

We define the intensity of the rest of capital as the ratio between non-ICT capital and labour (K_{NICT}/L) that is non-ICT capital intensity. Non-ICT capital is measured by conventional

---

[2] Encuesta de Estrategias Empresariales (ESEE) in Spanish.

[3] Similar depreciation rates close to 20% have been used, for example by Kafouros (2006), who assumes a depreciation rate for intangible technological resources of 20%, and Shin (2000, 2006) who uses a depreciation rate of 22.4 % for ICT investments. In any case, although rates used in prior studies vary widely, Bloom et al. (2006) show that the significance and the magnitude of the coefficient obtained for ICT is not affected by the exact choice of the alternative depreciation rate.

capital, calculated as the difference between the total net fixed assets (obtained from the balance sheets) minus that part of net fixed assets corresponding to ICT. The use of accounting depreciation, rather than economic depreciation, and historic values, rather than replacement values, in the determination of conventional capital may reduce measurement accuracy (Lichtenberg, 1995). Nevertheless, this is not a problem, since both ratios, historic value and real value, and accounting depreciation and economic depreciation, remain constant for all the firms in the same sector; including a sectorial dummy variable, as we do, removes any bias. The labour factor is measured by the total number of employees of the firm.

Interaction variables

We include in our regression two multiplicative dummy variables that reflect the interaction effect of ICT and worker qualifications, and ICT and quality management strategy, respectively. They are constructed as follows:

- ICT and worker qualifications (ICT_{QUALIF}): this variable captures the interaction between the availability of ICT capital in the firm and the intensity of qualified workers, measured as the existence of a higher than sector-average number of qualified employees in the firm. According to the educational level completed by the workers (OECD, 2001b), we classify as qualified those employees who have completed a university degree, or trained as technicians and experts. The multiplicative variable takes a value of 1 when the firm has more highly qualified workers than the sector average, along with positive ICT stock, and 0 otherwise.
- ICT and Quality management (ICT_{QUALIT}): the quality management variable reflects activities related to quality carried out by managers. The multiplicative variable takes a value of 1 if the firm has made efforts in standardization and quality control and, at the same time, it has ICT capital, and 0 otherwise.

Control Variables

Following prior studies, the following variables have been considered as independent control variables to proxy for industry characteristics, legal structure of the firm, and organizational size:

- Sector of activity (D_S): This variable is defined by the first two digits according to the Spanish National Classification of Economic Activities (CNAE) Code.
- Legal structure of the firm (D_{LS}): We include a set of dummy variables in order to control for the effect of the diversity of corporate structure in our sample (limited liability company, public limited liability company, cooperative, Employee-owned).
- Size (L): Firm size variable is measured by the total number of employees of the firm at the end of the year.

Table 1 presents the main statistics (mean and standard deviation) of the variables for the total sample:

As we can see, intensity of ICT is lower than non-ICT capital intensity. However, the standard deviation value is higher for ICT intensity, reflecting the dispersion among firms and it shows that there exist significant differences in the level of effort that Spanish manufacturing firms are expending on ICT investments.

	Mean	Standard Deviation
ICT intensity	2.0870	7.99527
Rest of capital intensity	10.0388	1.67310
Size	4.2065	1.45488
ICTQUALIF	0.3436	0,47509
ICTQUALIT	0.3034	0.45990
Limited Company	0.3278	0.46960
Labour Anonymous	0.6225	0.48494
Productivity (VA / N° workers)	0.0165	0.12762
Total number of firms	1,269	

Table 1. Main statistics of the sample

4.2 Main results and discussion

As noted previously, the dependent and independent variables are constructed and the empirical model estimated using the data set of 1,269 firms. First, we review the productivity differences between the group of firms that invest and the group that does not invest in ICT.

	$K_{ICT}=0$	$K_{ICT}>0$
Ln (Q/L)	10.2686***	10.6098***
N	531	738

*** Denote statistical significance at the 1% level (p<0,01)

Table 2. Productivity mean differences

As we can see in Table 2, firms with a positive value for the K_{ICT} variable achieve statistically higher values of productivity. Nevertheless, although it gives us a first idea about the positive relationship between ICT and performance, the empirical analysis offers a further understanding of the relationship. The most relevant results of the model of equation (2) are presented in Table 3.

We can see that the coefficient of the ICT intensity variable is positive and statistically significant. This allows us to accept Hypothesis 1, postulating a positive effect of ICT on the results. Moreover, the multiplicative variable representing ICT and qualifications interaction takes positive values and confirms the importance of qualifications. This allows us to corroborate Hypothesis 2. It is possible that the presence of more qualified workers can be related to the use of new workplace organisation such as team-work, decentralised decision-making, and flattering hierarchies, all practices that increase the possibilities of ICT. These results are in accordance with the studies that found that the use of equipment for data processing is mainly in the hands of workers with medium and high skills (Borghans and Ter Weel, 2007, 2011; Bresnahan et al., 2002).

	Original coefficients	Standardized coefficients
Constant	8.687***	---
Ln (K_{ICT}/L)	0.006***	0.068
Ln (K_{NICT}/L)	0.107***	0.267
Ln SIZE	0.058***	0.126
ICT$_{QUALIF}$	0.102**	0.073
ICT$_{QUALIT}$	0.075*	0.052
Limited Company	0.128	0.090
Anonymous Company	0.268***	0.194
Labour Anonymous Company	0.249*	0.047
D$_S$	Yes	Yes
R^2	35.6%	
Adjusted R^2	34.2%	
Total number of firms	1,269	

* Denotes statistical significance at the 10% level ($p<0,10$)
** Denotes statistical significance at the 5% level ($p<0,05$)
*** Denotes statistical significance at the 1% level ($p<0,01$)

Table 3. Results of the productivity regression analysis

As for the variable measuring ICT and quality strategy, a positive coefficient is obtained. Hence, we can conclude that Hypothesis 3 is accepted. These results are consistent with those obtained by Mata et al. (1995), who argue that ICT becomes really effective when it is handled proactively by management, and Bharadwaj (2000) who concludes that capacity for technical management allows firms to achieve better business results than their competitors.

The firm size variable presents a positive and statistically significant coefficient, consistent with those theories that establish a positive influence of firm size on performance, and with the presence of positive economies of scale. Regarding the legal structure of the firm, we can also highlight the positive and statistically significant effect of Anonymous Company and Labour Anonymous Company on firm productivity.

To sum up, the results obtained in the empirical work can be summarized according to the three hypotheses we have tested as follows:

Hypothesis	Result	Implications
Hypothesis 1	Accepted √	⇑ICT ⇒ ⇑ Productivity
Hypothesis 2	Accepted √	⇑ICT&Qualification ⇒ ⇑ Productivity
Hypothesis 3	Accepted √	⇑ICT&Quality ⇒ ⇑ Productivity

The model achieves an acceptable explanatory power – with a correlation coefficient of around 34%– and it confirms the importance of ICT and the interaction between those technologies and organizational complementary resources in improving firm performance. Our results reinforce those obtained by Aral et al. (2010), who concluded that one reason for the variations in the returns to ICT investments across firms may be differences in the adoption of complementary organizational practices.

5. Conclusion

The main goal of this chapter has been to gain a better comprehension of how information and communication technologies affect firm productivity, and of the importance of other complementary factors. Since there are few studies investigating the impact of ICT on firm productivity in Spain, the present study attempts to fill this gap. We have analyzed the role of the alignment of different elements in increasing the ICT impact, in a sample of Spanish manufacturing firms.

Our main contribution lies in the possibility of offering evidence supporting the existence of a statistically significant relationship, not only between ICT and productivity, but also between the multiplicative variables that represent ICT and other complementary factors.

The empirical evidence offered highlights the need to consider organizational aspects, such as human resources and strategic adjustment, in order to raise the potential benefits of ICT. According to this evidence, we conclude that investment in organisation and human resources skills is crucial in achieving higher levels of performance. Business management capabilities are growing in importance, rather than ICT alone. Thus, we can make certain practical recommendations that will be useful for all responsible agents in the management of ICT and other complementary factors.

Our findings suggest that differences in the use and impact of ICT across firms are probably due to the lack of complementary resources or the lack of fit between key organizational aspects. We can further postulate that the gap between Spanish and other European firms in realizing the potential of ICT may be due to the use of these technologies without adequate complementary resources at the firm level.

Factors related to the social, technological or legal environment may also play an important role in ensuring that the potential gains in productivity from the successful use of ICT are among the primary policy targets. For example, many governments provide ICT training or training support, some of them free of charge, or they offer financial support to cover part of training expenses.

Nonetheless, public policies need to be carefully designed to enhance their effectiveness and, in particular, it is necessary to take into account that without complementary firm investments and organisational change, the economic impact of ICT may be limited. Government programmes may sometimes fail because of a lack of organizational redesign at the firm level.

To conclude, it seems appropriate to indicate that this study has some limitations that point the way to further investigation. In future research, it would be interesting to include the time dimension to assess the impact of ICT investment on performance after a time lag, and compare this with earlier results. Another extension could be to distinguish among various types of information and communication technologies, testing possible differences on their impact. Unfortunately, the Survey on Business Strategies database does not offer this information, which prevents us from calculating separately the stock of different Information and Communication Technologies.

6. Acknowledgment

The authors would like to express their thanks for the financial support received under the Research Project ECO2009-13158 *"Entrepreneurship and enterprise development: analysis, implications for economic and social welfare and for the public policies"*, Spanish Ministry of Science and Technology. The authors also would like to thank the Fundación Empresa Pública for providing the data.

7. References

Acemoglu, D. (1998). Why Do New Technologies Complement Skills? Directed Technical Change and Wage Inequality. *The Quarterly Journal of Economics*, Vol.113, No.4 ,(November), pp. 1055-1089.

Aral, S. & Weill, P. (2007). IT Assets, Organizational Capabilities & Firm Performance: How Resource Allocations and Organizational Differences Explain Performance Variation. *Organization Science*, Vol.18, No.5, (September-October), pp. 1-18.

Aral, S.; Brynjolfsson, E. & Wu, L (2010). Assessing Three-Way Complementaries: Performance Pay, Monitoring and Information Technology. (August 25, 2010). *Management Science*, Forthcoming. Available at SSRN: http://ssrn.com/abstract=1665945

Arvanitis, S. (2005). Computerization, Workplace Organization, Skilled Labour and Firm Productivity: Evidence for the Swiss Business Sector. *Economics of Innovation and New Technology*, Vol.14, No.4, pp. 225-249.

Arvanitis, S. and Loukis, E. (2009). Information and Communication Technologies, Human Capital, Workplace Organization and Labour Productivity: A Comparative Study Based On Firm-Level Data for Greece and Switzerland. *Information Economics and Policy*, Vol.21, No.1, (February 2009), pp. 43-61.

Atzeni, G.E. and Carboni, O.A. (2006). ICT Productivity and Firm Propensity to Innovative Investment: Evidence from Italian Microdata, *Information Economics and Policy*. Vol.18, pp. 139–156.

Badescu, M. & Garcés, C. (2009). The Impact of Information Technologies on Firm Level Productivity: Empirical evidence from Spain. *Technovation*, Vol.29, No.2, pp. 122-129.

Barney, J.B. (1991). Firm resources and Sustained Competitive Advantage. *Journal of Management*, Vol. 17, pp. 99-120.

Barua, A.; Kriebel, H.C. & Mukhopadhyay, T. (1995). Information Technologies and Business Value: An Analytic and Empirical Investigation. *Information Systems Research*, Vol.6, No.1, pp. 3-23.

Bélanger, F. & Collins, R. (1998). Distributed Work Arrangements: A Research Framework, *The Information Society: An International Journal*, Vol.14, No.2, pp. 137-152.

Bharadwaj, A.S. (2000). A Resource-Based Perspective on Information Technology Capability and Firm Performance: An Empirical Investigation. *MIS Quarterly*, Vol.24, No.1, pp. 169-197.

Black, S.E. & Lynch, L.M. (2001). How to Compete: The Impact of Workplace Practices and Information Technology on Productivity. *Review of Economics and Statistics*, Vol.83, No.3, pp. 434-445.

Bloom, N.; Sadun, R. and Van Reenen, J. (2006). It Ain't What You Do, it's the Waythat You do IT. Investigating the Productivity Miracle Using Multinationals. *Centre for Economic Performance, London School of Economics*, (August 2006).

Borghans, L. & Ter Weel, B. (2007). The Diffusion of Computers and the Distribution of Wages. *European Economic Review*, Vol. 51, No.3, pp. 715-748.

Borghans, L. and Ter Weel, B. (2011). Computers, Skills and Wages. *Applied Economics*, Vol.3, No.29, pp. 4607-4622.

Bresnahan, T.; Brynjolfsson, E. & Hitt, L. (2002). Information Technology, Workplace Organization, and the Demand for Skilled Labor: Firm-Level Evidence. *Quarterly Journal of Economics*, Vol.117, No.1, pp. 339-376.

Bruque, S. and Moyano, J. (2007). Organizational Determinants of Information Technology Adoption and Implementation in SMEs: The Case of Family and Cooperative Firms. *Technovation*, Vol. 27, pp. 241-253.

Brynjolfsson, E. and Hitt, L. (1995). Information Technology as a Factor Of Production: The Role Of Differences Among Firms. *Economics of Innovation and New Technology* (Special No. on Information Technology and Productivity Paradox) Vol.3, No.4, pp. 183-200.

Brynjolffson, E. & Hitt L. (1996). Paradox lost? Firm-level Evidence on the Returns to Information Systems Spending. *Management Science*, Vol.42, No.4, pp. 541-558.

Brynjolfsson, E. & Hitt, L. (2003). Computing Productivity: Firm-level Evidence. *Review of Economics and Statistics*, Vol.85, No.4, pp. 793-808.

Brynjolfsson, E. & Yang, S. (1996). Information Technology and Productivity: Are View af the Literature. *Advances in Computers*, Academic Press, Vol.43, pp. 179-214.

Brynjolfsson, E. (1993). The Productivity Paradox of Information Technology. *Communications of the ACM*, Vol.36, No.12, pp. 66-77.

Brynjolfsson, E.; Hitt, L. & Kim, H.H. (2011). Strength in Numbers: How Does Data-Driven Decisionmaking Affect Firm Performance?. (April 22, 2011). Available at SSRN: http://ssrn.com/abstract=1819486.

Brynjolfsson, E.; Hitt, L. & Yang, S. (2002). Intangible Assets: Computers and Organizational Capital. *Brookings Papers on Economic Activity*, Vol.1, pp. 137-98.

Caroli, E. & Van Reenen, J. (2001). Skill-Biased Organizational Change? Evidence from a Panel of British and French Establishments. *The Quarterly Journal of Economics*, Vol.116, No.4, pp. 1449-1492.

Crespi, G.; Criscuolo, C. and Haskel, J. (2007). *Information Technology, Organisational Change and Productivity Growth: Evidence from UK Firms*. CEPDP, 783. Centre for Economic Performance, London School of Economics and Political Science, London, UK.

Davenport, T., (1994). Saving IT's Soul: Human Centered Information Management. *Harvard Business Review*, Vol.72, No.2, pp. 119-131.

Dewan, S. and Min, C.K. (1997). The substitution of Information Technology for Other Factors of Production: A firm level Analysis. *Management Science*, Vol.43, No.12, (December 1997), pp. 1660-1675.

Fariñas, J.C. & Jaumandreu, J. (1999). Diez años de Encuesta sobre Estrategias Empresariales (ESEE). *Economía Industrial*, Vol. 329, pp. 29-42.

Francalanci C. & Galal, H. (1998). Information Technology and Worker Composition: Determinants of Productivity in the Life Insurance Industry. *MIS Quarterly*, Vol.22, No.2, pp. 227-241.

Giuri, P., Torrisi, S. and Zinovyeva, N. (2008). ICT, Skills, and Organizational Change: Evidence From Italian Manufacturing Firms. *Industrial and Corporate Change*, Vol.17, No.1, pp. 29-64.

Gretton, P.; Gali, J. and Parham, D. (2004). The Effects af ICTs and Complementary Innovations on Australian Productivity Growth. In: OECD (2004). *The Economic Impact Of ICT Measurement, Evidence And Implications*. Organisation for Economic Co-operation and Development, París.

Gurbaxani, V.; Melville, N. & Kraemer, K. (1998). Disaggregating the Return on Investment to IT Capital. *Working Paper, Center for Research on Information Technology and Organizations*. Sept. 11, 1998.

Hitt, L. & Brynjolfsson, E. (1996). Productivity, Business Profitability and Consumer Surplus: Three Measures of Information Technology Value. *MIS Quarterly*, Vol.20, No.2, pp. 121-142.

Huerta, E (ed.). (2008). *La Innovación En La Empresa: Políticas Avanzadas De Gestión De Recursos Humanos*, Ed. Centro para la competitividad de Navarra, Navarra.

Kafouros, M. I. (2006). The Impact of the Internet on R&D Efficiency: Theory and Evidence. *Technovation*, Vol.26, pp. 827–835.

Kettinger, W.J.; Grover, V. ; Guha, S.; Segars, A.H. (1994). Strategic Information Systems Revisited: A Study in Sustainability and Performance. *MIS Quarterly*, Vol.18, No.1, pp. 31-58.

Krueger, A. (1993). How Computers Have Changed the Wage Structure: Evidence From Microdata. *Quarterly Journal of Economics*, Vol.107, No.1, pp. 33-60.

Lal, K. (1999). Determinants of the Adoption of Information Technology: A Case Study of Electrical and Electronic Goods Manufacturing Firms in India. *Research Policy*, Vol.28, No.7, pp. 667–680.

Lichtenberg, F. (1995). The Output Contribution of Computer Equipment and Personnel: A Firm Level Analysis. *Journal of Economics of Innovation and New Technology*, Vol.3, pp. 201-217.

Loveman, W. (1994). An Assessment of Productivity Impact on Information Technologies. In Scott Morton (ed.), *Information Technology and the Corporation of the 1990´s: Research Studies*, Cambridge, MA., Allen, T. J. and M.S. MIT Press.

Mahmood, M.A. & Mann, G.J. (2000). Special Issue: Impacts of Information Technology Investment on Organizational Performance. *Journal of Management Information Systems*, Vol.16, No.4, pp. 3–10.

Mata, F.; Fuerst, W.L. and Barney, J.B., (1995). Information Technology and Sustained Competitive Advantage: A Resource-Based Analysis. *MIS Quarterly*, Vol.19, No.4, (December), pp. 487-505.

OECD (2001a). *The New Economy beyond the Hype. The OECD Growth Project*. Organisation for Economic Co-operation and Development, Paris.

OECD (2001b). *OECD Science, Technology and Industry Scoreboard 2001*. Organisation for Economic Co-operation and Development, París.

OECD (2003). *ICT and Economic Growth - Evidence from OECD Countries, Industries and Firms* OECD, Paris. ISBN 92-64-10128-4.

OECD (2004). *The Economic Impact Of ICT Measurement, Evidence And Implications*. Organisation for Economic Co-operation and Development, Paris. Available at: http://www1.oecd.org/publications/e-book/9204051E.PDF

OECD (2005). *Guide to Measuring the Information Society*. Organisation for Economic Co-operation and Development, Paris. Available at:
http://www.oecd.org/dataoecd/41/12/35654126.pdf

OECD (2010). *OECD Information Technology Outlook 2010*, Organisation for Economic Co-operation and Development, Paris.

Peppard, J. and Ward, J. (2004). Beyond Strategic Information Systems: Towards an IS Capability. *Journal of Strategic Information Systems*, Vol.13, pp. 167-194.

Peteraf, M. (1993). The Cornerstones of Competitive Advantage: A Resource-Based View. *Strategic Management Journal*, Vol.14, pp. 179-191.

Pinsonneault, A. & Kraemer, K.L. (1997). Middle Management Downsizing: An Empirical Investigation Technology. *Management Science*, Vol.5, pp. 659-679.

Powell, T.C. & Dent-Micallef, A. (1997). Information Technology as Competitive Advantage: The Role of Human, Business, and Technology Resources. *Strategic Management Journal*, Vol.18, No.5, (May), pp. 375-405.

Ramírez, R.V.; Kraemer, K. L. & Lawler, E. (2001). The Impact of Organizational Improvement Efforts on the Productivity of Information Technology: A Firm-level Investigation, *UC Irvine, Center for Research on Information Technology and Organizations* (CRITO), Working Paper, 2001.

Roach, S.S., (1987). *American's Technology Dilemma: A Profile of the Information Economy*. Morgan Stanley Special Economic Study, (April, 1987).

Rumelt, R. (1984). Toward a Strategic Theory of the Firm, in R. Lamb (Ed.), *Competitive Strategic Management*. Englewood Cliffs, New Jersey Prentice Hall, 556-570.

Shin, I. (2000). Use of Information Network and Organizational Productivity: Firm-level Evidence in Korea. *Economics of Innovation and New Technology*, Vol.9, No.5, pp. 447-463.

Shin, I. (2006). Adoption of Enterprise Application Software and Firm Performance. *Small Business Economics*, Vol.26, pp. 241–256.

Soh, C. and Markus, M.L. (1995). How IT creates Business Value: A Process Theory Synthesis. In *Proceedings of the Sixteenth International Conference on Information Systems*, J. I. DeGross, G. Ariav, C. Beath, R. Hoyer, and C. Kemerer (eds), Amsterdam, 1995, pp. 29-41.

Solow, R.M. (1987). We'd better watch out. *New York Times Book Review*, July 12.

Strassmann, P.A. (1985). *Information Payoff. The Transformation of Work in the Electronic Age*. Free Press, New York.

Strassmann, P.A. (1990). *The Business Value Of Computers: An Executive's Guide*. New Canaan, CT, Information Economics Press.

Tachiki, D., Hamaya, S. and Yukawa, K. (2004). Diffusion and Impacts of the Internet and E-Commerce in Japan *Center for Research on Information Technology and Organizations* (CRITO). Globalization of I.T. Paper 339.

Tallon, P.P. and Kraemer, K.L. (2008). A Process-oriented Perspective on the Alignment of Information Technology and Business Strategy. *Journal of MIS* (JMIS), Vol.24, No.3, pp. 231-272.

Walton, R.E. (1989). *Up and Running. Integrating Information Technology and the Organization*. Harvard Business School Press. Boston, Massachusetts.

Weill, P. (1992). The Relationship between Investment in Information Technology and Firm Performance: A Study of the Valve Manufacturing Sector. *Information Systems Research*, Vol.3, No.4, pp. 307-333.

Wernerfelt, B. (1984). A Resource-Based View of Theof the Firm. *Strategic Management Journal*, Vol.5, pp. 171-80.

Wilson, D. (1993). Assessing the Impact of Information Technology on Organizational Performance. In *Strategic Information Technology Management*, Banker, R.; Kauffman, R. & M.A. Mahmood (ed.), Idea Group, Harrisburg, P.A.

Wilson, D. (1995). IT Investment and its Productivity Effects: An Organizational Sociologist's Perspective on Direction for Future Research. *Economic Innovation and New Technology*, Vol.4, No.3, pp. 235-251.

12

The Impact of Company Relationship and Institution Technology on R&D Activity and Innovation

Fredy Becerra-Rodríguez

Industrial Engineering Department, National University of Colombia-Manizales
Department of Business Administration, Manizales University
Colombia

1. Introduction

In today's competitive environment, the activities of a business cannot be imagined in an isolated manner, nor can they be exclusively associated with commercial-type relations (purchasing and sales of goods and services) as they were in the 1980s and 1990s (Buhman, et al., 2005; Más Ruíz, 2000). We are currently living in a networking society where strong interdependencies among actors are created through networks and inter-organizational alliances (Tikkanen & Parvinen, 2006), where today's companies are network-centric enterprises (Buhman et al., 2005), and where the external environment affects company behavior. At the same time, the companies, their culture and their learning processes influence the external stakeholders with whom they have relations (Minguzzi & Passaro, 2000). Therefore, companies must take into account not only their stage of development but also the development of the industry, in order to maximize their learning capability (Benson-Rea & Wilson, 2003) based on their relations with the other stakeholders located in the same region or country, and thus contribute to the development of competitive advantages at both business and regional levels (Bell & Albu, 1999; Carbonara, 2002; Dohse & Soltwedel, 2006; Feldman et al., 2005).

Section 1 presents some common proposals and perspectives based on the study of inter-company relations and of company-institution relations using the following theoretical approaches: agglomeration economics, clusters and industrial districts, and networking. Section 2 addresses theoretical aspects regarding R&D activities and innovation in a business network environment and proposes the hypotheses for this article. Section 3 explains the research methodology; based on the literature on the topic, it indicates the main methodologies employed in this article and defines analysis variables. Section 4 presents the general theoretical and empirical findings. And finally suggests future lines of research.

2. The study of inter-company relations and of company-institution relations

Inter-company relations as well as company-institution relations (public and private institutions) are determining factors for business competitiveness and contribute to the socioeconomical development of the area where they are located. They especially influence the companies' innovation and R&D capabilities. The study of these relations is based on the concepts of agglomeration economics, clusters and industrial districts, and networking.

These different concepts have theoretical aspects in common, such as:

- The existence of a company cluster in a clearly delimited territory (concentrated in one space) through which the companies take advantage of externalities resulting from their proximity (Alonso-Villar et al., 2004; Bell, 2005; Feser & Bergman, 2000; Mella et al., 2007; Nassimbeni, 2003)
- High levels of specialized goods and services, knowledge flows and innovations, and intense, constant technological change (Antonelli, 2006; Beesley, 2004; Callois, 2008; Carbonara, 2002; Groenewegen & Van Der Steen, 2006; Hagedoorn & Duysters, 2002; Hervás & Dalmau, 2006; OECD, 1999 a, 1999b)
- Building and sustaining vertical and horizontal relations of inter-company trust and of trust between companies and other stakeholders in and outside the territory (Brenner & Greif, 2006; Dohse & Soltwedel, 2006; Eraydin & Armatli-Köroglu, 2005; Hotz-Hart, 2002; McCann, 1995; Tracey & Clark, 2003; Walker et al., 1997)

And from a practical view, the empirical research conducted in the context of the above concepts has the following perspectives well worth mentioning.

- Local business networks and their role in the development of knowledge, of innovation processes, and in innovation results (Baptista, 1996; Baptista & Swam, 1998; Beaudry & Breschi, 2003; Beesley, 2004; Bell, 2005; Brenner & Greif, 2006; Hagedoorn & Duysters, 2002; Hervás & Dalmau, 2006; Muscio, 2006; Wolfe & Gertler, 2002; Yogel et al., 2000; Zhang, 2007)
- The importance of business networks in business strategy, productivity and competitiveness, and business infrastructure (Carbonara, 2002; Carrie, 1999; Feldman et al., 2005; Hervás et al., 2007; Lechner & Dowling, 2003)
- Business networks and their incidence *vis à vis* the job market, human resources and company development (Blasio & Di Addario, 2005; Hervás & Dalmau, 2006; Hu et al., 2005; Power & Lundmark, 2004; Pöyhönen & Smedlund, 2004)
- Business networks and public policies for regional economic development at State, regional, city or location level (Altengurg & Meyer-Stamer, 1999; Dohse & Soltwedel, 2006, Gibb, 2006; Jensen, 1996; Koch et al., 2006; Lai et al., 2005; McDonald et al., 2006; Oyelaran-Oyeyinka & Lal, 2006; Viladecans-Marsal, 2004).

All of the aspects mentioned above prove the importance of inter-company relations as well as of company-institution relations because of their positive influence not only on the economic and social development of specific areas but also on each company's development, especially in the development of technological capabilities for innovation. Likewise, a conclusion to be drawn is that the studies on inter-company relations are based on the idea of networks, which may allow inferring that "the business network" (Becerra, 2008) is the common unit of analysis, independently of the proposals presented in each approach.

3. Innovation as a process: Innovation activities

The idea of innovation stems from Schumpeter's ground-breaking work that proposes that innovation is achieved upon introducing a new product or a modified product, upon inventing a new method of doing something, upon entering a new market or finding a new source of provisioning, or upon creating a new organization (Schumpeter, 1997), which basically implies understanding innovation as a result. Along the same lines, Damanpour (1987) differentiates technological innovations and administrative innovations, and classifies innovations as radical innovations and as incremental innovations (Gopalakrishnan & Damanpour, 1997), according to the degree of innovation.

The Oslo Manual states that innovation "is the implementation of a new or significantly improved product (good or service), or process, a new marketing method, or a new organizational method, in business practices, workplace organization or external relations," (OECD/EUROSTAT, 2005, pg. 46). This definition comprises the different ideas discussed in the literature on the topic and highlights the external relations of a company, meaning that it alludes to the relational capital (Capello, 2002; Capello & Faggian, 2005) that can be found in business networks, which is the object of this study although the above is based on a perspective of innovation as a result.

Innovation as a process, not as a result, implies understanding the activities that take place in order for new ideas, objects and practices to be created, developed or reinvented (Slappendel, 1996). In that sense, the literature on the topic refers to stages that occur from the time when an invention is created to the time when it is commercialized. Such stages include research (basic and applied), R&D, the development of prototypes and models, the acquisition of technology, and some project engineering stages (OECD, 2002, OECD/EUROSTAT, 2005; Rammer & Schmiele, 2009). Nevertheless, the stages are recursive rather than sequential, in which knowledge is developed, communicated and transferred (Robertson et al., 1997 as cited in Edwards, 2003), resulting in an "interactive process" that is common in the field of innovation and that has been used to describe intra-company and inter-company innovation activities (Rothwell et al., 1990 as cited in Edwards, 2003). Innovation activities that use that process are the main input for obtaining innovations (results) and they are also essential in building knowledge and technological capabilities in the company.

4. Innovation activities (R&D and TKT) as determinants of business innovation and R&D activities

According to the Oslo Manual, "innovation activities are all those scientific, technological, organizational, financial and commercial steps, including investment in new knowledge, which actually lead to, or are intended to lead to, the implementation of innovations," (OECD/EUROSTAT, 2005, pg. 91), which implicitly includes R&D. According to the Frascati Manual, R&D includes basic research, applied research, and experimental development[1]. Other activities that are not technically R&D activities but that are carried out in R&D projects are also included in this category[2].

---

[1] This article focuses on experimental development because the characteristics proper to the Colombian economy and more specifically to the Provincial Department of Caldas, mostly comprising traditional manufacturing industry micro-companies and small and medium-sized companies (SME), do not allow assuming that the companies on their own or in association with other companies or institutions conduct basic or applied research (or if they do, it is solely residual).

[2] (See OECD, 2002, pg. 30-33).

Among the activities related to knowledge flow and technology (Oslo Manual, 2002 and Bogotá Manual, 2001) that are carried out in a business network environment, the activities related to machinery and equipment, specialized software, technical and technological information, and the dissemination of R&D and innovation results (Arvanitis et al., 2007) are included under technology and knowledge transfer (TKT) as determinants of the innovation and R&D proper to a company.

In summary, this article analyses the innovation activities (R&D and TKT) that stem from business network links and that are incentives for company R&D as well as for business innovation (product, process, and administrative innovations).

The influence of business network R&D on company innovation and R&D activities has been proposed in the empirical literature on the topic. By observing the systemic interactions that can favor or hinder innovation activities in the four regional innovation systems in Italy, Evangelista et al. (2002) found that there were differences in the level of importance given to R&D activities and to non-R&D activities at a company level. By studying the impact of relational capital on innovation in urban areas and non-urban areas and in the industrial districts of the Emilia Romagna, Capello (2002) found that relational capital had an impact on company innovation activities, which mostly benefited the large production companies in the district. Capello also found that relational capital had a positive impact on product innovation in small companies that operate in specialized sectors as they achieve synergies and cooperation with one another. The above was later reinforced by Capello & Faggian (2005) who established a positive relation between relational capital and company innovation activity.

Taking the R&D expenses of a company in a network context as a reference, Filatochev et al. (2003) found a positive significant relation between the intensity of R&D activities in firms and the industry concentration. Capello & Faggian (2005) affirm that the "physical proximity" between firms plays a crucial role in the increase of a firm's innovation capability, especially thanks to knowledge spillovers. Cassiman & Veugelers (2006) suggest that internal R&D activities and knowledge acquisition and innovation activities are complementary; regarding basic R&D, they suggest the importance of universities and research centers as sources of information for innovation processes.

Analyzing the determinants of R&D cooperation between innovative firms (foreign and domestic) and universities and public knowledge institutions, Van Beers et al. (2008) studied small economies (The Netherlands and Finland) and found that in The Netherlands foreign firms were less involved in cooperating with public knowledge institutions than domestic firms were whereas the opposite occurred in Finland. Both countries proved that spillovers have a positive effect on the probability of cooperating with universities and public knowledge institutions. That aspect was highlighted as having a positive impact on company innovation and R&D.

Taking the above into account, the following hypotheses are proposed:

H1: In a localized business network environment, as concerns R&D activities inter-company relations and company-institution relations have a positive impact on each company's innovation (product, process and administrative innovation).

H2: *In a localized business network environment, as concerns R&D activities inter-company relations and company-institution relations have a positive impact on each company's R&D activities (input for innovation).*

In business network exchanges, TKT is fundamental to disseminating and absorbing innovations (Banyte & Salickaite, 2008); therefore, it is fundamental to a company's innovation performance (Arvanitis et al., 2007; Capello, 2002; Evangelista et al, 1997; Lin & Chen, 2006). Furthermore, it contributes to each company increasing its R&D activities, creating its own innovations, and decreasing its dependence on ideas and technologies developed by others (Rammer & Schmiele, 2007). The literature on TKT refers to formal and informal exchanges (Allen et al., 2007) that take place among the personnel of the companies and institutions that are part of the network. That aspect constitutes relational capital (Capello, 2002; Capello & Faggian, 2005).

TKT has been studied as technology acquisition (Bin, 2008) or as the absorptive capability for technology (Fabrizio, 2009). Nevertheless, it is worth considering that TKT is not always the product of strictly commercial transactions; TKT occurs by employing diverse strategies in the form of joint research projects (or technology development projects), joint training projects (Arvanitis et al., 2007) or mutual support contracts for innovation in companies where public institutions play an important role. TKT also occurs in the joint use of technical infrastructure and laboratories, among others (Arvanitis et al., 2007; Filatochev et al., 2003).

Some empirical papers illustrate how inter-company TKT in localized networks influences company innovation and R&D activities. Capello (2002) found that scientific knowledge spillovers generated by universities and R&D centers influence business innovation activities. Small firms in non-urban areas use that knowledge in particular to the best advantage given such companies' production specialization. Lin & Chen (2006) evidenced that industry network knowledge integration has a positive effect on the process of developing new products.

By studying two Spanish clusters, Martínez & Céspedes (2006) found that companies used their relation with regional associations to obtain knowledge that contributes to their capabilities, even though there was no significant relation with innovation per se. Arvanitis et al. (2007) found that business innovation development indicators had a positive relation with TKT although at different levels of significance; process innovation and product innovation presented the best results in terms of the relation considering public research organizations. Knudsen (2007) found that inter-company relations had an effect on the success of new product development in the form of more frequent customer participation. He also found that companies had a tendency to make alliances with other companies in the same industrial sector and mentioned the danger that that involves. Indeed, he argued that the knowledge contributed was similar; therefore it hindered the possibility of developing radical product innovations. He further found that customer, university and competitor participation had a significant negative influence on innovation development, which he defined as an apparent paradox[3]. Finally, he concluded that the set of supplementary and

---

[3] There are two explanations for this apparent paradox: 1) average customers are incapable of articulating their needs regarding advanced technology products and 2) average customers are incapable of conceiving ideas beyond their own experience (Knudsen, 2007, pg.117).

complementary knowledge with external partners for new product development has a positive effect on innovative performance. Based on the above, the following hypotheses are proposed:

H3: In a localized company network environment, inter-company and company-institution technology and knowledge transfer has a positive effect on company innovation (product, process, administrative innovation).

H4: In a localized company network environment, inter-company and company-institution technology and knowledge transfer has a positive effect on company R&D activities (input for innovation).

5. Methodology

The empirical studies reviewed above regarding the topic of business networks show the wide variety of technologies and techniques available for use (Vom Hofe & Chen, 2006 and Wolfe & Gertler, 2004). Dhose and Soltwedel (2006) drew a similar conclusion upon reviewing the papers presented in the Workshop on Spatial Econometrics in April 2005. Those authors stated that the topics of the most important papers presented – which analyzed what they called "innovation clusters" – had very different points of view and employed very different methodologies.

Nevertheless, there are papers that classify research into certain typologies of techniques. One of them is the classification established by the OECD team specialized in industrial clusters; it identifies five categories of analysis techniques: input–output tables, the innovation interaction matrix, the graph theory (or network analysis), the correspondence analysis (quantitative studies that use statistics and econometrics techniques), and case studies (OECD, 1999b). Wolfe and Gertler (2004) proposed three methodology perspectives for studying clusters: statistic analysis tools with different levels of sophistication for measuring the degree of grouping in a local or regional economy, case studies for an comparative analysis of an individual cluster or a comparative analysis of a group of clusters, and the analysis of public policy and strategies specifically designed for promoting the establishment and growth of a cluster or of a group of clusters in a location or region. This last methodology is frequently combined with quantitative studies and case studies.

Based on the above, to establish the methodology to be used in this work, the author analyzed 64 empirical studies and found that the various authors had used one of the three methodologies mentioned above. Table 1 presents a summary of the methodologies most commonly used in the mentioned analysis.

The table above shows a general tendency towards using combined methodologies, seen in 47% of the studies reviewed, 20% of which used the combination CA– CS that implies combining the data (qualitative and quantitative) obtained through field work. Combining techniques can lead to a more in-depth analysis of the set of events and phenomena that take place in business networks and that cannot be successfully explained by data gathered from secondary sources, which is common in studies that solely use correspondence analysis (quantitative studies). To be consistent with the literature on the topic, the analysis in this article combines case studies with quantitative techniques to understand gathered data.

METHODOLOGY USED	THEORETICAL APPROACHES IN THE ANALYSIS OF BUSINESS NETWORKS						TOT.	%
	AGGLOMERATION ECONOMICS		CLUSTER AND INDUSTRIAL DISTRICTS		NETWORKING			
	#	AUTHORS	#	AUTHORS	#	AUTHORS		
ONLY CA	8	Alecke et al., 2006; Alonso-Villar et al.,2004; Blasio & Di Addario, 2005; Callejón, 1998; Ciccone, 2001; Feldman & Audretsch, 1999; Le Bass & Miribel, 2005; Viladecans-Marsal, 2004	3	Baptista & Swann, 1998; Beaudry & Breschi, 2003; McDonald et al., 2006	2	Hagedoorn & Duysters, 2002; Minguzzi & Passaro, 2000	13	20,3
ONLY CE	1	Mun y Hutchinson, 1995	11	Eraydin & Amartli-Köroglu, 2005; Feldman et al., 2005; Heath, 1999; Khan & Ghani, 2004; Lagendijk & Charles, 1999; Nadvi, 1999; Nadvi & Halder, 2005; Nassimbeni & Sartor, 2005; Perdomo & Malaver, 2003; Power & Lundmark, 20004; Vega-Rosado, 2006	8	Benson-Rea & Wilson, 2003; Biggiero, 2001; Carbonara, 2002; Huggins, 2000; Lechner y Dowling, 2003; Pöyhönen y Smedlund, 2004; Steinle et al., 2007; Yogel et al., 2000	20	31,2
OTHERS					1	Beesley, 2004	1	1,5
I/O – GT			1	Hauknes, 1999			1	1,5
I/O – CA	1	Trueba y Lozano, 2001	1	Vom Hofe & Bhatta, 2007			2	3,1
I/O – CE			4	Bishop et al., 2000; Chaminade, 1999; Oliveira & Fensterseifer, 2003 ; Roelandt et al., 1999;	1	Marceu, 1999	5	7,8
GT – CA					1	Giuliani, 2007	1	1,5
GT – CE					2	Macías, 2002; Carrasco & Albertos, 2006	2	3,1
CA – CE	4	Gordon & McCann, 2000; O'Donoghue y Gleave, 2004; Tuan & Ng, 2001; Zheng, 1998	6	Hervás et al., 2007; Hu et al., 2005; Lai et al., 2005; Mezquita & Lazzarini, 2006; Nassimbeni, 2003; Oyelaran-Oyeyinka & Lal, 2006	3	Bell, 2005; Malewicki, 2005; Mas-Ruíz, 2000	13	20,3
OTHER COMBINATIONS	4	Black et al., 2004; Davis & Weinstein, 1999; Frenken et al., 2005; Rosenthal y Strange, 2003			2	Johannisson & Ramirez-Pasillas, 2002; Reid et al., 2007	6	9,3
TOTAL	18		26		20		64	100

I/P: Input – output; GT: Graph theory; CA: Correspondence analysis; y CE: Case studies

Table 1. Summary of the Methodologies Used, by Theoretical Approach

5.1 Population and sample

The study consisted of 101 companies in the tools manufacturing sector (ISIC[4]); they represent the total population in the Provincial Department of Caldas in Colombia. The information was obtained from the Manizales Chamber of Commerce (CCM is the Colombian acronym), the Colombian National Industrialists Association (Asociación Nacional de Industriales - ANDI), the Colombian Small Industrialists Association (Asociación Colombiana de Pequeños industrials - ACOPI) and the Manizales telephone directories. Those databases were compared, to obtain the population under study. The population is distributed in three links of the tool cluster value chain: suppliers, core companies (tool manufacturers), and customers. Among those companies, 90% are micro-companies or small companies and 97% of them are part of forward and backward linkages, which enabled analyzing the networks and identifying their impact on the link nucleus (see Table 2).

FEATURES	(%)
Type of link	100
Suppliers	36
Main companies	3
Customers	61
Company size	100
Micro	64
Small	26
Medium	4
Large	6

Table 2. Tool Cluster General Data

5.2 Measuring

The study involved four company factors: innovation (product, process and administrative innovations), R&D activities (experimental development, innovation projects, use of infrastructure), links for inter-company and company-institution R&D activities (experimental development, innovation projects, use of infrastructure), and technology and knowledge transfer (machinery and equipment, specialized software and technical information, and dissemination of research results). To adjust the models, the control variable *company size* (number of employees) was included. The dependent variables as well as the independent variables are dichotomic variables (Jensen, et. al. 2007; Knudsen, 2007; Rammer & Schmiele, 2009). They are defined and operationalized in Chart 1.

5.3 Validity and reliability

The instruments used in this research have been adapted from prior research papers (Capello, 2002; Capello & Faggian, 2005; Cassiman & Veugelers, 2002; Earydin & Amartli-Köroglu, 2005; Filatochev et. al., 2003; Jensen et. al. 2007; Johansson & Karlsson, 2007; OCDE/EUROSTAT, 2005; OCDE; 2002). However, to establish the validity and

---

[4] International Standard Industrial Code. This group manufactures knifes, hand tools and hardware store items.

VARIABLE	OPERACIONALIZATION
Business innovation	The company has made innovations (product, process, administrative) in the last five years. Yes 1, No 0
R&D in the company	The company carries out R & D (experimental development, innovation projects, use of infrastructure). Yes 1, No 0
Links for R&D	The company partners or partnering with other local actors to perform R & D (experimental development, innovation projects, use of infrastructure). Yes 1, No 0
Inter-company TKT	The company has or has had technology and knowledge transfer (machinery and equipment, specialized software and technical information and dissemination of research results) with other local. Yes 1, No 0
Company size	Number of employees

Chart 1. Variable Operacionalization

reliability of the instrument, the following analyses were made: content validity by experts on the topic and a pilot test involving ten companies; Cronbach's alpha, to evaluate instrument consistency and proposed dimension consistency, obtaining $\alpha > 0.7$; and a correlation analysis, to establish convergent validity, obtaining significant correlations ($p < 0.05$) and a concordant theoretical value ($\rho > 0.7$) (Nunally, 1978). The above enabled establishing that the dimensions proposed for measuring inter-company business innovation, R&D activity links, and technology and knowledge transfer presented homogeneous measurements.

5.4 Comparison of hypotheses

First, a descriptive analysis was made in order to make an exploratory identification of the aspects related to the variables studied in the tool cluster. Also, a cluster analysis made with the maximum verisimilitude method using the criteria of "closest neighbor" and "unit of measure *lambda*, to identify the percentage of companies that develop innovations, by type of innovation (product, process and administrative). Based on the cluster analysis, a contingency analysis was made, to establish the realization indicator for the links defined in the independent variables, according to the cluster.

For comparing the hypotheses, a logistic regression analysis was made using the stepwise logistic regression method, considered appropriate for analyzing dichotomic variables, measuring impact, and adjusting variables. The Hosmer-Lemeshow test was run, to prove the specificity of the regression model. A linktest was run, to prove that the logit models are a linear combination of the dependent variables and independent variables. An analysis of estimated correlations was made, to determine that there was no correlation among the explanatory variables. And a marginal effect analysis was made, to establish incidence in terms of the probability of the independent variables / dependent variables.

6. Findings

First, the article presents a general profile of the sample companies according to study variables. The results indicate that 36% of the companies had developed innovations (product, process and administrative) and that 26% had carried out R&D activities; only 25% had links for the R&D activities and 38% had participated in technology and knowledge transfer (see Tables 3 and 4).

Business innovation	Inter-company TKT		Business innovation	Links for R&D	
	NO	YES		NO	YES
NO	54%	10%	NO	62%	2%
YES	8%	28%	YES	13%	23%

Table 3. Contingency Analysis among Business Innovation, Inter-company TKT, and Links for R&D

R&D in the company	Inter-company TKT		R&D in the company	Links for R&D	
	NO	YES		NO	YES
NO	56%	18%	NO	68%	6%
YES	6%	20%	YES	7%	19%

Table 4. Contingency Analysis among R&D in the company, Inter-company TKT, and Links for R&D

For the purpose of determining company behavior regarding the variable "innovation" considering the three types of innovation studied, a hierarchical cluster analysis was made, using as a reference the "closest neighbor" categories and *Goodman and Kruskal's lambda*[5] as measurement interval, the latter commonly used for dichotomic variable analysis. That led to identifying two types of business conglomerates in the population, according to the innovation.

The first type was comprised of 79 companies with low levels of process innovation and administrative innovation and with no product innovation; it was called low innovation level conglomerate. The second type was comprised of 22 companies with a high level of process innovation (77%), administrative innovation (50%), and product innovation seen in all of the companies; it was called high innovation level conglomerate (see Figure 1 and table 5).

TYPES OF BUSINESS INNOVATION	CONGLOMERATE 1	CONGLOMERATE 2
process Innovation	0,1	0,77
product innovation	0	1
organizational Innovation	0,08	0,5

Table 5. Business Innovation Conglomerates

---

[5] This type of interval is used taking into account that the analysis variables are dichotomic.

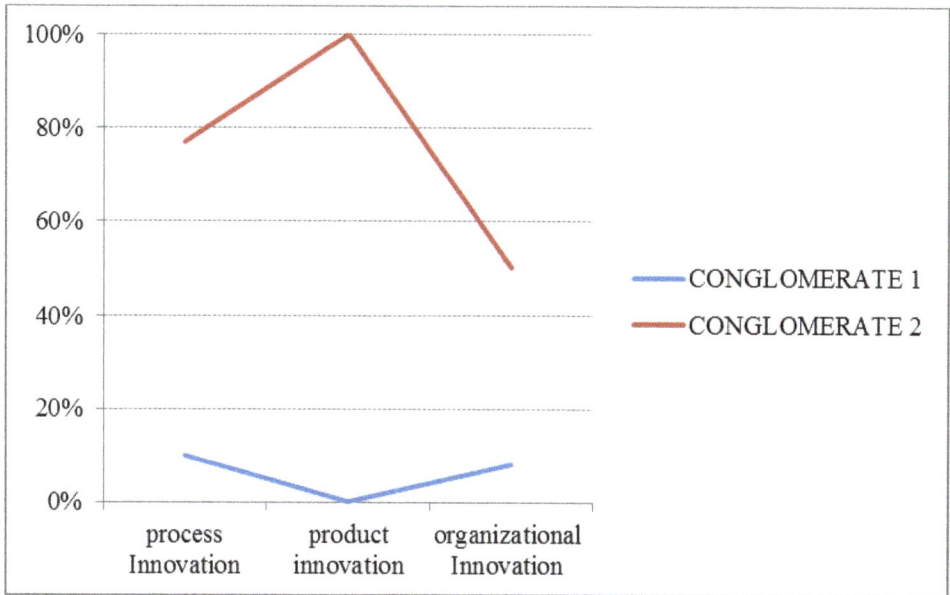

Fig. 1. Business Innovation Conglomerates

Using the data above and through a contingency analysis, the high innovation level conglomerate (Conglomerate 2) presents a higher inter-company association index for R&D activities (86%) and a higher TKT index (95.5%) than the low innovation level cluster (Conglomerate 1) (see Table 6).

INDEPENDENT VARIABLES	CONGLOMERATE 1	CONGLOMERATE 2
Links for R&D	7.6%	86.4%
Inter-company TKT	21.5%	95.5%

Table 6. Contingency Table of Innovation conglomerates and Links for R&D and inter-company TKT

To compare the hypotheses, a logistic analysis was made. For the first two hypotheses, the author studied the incidence of inter-company and company-institution cooperation on carrying out R&D activities for business innovation (H1) and on each company's R&D activities (H2). For the third and fourth hypothesis, the author studied the effect of inter-company and company-institution technology and knowledge transfer (TKT) on business innovation (H3) and on each company's R&D activities (H4).

The results showed that both H1 and H2 have a positive significant effect (p $value$ < 0.001); such findings validate accepting both of those hypotheses. Likewise, both H3 and H4 have a positive significant effect (p $value$ < 0.001), which means that they can be accepted (see Table 7 and 8).

Variables	BUSINESS INNOVATION								
	Model 1			Model 2			Model 3		
	β	Z	Sig.	β	Z	Sig.	B	Z	Sig.
Control Var.									
Constant	-1,440441	-4,36	***	-1,990311	-5,31	***	-2,167	-5,09	***
Size	0,0611753	2,53		0,025925	1,48		0,03373	1,38	
Dependent Variable									
Links for R&D				3,748522	4,54	***			
Inter-company TKT							2,39628	4,28	***
Lr χ2	26,5		***	58,77		***	47,05		***
Pseudo R2	0,2014			0,4467			0,3576		
Goodnes fit				57,4			76,86		

Table 7. Logistic Regression Analysis for the Tool Business Network – Business innovation

Variables	R&D IN THE COMPANY								
	Model 1			Model 2			Model 3		
	β	Z	Sig.	β	Z	Sig.	B	Z	Sig.
Control Var.									
Constant	-1,852675	-5,52	***	-2,847551	-5,63	***	-2,4315	-5,46	***
Size	0,0401128	2,42	*	0,020963	1,5		0,02583	1,71	…
Dependent Variable									
Links for R&D				3,318111	4,85	***			
Inter-company TKT							1,64242	2,77	**
Lr χ2	27,31		***	55,41		***	35,33		***
Pseudo R2	0,2371			0,481			0,3067		
Goodnes fit				46,18			41,78		

Table 8. Logistic Regression Analysis for the Tool Business Network – R&D in the company

To complement the analyses above, a marginal effect analysis was made, to establish to what degree association for R&D and TKT activities affects business innovation and company R&D activities. On one hand, the analysis showed that a percentile increase in inter-company and company-institution cooperation for carrying out R&D activities generates an increase of 0.68% in innovation and an increase of 0.66% in each company's R&D activities. On the other hand, a percentile increase in inter-company and company-institution TKT generates an increase of 0.53% in innovation and an increase of 0.35% in each company's R&D activities. Upon comparing those results with the pseudo-coefficient of determination, the conclusion may be drawn that the models in which there is inter-company and company-institution cooperation for carrying out R&D activities present a greater fit than the models in which the variable inter-company and company-institution technology and knowledge transfer (TKT) is present (see Table 9).

VARIABLES	Business innovation		R&D in the company	
	$\delta y/\delta x$	Sig.	$\delta y/\delta x$	Sig.
Links for R&D	0.68%	***	0.66%	***
Inter-company TKT	0.53%	***	0.35%	**

** $p\ value$ <0,01; *** $p\ value$ <0,001

Table 9. Marginal Effects of the Independent Variables on Company R&D Activities and on Business Innovation

The results obtained from applying the logit model as discussed above were tested, to verify the goodness of fit. To do so, the level of significance[6] was identified for the model and Wald's linearity analysis parameters[7] were applied, plus an error term distribution analysis[8] was made. Likewise, the non-existence of multicolinearity and heteroskedacity was identified.

7. Discussion and conclusions

Upon reviewing the literature that studies inter-company and company-institution relations, three main theoretical approaches were studied (agglomeration economies, clusters and industrial districts, and networking), which have guided the research on such relations and that, given the convergence in the theoretical orientation of such approaches, as well as the perspective employed in each research paper, the notion of "business networks" can be understood as the common unit of analysis. In that sense, the conclusion can be drawn that there is a consensus in the literature on the topic regarding the positive effect of inter-company and company-institution relations in a network environment on business performance, especially because they result in greater possibilities of knowledge development, broadened company's innovation process capabilities, and better results in process, product, and administrative innovations.

There was evidence in the tool cluster in the Provincial Department of Caldas in Colombia that the percentage of companies that have developed innovations and carried out R&D

---

[6] The Lr χ^2 *(p value < 0.001)* test was run.
[7] The parameters were identified as being linear and consistent *(p value>0.05)*.
[8] The Roc Curve Graph was prepared and the sensitivity and specificity graphic analysis was made, as well as the Hosmer-Lemeshow test *(p value>0.001)* that proves the non-existence of stochastic perturbations.

activities is relatively low and, therefore, the author observes that the percentage for establishing relations for R&D and TKT is also low. That fact can be explained by the composition of the companies. Indeed, most fall into the category of SMEs with limited capabilities for carrying out those types of activities, as well as for establishing contacts with other regional stakeholders. Mahemba & De Bruijn (2003) found a similar situation in Tanzania; their explanation was that SMEs are not aware of opportunities in their midst, such as collaboration with research institutes, universities, technology centers, and the government. Nevertheless, that contradicts the findings of Barge-Gil (2009) who proposes that small firms and firms in low, medium-low, and medium sized sectors are more prone to innovation-based cooperation; he further highlights the role that suppliers play in innovation development.

The companies that stated that they had relations with other stakeholders enabled comparing the proposed hypotheses and, hence, enabled proving that inter-company relations and relations between companies and other regional stakeholders for R&D and TKT have a positive effect on the company's R&D and innovation. Therefore, the proposed hypotheses may be accepted. That fact corroborates other findings in the literature on the topic.

In general, the results obtained are consistent with the idea of "relational capital" (Capello, 2002; Capello and Faggian, 2005); nevertheless, they leave unanswered questions that may be addressed in future research. One practical suggestion for future research would be to study, in the business network under study herein, the determinants that influence the companies' low propensity to establish relations with other regional stakeholders in order to improve business innovation and even to observe other variables associated with the innovation process (Arvanitis et al., 2007) that may affect business innovation. Another future line of research may involve conducting similar inquiries in other business networks and in other geographical environments and then carrying out comparative analyses. It would be particularly interesting to study the relation between companies and support institutions (governmental and non-governmental) as that is a fundamental aspect for public policy regarding business competitiveness and regional social and economic development in countries such as Colombia.

Finally, the author proposes some other future lines of research to complement this article. Studies may be conducted inquiring what variables are the most determinant for the integral development of business networks, for example, employment is often used to evaluate productive specialization, competences or technological innovation. Yet another suggestion would be to identify the geographical contexts and sectors in which the research is focused and carry out comparative analyses.

8. References

Alecke, B.; Alsleben, C.; Scharr, F. & Untiedt, G. (2006). Are there really high-tech clusters? The geographic concentration of German manufacturing industries and its determinants, *Annals Regional Science*, Vol.40, pp. 19 – 42.

Allen, J.; James, A. & Gamlen, P. (2007). Formal versus informal knowledge networks in R&D: a case study using social network analysis. *R&D Management*, Vol.37, No3, pp. 179 – 196.

Alonso-Villar, O.; Chamorro, J. & Gonzáles, X. (2004). Agglomeration economies in manufacturing industries: The case of Spain, *Applied Economics*, Vol.36, pp. 2103–2116.

Altenburg, T. & Meyer-Stamer, J. (1999). How to promote clusters: Policy experiences from Latin America, *World Development*, Vol.27, No.9, pp. 1693-1713.

Antonelli, C. (2006). The bussines governance of localized knowledge: An information economics approach for the economics of knowledge, *Industry and Innovation*, Vol.13, No. 3, pp. 227 – 261.

Arvanitis, S.; Kubli, U.; Sydow, N. & Woerter, M. (2007). Knowledge and technology transfer (KTT) activities between universities and firms in Switzerland – the main facts: an empirical analysis based on firm-level data. *The Icfai Journal of Knowledge Management*, Vol.5, No.6, pp. 17 – 75.

Banyte, J. & Salickaite, R. (2008). Successful diffusion and adoption of innovation as a means to increase competitiveness of enterprises. *Engineering Economics*, Vol.1, pp. 48 – 56.

Baptista, R. (1996). Research round up: industrial cluster and technological innovation, *Business Strategy Review*, Vol.7, No.2, pp. 59 – 64.

Baptista, R. & Swann, P. (1998). Do firms in clusters innovate more?, *Research Policy*, Vol.27, pp. 525–540.

Barge-gil, A. (2009). *Cooperation-based innovators and peripheral cooperators: an empirical analysis of their characteristics and behavior.* Paper presented at the conference Danish Research Unit for Industrial Dynamics – DRUID. Denmark.

Beaudry, C. & Breschi, S. (2003). Are firms in clusters really more innovative?, *Economy Innovation New Technology*, Vol.12, No.4, pp. 325–342.

Becerra, F. (2008). Las redes empresariales y la dinámica de la empresa: aproximación teórica. *Innovar*, Vol.18, No.32, pp. 27 – 46.

Beesley, L. (2004). Multi-level complexity management of knowledge networks, *Journal of Knowledge Management*, Vol.8, No.3, pp. 71 – 88.

Bell, M. & Albu, M. (1999). Knowledge systems and technological dynamism in industrial clusters in developing countries, *World Development*, Vol.27, No.9, pp. 1715-1734.

Bell, G. (2005). Clusters, networks, and firm innovativeness, *Strategic Management Journal*, Vol.26, pp. 287 – 295.

Benson-Rea, M. & Wilson, H. (2003). Networks, learning and the lifecycle, *European Management Journal*, Vol.21, No.5), pp. 588–597.

Biggiero, L. (2001). Self-organizing processes in building entrepreneurial networks: A theoretical and empirical investigation, *Human Systems Management*, Vol.20, No.3, pp. 209 – 222.

Bishop, P.; Brand, S. & McVittie, E. (2000). The use of input – output models in local impact analysis, *Local Economy*, Vol.15, No.3, pp. 238 – 250.

Black, G.; Chucrch, H. & Holley, D. (2004). Empirical estimation of agglomeration economies associated with research facilities, *Atlantic Economic Journal*, Vol.32, No.4, pp. 320 – 328.

Blasio, G. & Di Addario, S. (2005). Do workers benefit from industrial agglomeration?, *Journal of Regional Science*, Vol.45, No.4, pp. 797–827.

Brenner, T. & Greif, S. (2006). The dependence of innovativeness on the local firm population: An empirical study of German patents, *Industry and Innovation*, Vol.13, No.1, pp. 21–39.

Buhman, C.; Kekre, S. & Singhal, J. (2005). Interdisciplinary and interorganizational Research: Establishing the science of enterprise networks, *Production and Operations Management*, Vol.14, No.4, pp. 493–513.

Callejón, M. (1998). *Concentración geográfica de la industria y economías de aglomeración.* Barcelona: Universidad de Barcelona Facultad de Ciencias Económicas y Empresariales. Free Press.

Callois, J. (2008). The two sides of proximity in industrial clusters: The trade-off between process and product innovation, *Journal of Urban Economics*, Vol.63, pp. 146–162.

Capello, R. (2002). Spatial and Sectoral Characteristics of Relational Capital in Innovation Activity. *European Planning Studies*, Vol.10, No.2, pp. 177 – 200.

Capello, R. & Faggian, A. (2005). Collective Learning and Relational Capital in Local Innovation Processes. *Regional Studies*, Vol.39, No.1, pp. 75–87.

Carbonara, N. (2002). New models of inter-firm networks within industrial districts, *Entrepreneurship & Regional Development*, Vol.14, pp. 229 – 246.

Carrasco, J. & Albertos, J. (2006). Redes institucionales y servicios a las empresas en el cluster cerámico de Castellón, *Scripta Nova*, X(213).

Carrie, A. (1999). Integrated clusters – the future basis of competition, *International Journal of Agile Management Systems*, Vol.1, No.1, pp. 45 – 50.

Cassiman, B. & Veugelers, R. (2006). In search of complementarity in innovation strategy: internal R&D and external knowledge acquisition. *Management Science*, Vol.52, No.1, pp. 68-82.

Chaminade, C. (1999). Innovation processes and knowledge flows in the information and communication technologies (ICT) cluster in Spain. En OECD (Edit.), *Boosting Innovation:The Cluster Approach* (pp. 219 – 242). París: OECD.

Cicconee, A. (2001). Efectos de aglomeración en Europa y en EE.UU. *Els Opuscles del Crei*, Vol.9, pp. 1 – 28.

Damanpour, F. (1987). The adoption of technological, administrative, and ancillary innovations: impact of organizational factors. *Journal of Management*, Vol.13, No.4, pp. 675-688.

Davis, D. & Weinstein, D. (1999). Economic geography and regional production structure: An empirical investigation, *European Economic Review*, Vol.43, pp. 379 – 407.

Dohse, D. & Soltwedel, R. (2006). Recent developments in the research on innovative cluster, *European Planning Studies*, Vol. 14, No.9, editorial.

Edwards, T. (2003). Innovation and organizational change: developments towards an interactive process perspective. *Technology Analysis & Strategic Management*, Vol.12, No.4, pp. 445 – 464.

Eraydin, A. & Armatli-Köroglu, B. (2005). Innovation, networking and the new industrial clusters: The characteristics of networks and local innovation capabilities in the Turkish industrial clusters, *Entrepreneurship and regional development*, Vol.17, pp. 237 – 266.

Evangelista, R.; Perani, G.; Rapiti, F. & Archibugi, D. (1997). Nature and impact of innovation in manufacturing industry: some evidence from the Italian innovation survey. *Research Policy*, Vol.26, pp. 521-536.

Evangelista, Rinaldo; Iammarino, Simona; Mastrostefano, Valeria & Silvani, Alberto (2002). Looking for regional systems of innovation: evidence from the Italian innovation survey. *Regional Studies*, Vol.36, No.2, pp. 173–186.

Fabrizio, K. (2009). Absorptive capacity and the search for innovation. *Research Policy*, Vol.38, pp. 255 – 267.

Feldman, J. & Audretsch, D. (1999). Innovation in cities: Science-based diversity, specialization and localized competition, *European Economic Review*, Vol.43, pp. 409 – 429.

Feldman, J.; Francis, J. & Bercovitz, J. (2005). Creating a cluster while building a firm: Entrepreneurs and the formation of industrial clusters, *Regional Studies*, Vol.39, No.1, pp. 129 – 141.

Feser, E. & Bergman, E. (2000). National industry cluster templates: A framework for applied regional cluster analysis, *Regional Studies*, Vol.34, No.1, pp. 1 – 19.

Filatochev, I.; Piga, C. & Dyomina, N. (2003). Network positioning and R&D activity: a study of Italian groups. *R&D Managemente*, Vol.33, No.1, pp. 37 – 48.

Frenken, K.; Van Oort, F. & Verburg, T. (2005). *Variety and regional economic growth in the Netherlands*. Paper presented to the Regional Studies Conference on Regional Growth Agenda's Aalborg, Aalborg, Denmark, may 28 – 31 .

Gibb, A. (2006). Making markets in business development services for SMEs: Taking up the Chinese challenge of entrepreneurial networking and stakeholder relationship management, *Journal of Small Business and Enterprise Development*, Vol.13, No.2, pp. 263-283.

Giuliani, E. (2007). Towards an understanding of knowledge spillovers in industrial clusters, *Applied Economics Letters*, Vol.14, pp. 87–90.

Gopalakrishnan, S. & Damanpour, F. (1997). A review of innovation research in economics, sociology and technology management. *Omega*, Vol.25, No.1, pp. 15 – 28.

Gordon, R. & McCann, P. (2000). Industrial Clusters: Complexes, agglomeration and/or social networks? *Urban Studies*, Vol.37, No.3, pp. 513 – 532.

Groenewegen, J. & Van Der Steen, M. (2006). The evolution of national innovation systems, *Journal of Economic Issues*, Vol.40, No.2, pp. 277 – 285.

Hagedoorn, J. & Duysters, G. (2002). Learning in dynamic inter-firm networks: The efficacy of multiple contacts, *Organization Studies*, Vol.23, No.4, pp. 525 – 548.

Hauknes, J. (1999). Norwegian input-output clusters and innovation patterns. En OECD (Edit.), *Boosting innovation: The cluster approach* (pp. 61 – 90). París: OECD.

Heath, R. (1999). The Ottawa high-tech cluster: Policy or luck? En OECD (Edit.), *Boosting innovation:The cluster approach* (pp. 175 – 192). París: OECD.

Hervás, J. & Dalmau, J. (2006). How to measure IC in clusters: Empirical evidence, *Journal of Intellectual Capital*, Vol.7, No.3, pp. 354 – 380.

Hervás, J.; Dalmau, J. & Canales, C. (2007). Localización y estrategia empresarial: Contraste empírico de un modelo de indicadores de competitividad en un sector industrial. Retrived from http://www4.usc.es/Lugo-XIII-Hispano-Lusas/pdf/01_ESTRATEGIA/30_hervas_dalmau_canales.pdf.

Hotz-Hart, B. (2000). Innovation networks, regions and globalization. En Clark, G.; Feldman, M. & Gertler, M. (Eds.), *The Oxford Handbook of Economic Geography* (pp. 432 – 450). Oxford: Oxford University Press.

Hu, T-S.; Lin, C-Y. & Chang, S-L. (2005). Role of interaction between technological communities and industrial clustering in innovative activity: The case of Hsinchu district, Taiwan, *Urban Studies*, Vol.42, No.7, pp. 1139–1160.

Huggins, R. (2000). The success and failure of policy-implanted inter-firm network initiatives: Motivations, processes and structure, *Entrepreneurship & Rregional Development*, Vol.12, pp. 111-135.

Jensen, R. (1996). Social issues in spatial economics, *International Journal of Social Economics*, Vol.23, Nos.4/5/6, pp. 297 – 309.

Jensen, M.; Johnson B.; Lorenz, E. & Lundvall, B. (2007). Forms of knowledge and modes of innovation, *Research Policy*, Vol.36, pp. 680–693.

Jhojannisson, B. & Ramirez-Pasillas, M. (2002). The institutional embeddedness of local inter-firm networks: A leverage for business creation, *Entrepreneurship & Regional Development*, Vol.14, pp. 297 – 365.

Johansson, S, & Karlsson, C. (2007). R&D accessibility and regional export diversity. *Annals Regional Science*, Vol.41, pp. 501 – 523.

Khan, J. & Ghani, J. (2004). Clusters and entrepreneurship: Implications for innovation in a developing economy, *Journal of Developmental Entrepreneurship*, Vol.9, No.3, pp. 221 – 238.

Knudsen, M. (2007). The relative importance of interfirm relationships and knowledge transfer for new product development success. *Journal of Product Innovation Management*, Vol.24, pp. 117 – 138.

Koch, T.; Kautonen, T. & Grünhagen, M. (2006). Development of cooperation in new venture support networks: The role of key actors, *Journal of Small Business and Enterprise Development*, Vol.13, No.1, pp. 62 – 72.

Lagendijk, A. & Charles, D (1999). Clustering as a new growth strategy for regional economies? A discussion of new forms of regional industrial policy in the United Kingdom. En OECD (Edit.), *Boosting innovation: The cluster approach* (pp. 127 – 154). París: OECD.

Lai, H-C.; Chiu, Y-C. & Leu, H-D. (2005). Innovation capacity comparison of China's information technology industrial clusters: The case of Shanghai, Kunshan, Shenzhen and Dongguan, *Technology Analysis & Strategic Management*, Vol.17, No.3, pp. 293–315.

Le Bas, C. & Miribel, F. (2005). The agglomeration economies associated with information technology activities: An empirical study of the US economy, *Industrial and Corporate Change*, Vol.14, No.2, pp. 343–363.

Lechner, C. & Dowling, M. (2003). Firm networks: External relationships as sources for de growth and competitiveness of entrepreneurial firms, *Entrepreneurship & Regional Development*, Vol.15, pp. 1- 26.

Lin, B-W. & Chen, C-J. (2006). Fostering product innovation in industry networks: the mediating role of knowledge integration. *International Journal of Human Resource Management*, Vol.17, No.1, pp. 155–173.

Macías, A. (2002). Redes sociales y clusters empresariales, *Revista Hispana para el Análisis de Redes Sociales*, Vol.1, No.6, pp. 1 – 20.

Mahemba, C. & De Bruijn, E. (2003). Innovation Activities by Small and Medium-sized Manufacturing Enterprises in Tanzania. *Creativity and Innovation Management*, Vol.12, No.3, pp. 162 – 172.

Malewicki, D. (2005). Member involvement in entrepreneur network organizations: The role of commitment and trust, *Journal of Development Entrepreneurship*, Vol.10, No,2, pp. 141 – 166.

Marceu, J. (1999). The disappearing trick: Clusters in the Australian economy. En OECD (Edit.), *Boosting innovation: The cluster approach* (pp. 155 – 174). París: OECD.

Martínez, J.& Céspedes, J. (2006). *Generación y difusión de la innovación en distritos industriales. Mi+d*, Free press.

Más-Ruíz, F. (2000). The supplier-retailer relationship in the context of strategic groups, *International Journal of Retail & Distribution Management*, Vol.28, No.2, pp. 93-106.

McCann, P. (1995). Rethinking the economics of location and agglomeration, *Urban Studies*, Vol.32, No.3, pp. 563 – 577.

McDonald, F.; Tsagdis, D. & Huang, Q. (2006). The development of industrial clusters and public policy, *Entrepreneurship & Regional Development*, Vol.18, pp. 525–542.

Mella, J.; López, A. & Yrigoyen, C. (2005). Crecimiento económico y convergencia urbana en españa. Investigación, nº 6, España: Instituto de Estudios Fiscales.

Mesquita, L. & Lazzarini, S. (2006). Vertical and horizontal relationships in an industrial cluster: Implications for firms access to global markets. Paper presented at the *Academy of Management Proceedings*, Atlanta, U.S., august 11 – 16.

Minguzzi, A. & Passaro, R. (2000). The network of relationships between the economic environment and the entrepreneurial culture in small firms, *Journal of Business Venturing*, Vol.16, pp. 181–207.

Mun, S-I. & Hutchinson, B. (1995). Empyrical analysis of office rent and agglomeration economies: A case study of Toronto, *Journal of Regional Science*, Vol.35, No.3, pp. 437 – 455.

Muscio, A. (2006). Patterns of innovation in industrial districts: An empirical analysis, *Industry and Innovation*, Vol.13, No.3, pp. 291 – 312.

Nadvi, K. (1999). Collective efficiency and collective failure: The response of the Sialkot surgical instrument cluster to global quality pressures, *World Development*, Vol.27, No.9, pp. 1605 – 1626.

Nadvi, K. & Halder, G. (2005). Local cluster in global value chains: Exploring linkages between Germany and Pakistan, *Entreprenerurship & Regional Development*, Vol.17, pp. 339 – 363.

Nassimbeni, G. (2003). Local manufacturing systems and global economy: Are they compatible? The case of the Italian eyewear district, *Journal of Operations Management*, Vol.21, No.2, pp. 151–171.

Nassimbeni, G. & Sartor, M. (2005). The internationalization of local manufacturing systems: evidence from the Italian chair district, *Production Planning & Control*, Vol.16, No.5, pp. 470–478.

Nunally, J. C. (1978). *Psychometric theory*. New York: McGraw Hill.

OECD (1999a). *Boosting Innovation:The Cluster Approach*. París: OECD.

OECD (1999b). *Managing National Innovation Systems.*París: OECD.

OECD/EUROSTAT (2005). *Oslo manual*: Guidelines for collecting and interpreting innovation data. París: OECD.

OECD (2002). *Frascati Manual: Proposed Standard Practice for Surveys on Research and Experimental Development*. París: OECD.

O'Donoghue, D. & Gleave, B. (2004). A note on methods for measuring industrial agglomeration, *Regional Studies*, Vol.38, No.4, pp. 419 – 427.

Oliveira, E. & Fensterseifer, J. (2003). Use of resource-based view in industrial cluster strategic analyisis, *International Journal of Operations & Production management*, Vol.23, No.9, pp. 995 – 1009.

Oyelaran-Oyeyinka, B. & Lal, K. (2006). Institutional support for collective learning: Cluster development in Kenya and Ghana, *African Development Review*, Vol.18, No.2, pp. 258 – 278.

Perdomo, J. & Malaver, F. (2003). *Metodología para la referenciación competitiva de clusters estratégicos regionales*. Bogotá: Centro de Investigaciones para el Desarrollo – CID, Universidad Nacional de Colombia.

Power, D. & Lundmark, M. (2004). Working through knowledge pools: Labour market dynamics, the transference of knowledge and ideas, and industrial clusters, *Urban Studies*, Vol.41, Nos.5/6, pp. 1025–1044.

Pöyhönen, A. & Smedlund, A. (2004). Assessing intellectual capital creation in regional clusters, *Journal of Intellectual Capital*, Vol.5, No.3, pp. 351 – 365.

Rammer, C. & Schmiele, A. (2009). Drivers and effects of internationalizing innovation by SMEs. *The Icfai University Journal of Knowledge Management*, Vol.7, No.2, pp. 18 – 61.

Red de Indicadores de Ciencia & Tecnología (RICyT), Organización de los Estados Americanos (OEA), Programa de Ciencia y Tecnología para el Desarrollo (Cyted) de Colciencias y Observatorio Colombiano de Ciencia y Tecnología (OCyT), (2001). *Manual de Bogotá: normalización de indicadores de innovación tecnológica en América Latina y el Caribe*. Bogotá: autores

Reid, N.; Carroll, M. & Smith, B. (2007). Critical steps in the cluster building process, *Economic Development Journal*, Vol.6, No.4, pp. 44 – 52.

Roelandt, T.; Den Hertog, P.; Van Sinderen, J. & Van Den Hove, N. (1999). Cluster analysis and cluster policy in the Netherlands. En OECD (Edit.), *Boosting innovation: The cluster approach* (pp. 315 – 338). París: OECD.

Rosenthal, S. & Strange, W. (2003). Geography, industrial organization, and agglomeration, *The Review of Economics and Statistics*, Vol.85, No.2, pp. 377–393.

Shumpeter, J. (1997) *Teoría del desenvolvimiento económico*. México: Fondo de Cultura Económica.

Slappendel, C. (1996). Perspectives on innovation in organizations. *Organization Studies*, Vol.17, No.1, pp. 107 – 129.

Steinle, S.; Holger, S. & Mietzner, K. (2007). Merging a firm-centred and a regional policy perspective for the assessment of regional clusters: Concept and application of a "dual" approach to a medical technology cluster, *European Planning Studies*, Vol.15, No.2, pp. 235 – 251.

Tikkanen, H. & Parvinen, P. (2006). Planned and spontaneous orders in the emerging network society, *Journal of Business & Industrial Marketing*, Vol.21, No.1, pp. 38–49.

Tracey, P. & Clark, G. (2003). Alliances, networks and competitive strategy: Rethinking cluster of innovation, *Grow and Change*, Vol.34, No.1, pp. 1 – 16.

Trueba, M. & Lozano, P. (2001). Las pautas de localización industrial en el ámbito municipal: Relevancia de las economías de aglomeración, *Economía Industrial*, Vol.337, pp. 177 – 188.

Tuan, C. & Ng, L. (2001). Regional division of labor from agglomeration economies' perspective: Some evidence, *Journal of Asian Economies*, Vol.12, pp. 65 – 85.

Van Beers, C.; Berghäll, E. & Poot, T. (2008). R&D internationalization, R&D collaboration and public knowledge institutions in small economies: evidence from Finland and the Netherlands. *Research Policy*, Vol.37, pp. 294–308.

Vega-Rosado, L. (2006). The international competitiveness of Puerto Rico using the Porter's model, *Journal of Global Competitiveness*, Vol.14, No,2, pp. 95 – 111.

Viladecans-Marsal, E. (2004). Agglomeration economies and industrial location: City-level evidence, *Journal of Economic Geography*, Vol.4, pp. 565–582.

Vom Hofe, R. & Chen, K. (2006). Whither or not industrial cluster: Conclusions or confusions? *The Industrial Geographer*, Vol.4, No.1, pp. 2 – 28.

Vom Hofe, R. & Bhatta, D. (2007). Method for identifying local and domestic industrial clusters using interregional commodity trade data, *The Industrial Geographer*, Vol.4, No.2, pp. 1 – 27.

Walker, G.; Kogut, B. & Shan, W. (1997). Social capital, structural holes and the formation of an industry network, *Organization Science*, Vol.8, No.2, pp. 109 – 125.

Wolfe, D. & Gertler, M. (2004). Clusters from the inside and out: Local dynamics and global linkages, *Urban Studies*, Vol.41, Nos.5/6, pp. 1071–1093.

Yoguel, G.; Novick, M. & Marin, A. (2000). Production networks: Linkages, innovation processes and social management technologies. A methodological approach applied to the volkswagen case in Argentina, *Danish Research Unit for Industrial Dynamics – DRUID working paper*, Vol.11, pp. 1 – 36.

Zhang, Y. (2007). Inter-firm networks and innovation: The difference between the horizontal and vertical type. Paper presented at the *Academy of Management Proceedings*, Philadelphia, U. S. , august 3 – 8.

Zheng, X-P. (1998). Measuring optimal population distribution by agglomeration economies and diseconomies: A case study of Tokio, *Urban Studies*, Vol.35, No.1, pp. 95 – 112.

13

Linking Process Technology and Manufacturing Performance Under the Framework of Manufacturing Strategy

Hongyi Sun
Department of Systems Engineering and Engineering Management
City University of Hong Kong
China

1. Introduction

Performance improvement is the goal of any manufacturing firms. A bunch of manufacturing practices are involved as suggested in the manufacturing strategy model. These include technologies, human resources and comprehensive programmes such as total quality management (TQM) and pull production. As a result, the linkage among various practices and performance are very complicated. Previous research in this field may have some limitations. The following part will review these limitations using TQM and AMT (Advanced Manufacturing Technology) as an example and argue the necessity of using structural equation modeling to deal with multiple variables.

First, most previous research on practice-performance linkage assumes that all practices directly contribute to the performance. Therefore, the conceptual models are mostly a one-layer model. The data analysis methods are mostly simple correlation or multiple correlation. The methodology is basically exploratory. The assumption of this research argues that practices may not all be directly correlated with performance. There may be several layers from practices to performance. Therefore, a comprehensive model based on path analysis or structural equation modeling is needed to investigate the practice-performance relationship. To specify the path-analysis model, a conceptual model is needed. In this research the conceptual framework from manufacturing strategy will be used.

Second, in previous research, the measures of practices vary from one single question to a set of questions which are grouped into a construct. It is not so common to develop constructs in AMT-performance research yet. The definition and classification of AMT are not consistent. Beaumont et al (2002) measure AMT in terms of direct (fabrication and assembly), indirect (engineering and design) and administrative (information management). Dasa and Narasimhan (2001) divided AMT into manufacturing technologies and design technologies. However, the classification of AMT is not consistent with technical definition (Groover, 1987; Goetsch, 1990; Singh, 1996; Kotha and Swamidass, 1998). In this research, AMT will be classified according to technical definition of computer integrated manufacturing (CIM).

In summary, practice-performance linkage has been mostly studied by simple or multiple correlation analysis in single areas such as technology or quality. In modern manufacturing companies, both practices as input and manufacturing performance as output are getting more and more complicated. Therefore, the relationship must be a complex one. This paper reports the research which aims to investigate this complex practice-performance linkage in a path-analysis model. The research is based on the manufacturing strategy framework. The idea is consistent with complex performance. Complex performance is described by Lewis and Roehrich (2009) in terms of the interaction between infrastructural complexity (e.g. buildings, enabling facilities, hardware) and transactional complexity (e.g. performance involving high degrees of embedded knowledge).

The paper is structure in five sections. In section two, literature on all types of practice and performance will be reviewed under the framework of manufacturing strategy and a set of hypotheses will be formulated. In section three, methodological issues such as data collection, operationalisation, validity and reliability tests and data analysis method will be described. In section four, the results will be presented. In section five, the results will be discussed and implications for practice and future research will be explored. In the final section, the research will be concluded; limitation and future research will be discussed.

2. A conceptual model and hypotheses formulation

2.1 The conceptual framework under manufacturing strategy

Manufacturing strategy is regarded as the manner in which the business unit deploys its manufacturing resources (Hayes and Wheelwright, 1984) and effectively uses its manufacturing strengths (Swamidass and Newell, 1987; Riis, 1992) to complement the business strategy. One of the themes in manufacturing strategy deals with various linkages or alignment among business objectives, manufacturing missions, manufacturing practices and performance. This paper aims to explore the relationship between manufacturing practices and performance. The key variables are practices and performance. The related variables include performance, structural decisions, infrastructural decisions, technology, and organization. The contents and possible relationships among the variables are illustrated in figure 1 and will be elaborated below.

2.2 Manufacturing performance

Under manufacturing strategy theory, manufacturing practices may not directly contribute to business performance such as market share and profitability. Their immediate contribution should be those at manufacturing levels such as cost reduction, quality improvement and shortening throughput time. Therefore, in manufacturing strategy research, business performance and manufacturing performance are distinguished (Tunalv 1991, McDermott and Stock, 1999, Sun and Cui 2002, Beaumont et al, 2002). These manufacturing performance dimensions, if being well aligned with business competitive objectives, will contribute to the achievement of business performance (Dasa and Narasimhan, 2001, Sun and Cui 2002). Therefore, there should be a corresponding relationship between manufacturing performance, manufacturing missions and business objectives. So in this research on practice-performance linkage, the performances refer to manufacturing performance. In manufacturing strategy research, manufacturing

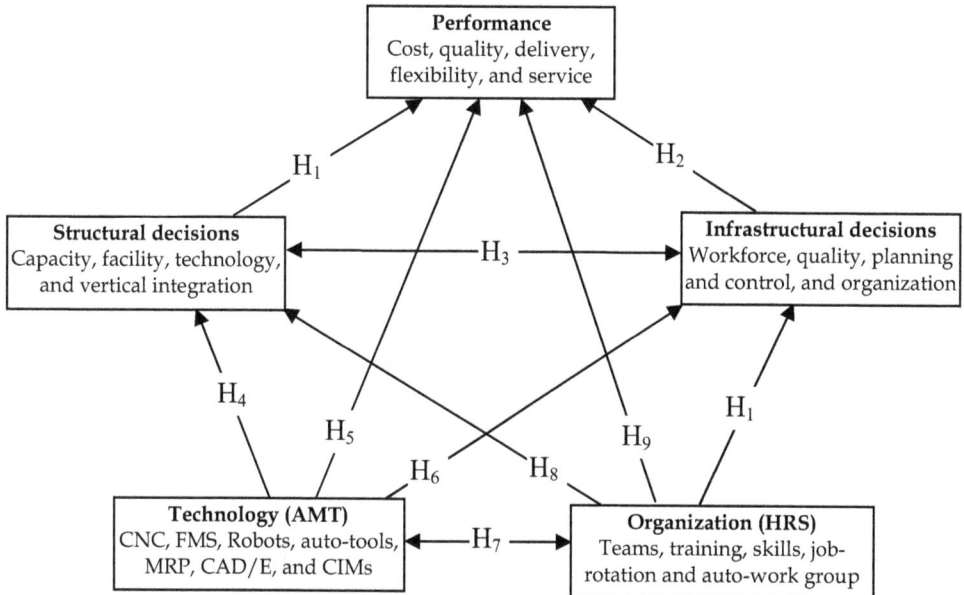

Fig. 1. A conceptual model for studying practice-performance linkage

performance should be corresponding to manufacturing missions/tasks which cover cost, quality, delivery, flexibility and service (Skinner 1969, Wheelwright 1984, Kim and Arnold 1996 etc.). The service often refers to customer satisfaction. Based on the alignment and corresponding theory, manufacturing performance can also be divided on into these five categories.

2.3 Action programmes based on structural and infrastructural decisions

Manufacturing action programs are often regarded as sets of decisions, that derive from the experience of a number of leading companies and that have proved to be successful (Schonberger, 1982; 1986; Hanson and Voss, 1993, Hanson *et al.*, 1994). They are the resources or functions that must be performed by manufacturing (Schroeder *et al.*, 1986). Because of the diversity of manufacturing decisions that must be made over time, Hayes and Wheelwright (1984) developed an organizing framework that groups them into two major categories, structural and infrastructural decisions. There is an essential agreement on this structure-infrastructure dichotomy in the literature (e.g., Leong *et al.*, 1990; Hill 1995, Tseng et al, 1999, Ng and Hung 2001). Structural decision category addresses the "bricks and mortar" decisions of capital spending. Examples of structural decisions include decisions on capacity, facility, the investment in technology, and vertical integration (Hayes and Wheelwright, 1984). Infrastructural decision category addresses more "tactical" issues, which affect the people and systems that make manufacturing work (Leong *et al.*, 1990). The infrastructural decisions may include decisions on workforce, quality, production planning and organization.

Corresponding to the above two decision areas, there are two types of action programmes. Those programmes supporting structural decisions such as increasing equipment and capacity are named as structural programmes. The programmes to support infrastructural decisions and choices are named as infrastructural programmes. Regarding contribution to performance, Hayes *et al.* (1988) suggested that infrastructure decisions were equally important as structure decisions. Performance improvement has been found positively correlated with infrastructural programs such as quality management programs, pull production systems, total productive maintenance (Cua, McKone and Schroeder, 2001), and supply chain management. Structural and infrastructural decisions are the two sides of the same manufacturing process. So they must be related to each other. Hayes *et al.* (1988) suggested that the distinction between structure and infrastructure was analogous to the distinction between computer hardware and software. The fixed, long-term and often unrecoverable investments of the firm in durable or facilities are analogous to computer hardware, while those that are more controllable by management are analogous to software. Based the contents and analysis of the two types of programmes, the following hypotheses are formulated.

H_1: Structural programs directly contribute to Performance.

H_2: Infrastructural programs directly contribute to Performance.

H_3: Infrastructural programs are positively related to Structural programs.

2.4 Technology

Under manufacturing strategy framework, technology is part of structural decisions. However, since technology has changed dramatically in the past decades years and it has very different features compared with other items such as capacity and facility etc, technology is treated separately and refers to Advanced Manufacturing Technologies (AMT).

AMT refers to those computer-aided technologies in information management, design, engineering and fabrication processes such as Computer Aided Manufacturing (CAM), Computer Aided Design (CAD) and Computer Aided Process Planning (CAPP). AMTs are the main technical components of Computer Integrated Manufacturing (CIM) systems. It is more than a group of advanced and automated technologies (Haywood, 1990). The main feature of CIM is the total integration of all manufacturing functions, including design, engineering, planning, control, fabrication, and assembly etc. through the use of computers. According to the CIM wheel model of the Society of Manufacturing Engineer (SME), there are one business and four technical components of a CIM system (Goetsch 1990). The four technical components are *planning and controlling, information resources management, product and process definition,* and *factory automation.* The four components and relevant AMTs involved have been described in details in literature (Groover, 1987; Goetsch, 1990; Singh, 1996; Kotha and Swamidass, 1998). The contents of the four components as well as their relationship with other variables will be analyzed below.

The factory automation component contains will directly influence the structural decision on the manufacturing process, especially the level of automation, new equipment implementation, capacity incensement and facility investment (Goetsch 1990, Bessant and Haywood 1988). In fact, the structural decision is called process choices in Hill's model (Hill

1995). Regarding the relationship between processes and AMT selection, there have been many similar models reported (Fix-Sterz et al 1987, p.11, Greenwood 1988, Lindberg 1990 p.12, Noori 1990, Ayres 1991, Parthasarthy & Sethi 1992). In general, for small batch and large variety job shop processes, standalone NC and MC will be suggested. For medium batch and variety, FMS is recommended. For large volume and few varieties, dedicated and automated lines are suggested. All these suggest that different processes may use different type of AMTs. In either case, the changes in process will require the changes in the technological dimensions. In other words, AMT is needed to support the implementation of structural programs for the purpose of updating manufacturing processes. The above reference leads to the fourth hypothesis.

H_4: The implementation of structural programs will be positively correlated with the utilization of manufacturing technologies.

The planning and controlling component includes such elements as planning/scheduling and controlling of facilities, materials, tools and shop floor activities. Hardware and software are available to automate each of the elements. Material Requirement Planning (MRP), as well as Manufacturing Resources Planning (MRP II), is an important concept with a direct relationship to CIM. Information resources management is the nucleus of CIM. Information, updated continually and shared instantaneously, is what CIM is all about. One of the major goals of this nucleus is to overcome the barriers that prevent the complete sharing of information among all other CIM components. The AMTs used for this purpose include Shared Databases (Shared DB), Wide Area Network (WAN), and Local Area Network (LAN). Planning and control is one of the key issues in infrastructural decision. However, it needs the support of technologies such as MRP and IT system. Re-engineering program is especially based on IT system implementation. The implementation of IT systems also needs the support of the relevant infrastructural changes. The above analysis leads to H_5.

H_5: The utilization of AMT is positively correlated with the implementation of infrastructural programs.

The need to achieve cost efficiency, quality, and flexibility is necessary, and has imposed a major challenge to the manufacturing industry in the nineties and beyond. AMT has been widely regarded as a new and valuable weapon to rise to the challenge proposed by the new market situation to manufacturing industries (Hunt, 1987; Noori, 1990). Therefore, AMT is widely regarded as the new weapon to improve manufacturing performance. This leads to the following hypothesis.

H_6: There is a positive relationship between technology utilization and manufacturing performance.

2.5 Organisational dimension

Workforce and organisation are part of the infrastructural decisions. However, the issue is different to other items such as quality, planning and control. Additionally, HRS and organisational issues have been studied intensively from AMT perspective. So the organisational issue is separated in the research. Since the scope of study is in manufacturing function, the organisation refers to work organisation on the shop floor.

Plenty of previous research was found on the changes in human resources in association with single AMTs. Lee and Leonard (1990) discovered that the Automated Guided Vehicle (AGV) in a small batch-manufacturing environment altered the nature of human work. Saraph and Sebastian (1992) reviewed many previous studies and concluded that the failure of AMT is mainly due to the implicit or explicit neglect of critical human resource factors. Gerwin and Kolandy (1992, p.215) said that AMT invites a wide range of changes in human resources management and practices. They further suggested that human resources development should be integrated with the design of new technologies in the manufacturing environment. Samson, Sohal and Ramsay (1993) argue that human resources issues such as commitment, involvement, the acceptance of changes, culture, work and skills should be considered for the successful implementation of AMT. According to these previous studies, the human resources suitable for AMT are characterised by lower division of labour, frequent job rotation, stable employment, active employees' participation, loose first-line supervision, more training, team-based work organisation, group-based incentive system (Sun 2001). Based on the requirement of the development in HRS and organisational dimension for AMT implement, the following hypotheses can be formulated.

H_7: The utilization of AMT is positively correlated with the adoption of new form of work organization.

The most influential research on organizational structure and technology was made by Woodward (1965) at Imperial College in England. The very original research was conducted through a survey of 203 British manufacturing firms (p.8). Woodward's research was carried out at the level of the work organization in the production department. The samples are purely industrial companies. Woodward found that type of production, i.e., the structural decision area, was related to a specific type of organizational structure. The found that production process was the most important factor deciding the organizational structure. The number of levels in the management hierarchy, the span of control of first-line supervisors, and the ratio of managers and supervisors to other personnel were all affected by the type of the employed production technology. Besides, the success or effectiveness of the organizations was related to the "fit" between processes and organizational structure. The successful firms of each type were those that had the appropriate structured technical systems. The theory leads to hypothesis H9.

H_8: The adoption of new work organization is positively correlated to the implementation of structural action programs.

HRS and organization is part if infrastructural decision area, there it is of course related to the infrastructural decisions and relevant action programmes to support the decision. For example, teams work, employee involvement and suggestions have been proved to be a necessary part of quality management program. Employee involvement in terms of suggestions and participation are associated with quality management activities such as quality circles and communication. Research has shown that job enrichment and task characteristics such as skill variety and autonomy are directly associated with higher work quality and employee satisfaction (Kopelman, 1986). Self-managing work teams typically produce positive results in terms of quality and costs (Beekun, 1989; Sundstrom, 1990). Teams are also proved to be useful for new product development (Sobek II et al, 1998). Therefore, it is natural to formulate hypothesis H_9 and H_{10}.

H_9: The autonomous working organization is positively correlated with performance.

H_{10}: The adoption of autonomous working organization is positively correlated with the implementation of infrastructural programmers.

The relevant variables and would-be relationships are illustrated in the conceptual model as shown in figure 1. The ten hypotheses will be tested in several models.

3. Empirical data

3.1 Questionnaire and data collection

The data for this research are from the International Manufacturing Strategy Survey (IMSS). The project was initiated by London Business School and Charlmes University of Technology in 1992. IMSS is an international research network consisting of 20 countries and 600 companies around the world, including developed countries, i.e. USA, Japan, British, Germany, and developing countries, i.e. China, Argentina, Mexico. The participant companies are in the metal products, machinery and equipment industry, i.e. the international Standard Industry Classification (ISIC) 38. For details regarding IMSS project, please refer to the book by Lindberg et al., 1998.

The research reported in this paper is based on the data from the third round of IMSS survey. Data collection methods varied from country to country. In some countries, sample selection was at the coordinators' convenience, and others used random sampling. Phone contact was followed in most of the participating countries, except for the Netherlands. The questionnaires were forwarded to participating companies via mailing, fax or on-site interview. In those countries where English is not used, the questionnaire was translated into local native languages. Participating countries sent their data to the coordinator who forwarded the final database to all participants. When this research is conducted, 282 sets of data are available.

IMSS questionnaire covers four aspects of manufacturing practices and strategies. In this research those questions that are related to practice and performance are selected. In the practice part, there are three sections, namely, technology, organization and improvement programs. The section on organization contains questions on suggestions, training, skills, teams, and job rotations. The performance section contains questions related to quality, flexibility, delivery, cost and customer satisfaction. These questions are listed in the Appendix.

3.2 Method for validity and reliability tests

Validity and reliability tests cover content validity, construct validity and reliability. *Content validity* refers to whether the items in a scale represent the contents of a theoretical construct. The content validity is based on literature review, research experiences, and case studies. The contents of technology, organization, improvement programmes and performance have all been reviewed and discussed in literature review section.

Reliability refers to the internal consistent of the items within a scale that aims to measure a theoretical construct. The most commonly used test method is internal consistency (Saraph, Benson, and Schoeder, 1989; Flynn, Schroeder and Sakakibala, 1994; Nunnally, 1978). It is estimated by using Cronbach's alpha. Peterson's (1994) summary of Cronbach coefficient shows that a value above 0.7 was thought to be sufficient in most of the situations. However, in the early stage of a research where the construct had not been well tested in previous studies, Nunnally (1967) recommended a level above 0.5 be acceptable.

Construct validity refers to whether a scale is an appropriate operational definition of an abstract variable or a construct (Nunnally 1978). It is established through the use of principal factor analysis. Factor analysis (de Vaus 1993) groups variables (i.e., single questions) into factors based on their common correlation. Those variables that are correlated with each other will be grouped together. Such a group of variables is called a factor. The grouping is based on the rotated loading coefficients. The threshold of the loading coefficients is related to the size of the sample. For example, Flynn, Schroeder and Sakakibala (1994) claim that for a sample of 100, the loading of 0.19 and 0.26 indicate significance at the 0.05 and 0.01 levels, respectively. This is based on the seminal work by Cohen (1988), who suggested that in 'soft' behavioral and management research, an effect size of 0.3 is often encountered (p.95). Based on Cohen's argument, de Vaus (1993) suggested a rule of thumb as follows: if its rotated loading coefficient is more than 0.30, then a variable will be included in the corresponding factor; if the loading coefficients for all the factors are more than 0.3, then the variable will be grouped according to the largest coefficient and conceptual analysis. As the sample size of this study is 250 (180 plus 71), with a 95% confidence level and an effect size of 0.3, the statistical power of this sample is larger than 0.95 (Cohen, 1988, p.102), which is high enough to identify inherent statistical relationships.

3.3 Construct measurement

All the questions used in this research are coded and corresponding to the questionnaire in the appendix.

3.3.1 Manufacturing performance and the latent variable

Manufacturing performance is directly measured by asking the respondents to indicate the amount of change of the performance dimensions over the past three years, with 1=strongly deteriorated and 5=strongly improved. According to the classification of manufacturing mission and performance under manufacturing strategy, five constructs/dimensions are formulated as shown in table 1. All the constructs passed validity and reliability tests. Additionally, a second level factor analysis of the five performance dimensions produces a valid and reliable performance scale. This means that a latent variable of performance exists.

3.3.2 Technology constructs and the latent variables

Based on the classification in literature, AMT is divided onto four constructs, namely, fabrication (NC, MC and FMS) assembly, design (CAD/E), information technology (IT) and integrated manufacturing with automated materials transportation and inspection. Confirmative factor analysis revealed that the FMS and NC, MC are separated into two factors which are named standalone automation and FMS, respectively. Other items passed the factor analysis. Finally five AMT constructs are identified. Their validity and reliability tests are list in table 2. Additionally, a second level factor analysis of the five technological dimensions produces a valid and reliable technology scale. This implies that there exist a latent variable of technology.

Code	Factors and items	1	2	3	4	5	Performance
	1. Quality:						0.64
D21	Manufacturing conformance	0.74					
D22	Product quality and reliability	0.72					
	2.Flexibility:						0.39
D24	Volume flexibility		0.88				
D25	Mix flexibility		0.63				
	3. Delivery:						0.78
D28	Delivery speed			0.73			
D29	Delivery reliability			0.88			
	4. Cost:						0.55
D213	Labor productivity				0.67		
D214	Inventory turnover				0.52		
D215	Capacity utilization				0.62		
D27	5. Service (customer satisfaction):					/	0.69
	Extraction Sums of Squared Loadings						
	Total	1.73	1.94	2.67	1.89	/	1.9
	% of Variance	38.64	54.23	43.46	39.45	/	40
	(Cronbach's α)	0.70	0.77	0.71	0.63	/	0.60

Table 1. Manufacturing performance constructs

Factors and items	1	2	3	4	5
1. Integrated manufacturing:					
BT15 Robots	.712				
BT16 Automated guided vehicles (AGVs)	.602				
BT17 Automated storage-retrieval systems (AS/RS)	.721				
BT19 Computer-aided in inspecting/testing/ tracking	.666				
2. CAD/E:					
BT110 CAD; CAE		.817			
BT111 CAD-CAE-CAM-CAPP		.807			
BT112 Eng'g DB, Product Data Management systems		.654			
3. IT and MRP:					
BT23 Purchasing and supply management			.884		
BT21 Material management			.867		
BT22 Production planning and control			.786		
BT24 Sales and distribution management			.760		
BT25 Accounting and finance			.730		
BT113 LAN-WAN/ Intranet / Shared databases/Internet			.551		
4. Standalone automation:					
BT13 CNC-DNC				0.80	
BT12 Machining centers				0.77	

	Factors and items	1	2	3	4	5
BT14	Automated tool change - parts loading/unloading				0.75	
BT11	Stand-alone/NC machines				0.66	
	5. FMS:					0.63
	Extraction Sums of Squared Loadings					
	Total	1.83	1.75	3.66	2.23	/
	% of Variance	45.79	58.25	59.43	55.76	/
	% of Variance	45.79	58.25	59.43	55.76	/
	(Cronbach's α)	0.60	0.64	0.86	0.74	/

Table 2. Factor analysis of technologies by CFA

3.3.3 Organsiation construct and a representative partial model

The organization part contains ten questions. Some of them were deleted since they are not relevant. Corresponding to literature review on HRS development, questions on training, skills, working in teams and job rotation are selected. Since the constructs for HRS development as discussed in the paper are not as common as AMT constructs, explorative factor analysis is used to explore all the items. It is found that the two questions related to training do not significantly related to other items. Scanning the data revealed that the data on training may have something wrong. Maybe due to different training systems, there are quite many data that are not explainable at all. For example, annual training hours are more than 10,000 hours. So questions on training are neglected. A question on labour union cooperation is also deleted since it is not a common question for all participating countries. The rest questions are analyzed and produce 3 factors which are named, working in teams, autonomous working group and suggestions, and skills and job rotation. The validity and reliability tests are shown in table 3. The construct "auto work org. & suggestions" does not pass the reliability test. Its Cronbach alphas is only 0.39, less than the minimum threshold of

Code		F1	F2	F3
		Team	Skills & rotation	Auto work org. & suggestions
B06a	Team in fabrication	**0.90**	0.15	0.06
BO6b	Team in assembly	**0.90**	-0.01	0.12
BO9	Multiple skills	0.03	**0.86**	0.12
BO10	Job rotation	0.10	**0.86**	-0.01
BO5	Suggestions	-0.06	0.00	**0.87**
C512a	Auto work org.	0.30	0.13	**0.67**
	Rotation Sums of Squared Loadings:			
	Total	1.72	1.51	1.24
	% of Variance	28.59	25.10	20.60
	Cumulative %	28.59	53.69	74.30
	Cronbach's α	0.80	0.67	*0.39<0.5*

Note: ** significant at the level of p=0.01, * significant at the level of p=0.05

Table 3. Factor analysis of human resources items by EFC

0.05. The construct is not accepted. Instead, the two items "auto work org." and "suggestions" are treated as separate variables. So there are four variables in transitional dimension, namely, autonomous working organization, suggestions, working in team and skills and rotation.

The second level factor analysis of the four variables does not produce a valid and reliable scale. Therefore these four factors cannot be treated as a latent variable in data analysis. Based on the correlation analysis, it is found that "autonomous working organization" is correlated with all other three variables and no other correlation relationships exist. So this variable will be used as a representative variable of organizational dimension while other three are linked to the representative one. In fact, the measure of autonomous working organization is a quite representative since it covers knowledge of employees, delegation, training, improvement and autonomous teams. Details will be shown it the specified models in figure 2, 3, and 4.

3.3.4 Structural and infrastructural programmes

The programmes used in this research refer to a major project aimed at producing considerable changes in the company's management practice and organization. There are fourteen improvement action programs listed in the questionnaire. These programmes cover many aspects of manufacturing improvement. However, based on manufacturing strategy framework, improvement activities can be divided into structural and infrastructural areas. Based on this concept, the programmes are divided into two groups, namely structural and infrastructural programmes as shown in table 4. These two groups of programmes both pass the validity and reliability tests as shown in table 4. This indicates that companies do no implement action programme individually, rather in a coherent and systematic way. The validity and reliability tests imply that there exist a latent variable of structural programs and a latent variable of infrastructural programs.

Code		Component	Component
	Structural programmes:		
C53A	Process automation	.767	
C51A	Updating process equipment	.763	
C511A	Equipment productivity	.667	
C58A	Process focus	.634	
C52A	Expanding manufacturing capacity	.528	
	Infrastructural programmes:		
C59A	Pull production		.717
C513A	New product development		.713
C510A	Quality improvement		.687
C56A	Restructuring supply strategy		.623
C57A	Outsourcing		.582
C514A	Environmental compatibility		.490
	Extraction Sums of Squared Loadings:		
	Total	2.296	2.46
	% of Variance	45.928	41.06
	Cumulative %	45.928	41.06
	Cronbach's α	0.69	0.77

Table 4. Factor analysis of action programmes by CFA

3.4 Structural Equation Modeling (SEM) and model fitness test

In this study, structural equation modeling (SEM) is used to test the hypothesis as well as the fitness of the whole model. SEM is a method that can be used to establish relationships among multiple variables. It has several advantages over simple correlation, such as considering the collinearity effect. It can also include any possible relationships among a set of variables. SEM is applied in the following procedures.

An initial model is specified and assessed by examining the whole model fit and individual parameter significance. Multiple criteria will be used to evaluate the whole model fitness (Hu and Bentler, 1999; Kaplan, 2000; Byrne, 2001), goodness of Fit Index (GFI) (Jöreskog & Sörbom, 1984), comparative fit index (CFI) (Bentler, 1990) and root mean square error of approximation (RMSEA) (Hu and Bentler, 1998; MacCallum and James, 2000). Rule of thumb recommended by scholars regarding the fit indexes is used to evaluate the model fit. Generally, GFI and CFI value above 0.9 are regarded as a good fit; RMSEA value less than 0.05 indicates good fit and value between 0.05-0.08 (Browne and Cudeck, 1993) represents reasonable fit. For normed Chi Square, Carmines and McIver (1981) recommended the value be below 3, but a value up to 5 also represents a reasonable fit (Wheaton et al., 1977; Marsh and Hocevar, 1985). If the model doesn't fit well, it should be re-specified. Those items whose path loading coefficients are insignificant ($\alpha>0.05$) should be deleted for further test. In case all the measure coefficients are significant ($\alpha<=0.05$), the item with smallest coefficient is deleted. The process should be one by one gradually. The process ends when the whole model satisfies all the fitness criteria and all individual measurement coefficients are significant. The evaluation criteria and standards are summarized below:

- Coefficients for all paths are significant at 0.05 level
- χ^2/df: <3 good fit, 3- 5 reasonable fit
- GFI and/or CFI: 0.9-0.95 good fit, > 0.95 superior fit
- RMSEA: <0.05 good fit, 0.05- 0.08 reasonable

4. Results

The data analysis includes the test of four models. The first model (model-1) is based the conventional simple correlation. The second model (model-2) is based on multiple correlation with performance as dependent variable and four practices as independent variable. The third model (model-3) is based on the conceptual model in figure 2, i.e., all the hypotheses paths being included. The last model (model-4) will be the model deleting the no-significant paths gradually, if any. The testing results of the four models are summarized in table 5 and presented in details below.

4.1 Model 1 based on simple correlation

In model-1, each pair of the five variables are linked separately and simple bivariate correlation is calculated. The result is shown in table 5, the column of model-1. The result shows that all the correlation coefficients are significant. Based on the results from model-1, all the hypotheses should be accepted.

Hypotheses and paths		Model-1 Simple correlation		Model-2 Multiple correlation		Model-3 SEM (Initial)		Model-4 SEM (Final)	
H_1	Structural programs → Performance	0.00	✓	0.01	✓	0.44	×	0.01	✓
H_2	Infrastructural programs → Performance	0.00	✓	0.01	✓	0.92	×	/	
H_3	Infrastructural programs → Structural programs	0.00	✓	/		0.00	✓	0.00	✓
H_4	Technology → Structural programs	0.00	✓	/		0.04	✓	0.00	✓
H_5	Technology → Infrastructural programs	0.00	✓	/		0.00	✓	0.00	✓
H_6	Technology → Performance	0.00	✓	0.35	×	0.64	×	/	
H_7	Auto work org. ⬚ Technology	0.00	✓	/		0.00	✓	0.00	✓
H_8	Auto work org. → Structural programs	0.00	✓	/		0.89	×	/	
H_9	Auto work org. → Performance	0.00	✓	0.71	×	0.28	×	/	
H_{10}	Auto work org. → Infrastructural programs	0.00	✓	/		0.00	✓	0.00	✓
SEM Model fitness indexes		n/a		$X^2=858$ $X^2/df=3.15$ CFI=0.95 RMSEA=0.088		$X^2=516$ $X^2/df=1.94$ CFI=0.98 RMSEA=0.06		$X^2=518$ $X^2/df=1.92$ CFI=0.98 RMSEA=0.057	
Model fitness test (Figure)		n/a n/a		Not (Cf., Fig.2)		Not (Cf., Fig.3)		Yes (Cf., Fig.4)	

Note: ✓: significant with p<0.05, ×: not significant with p>0.05, /: the path was not specified or deleted due to insignificance

Table 5. The path significance (p) and model fitness tests of the four models

4.2 Model 2 based on multiple correlation

However, simple correlation does to take collinearity into consideration. This is proved by the test of model-2, which is based on multiple correlation. Model-2 is specified with performance as dependent variable and the four practice variables as independent simultaneously. The SEM model fitness test shows that only two paths are significant while two others are not significant as shown in table 5, the column of model-2. Different results can be observed in the two models. According to the SEM principle, as long as there is a non-significant path, the whole model does not fit well and no conclusion should be drawn. The reason is that the interrelations among the four practice variables have not been considered yet. This interrelationship may influence the relationship among practice and performance, as will be illustrated in the model-3 and 4.

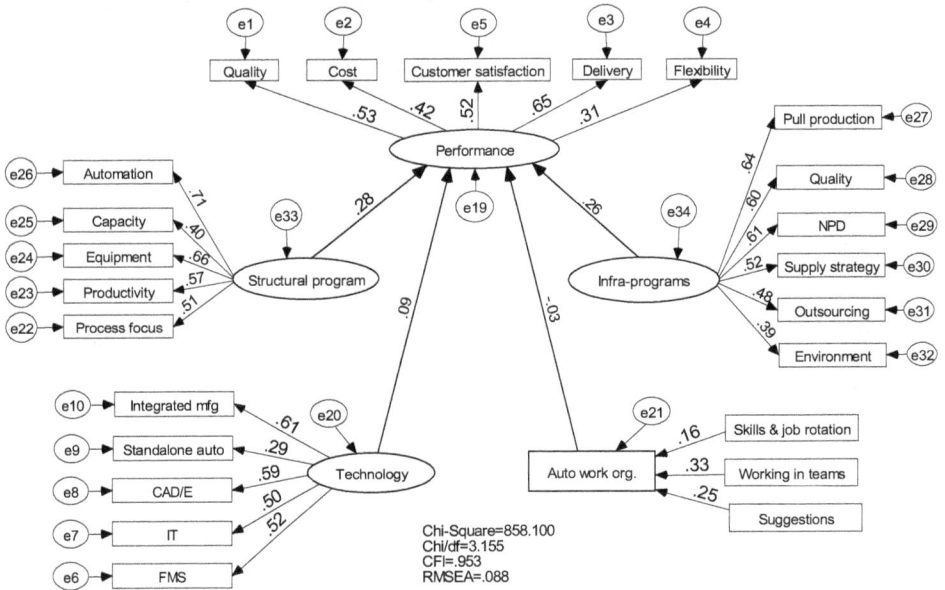

Fig. 2. The test of model-2 based on the multiple correlation principle

4.3 Mode-3 & 4 based on SEM

Model-3 is specified based on the conceptual model (cf., figure 1) of manufacturing strategy and incorporates all the possible hypotheses among the five variables. It is the initial specified model for testing. The test result of model-3 is shown in the column of model-3 in table 5. The details are shown in figure 3. The test shows that five paths are not significant. Obvious differences can be found between model-2 and model-3. In model-2, the paths for H1 and H2 are significant but not significant in model-3. According to the SEM principle, as long as there are non-significant paths, the whole model does not fit well and no conclusion can be drawn.

In the next step, the non-significant paths are removed one by one GRADUALLY and the model is tested again. The principle for removing non-significant paths should follow the principle from the least non-significant to the next least non-significant each by each. The reason is that removing one of the paths may change the path significance of other remaining paths. In this case, the path for H2 (p=0.92) should be removed from the model first. Then the path for H8 (p=0.89) is removed. The process continues until all the remaining paths are significant and the whole model fits well. Finally a model-4 is obtained as shown in figure 4. In this model, all the paths are significant and the whole model passes the fitness test as well. Therefore, conclusion can be drawn from model-4.

According to the results from model-4, it can be found that among the 10 hypotheses, four hypotheses are rejected and six are accepted, as shown the column of model-4 in table 5 as well as figure 5. Hypotheses 1, 3, 4, 5, 7 and 10 are accepted, while hypotheses 2, 6, 8 and 9 are rejected.

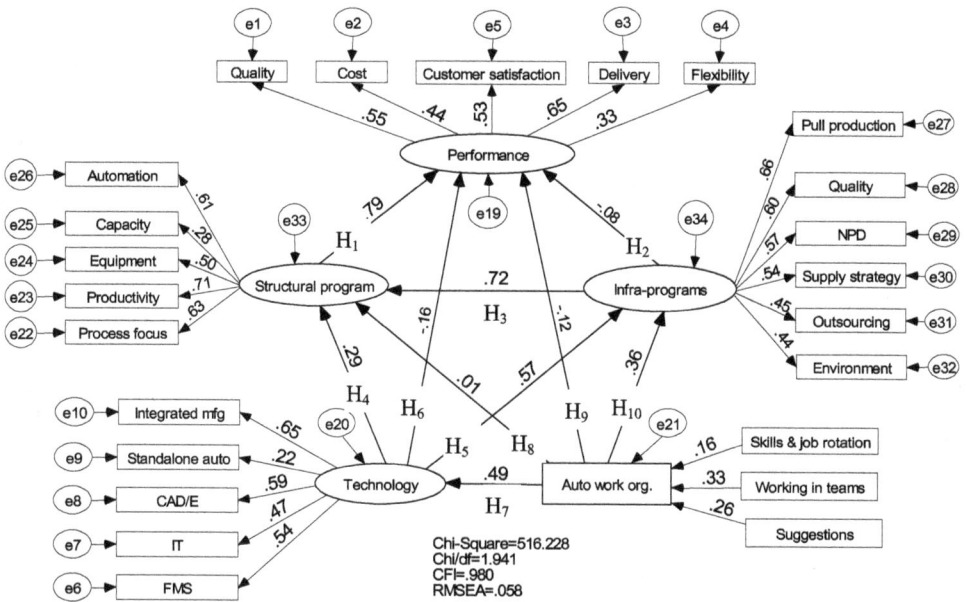

Fig. 3. The specified model (model-3) and test result

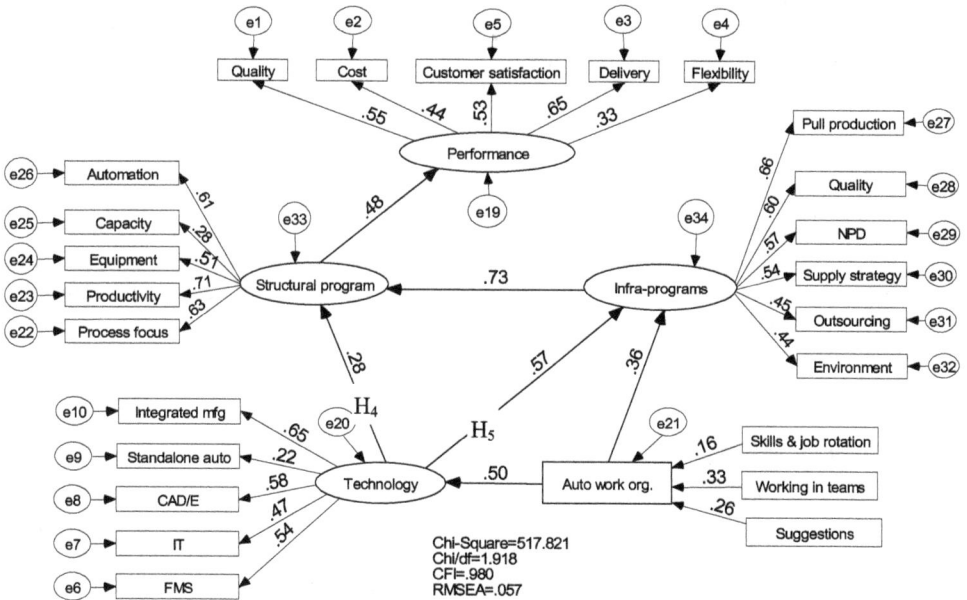

Fig. 4. The modified model (model-4) by gradually deleting no-significant paths

5. Discussions and implications

The research finds that structural programs are the only practice that directly contributes to manufacturing performance, while other three dimensions such as infrastructural programs, technology and organization contribute indirectly through structural programs. The research results trigger the following discussions.

5.1 Manufacturing process is the core

This research reveals that the improvement programs that are related to the physical process directly contribute to manufacturing performance. The structural programs work on the manufacturing process. Therefore, the process is the core and direct factor that explains manufacturing performance. This can be supported by another stream of structural research on quality management. The research based on the USMBQA framework is also a structural model and produced very valid and reliable research results. For most of the research based this model, process management is directly correlated with performance (Kaynak, 2003, Meyer and Collier 2001, Pannirselvam and Ferguson 2001, Wilson and Collier 2000). The implication is very clear. To improve the manufacturing performance, it is critical to improve the manufacturing process.

5.2 Infrastructure is the basis

It is surprising that infrastructural programs like quality management, full production etc do not directly contribute to manufacturing performance. This is opposite to many previous studies on the relationship between quality management and performance. However, if looking at the research models in previous research, it will be possible to explain the difference. In previous research, only part of the programs is investigated and other relevant factors such as structural programs are ignored. When simple correlation or multiple correlation is used in this research, the infrastructural programs are found to be positively correlated with performance (cf., model-1 in table 5). Then the conclusion will be different.

The explanation is that infrastructural programs are useful. However, they do not contribute directly to performance but through the structural programs. The path loading between infrastructural to structural programs is very high (0.73) and very significant (p=0.001). These infrastructural programs are for the establishment of infrastructure. They support the manufacturing structural technical process. The finding implies that whatever infrastructural programs to be implemented, the evaluation may not be whether it directly contributes to performance, but the requirement of the process or programs related to the structural side of the process.

5.3 Technology and organization are useful, but not directly contributing

Technology is not found to be directly correlated with performance. In the past 20 years, AMT has been widely used by manufacturing companies all-over the world. However, world-wide research found that not all AMT perform as expected. Some AMTs performs "satisfactory", but did not produce the full benefits. Other AMTs perform well on the shop floor level, while the business performances of the companies were not improved (Voss, 1988). All these problems have caught the attention of both researchers and practitioners.

Since the beginning of the 1980s, management of technology, especially implementation of AMT, has been a hot topic (Gerwin, 1982; Voss, 1988). The relationship between AMT and performance was investigated conceptually (Macbeth, 1989, p.71; Bishop and Schofield, 1989, p.44), by case studies (Sohal, 1996; Sun, Hjulstad and Frick, 1997) and by survey (Sun 2000, Small, 1998). Recent empirical research does not found that the use of AMT has direct impact on business or manufacturing performance (Swamidass and Kotha, 1998). The research by Beaumont et al (2002) intents to investigate AMT investment and performance in foreign-owned and Australian domestic companies. They did not conclude whether the AMT is significant related to performance. Sun (2000) found that little linear relationship exists between AMT and performance. The result from this research provides a reasonable explanation. Future research is needed to investigate the detail relationship between AMT and structural infrastructural action program. For example IT and supply chain management is one of the topic recently attracts researchers' attention.

5.4 Methodological implications

In this research, four different models are tested for the same set of hypotheses tests. Obvious differences are found among the four models. The differences have significant implications for selection of research methods on relationships among multiple variables. Simple correlation is simple and visual. However, its main limitation is the ignorance of the collinearity effects among variables. It can be used for identity or specify the preliminary model or explorative research at preliminary stage. Multiple correlation has the advantage of taking collinearity into consideration. However, it does not cover the interactions among the independent variables. If there are such interactions, multiple correlation results may not be reliable. Structural Equation Modeling (SEM) method is a good method since it covers collinearity effects and interactions among all the variables. As a result, it is more reliable for investigating relationships among multiple variables. More research on operations management, technology management and quality management are using more SEM to investigate multi-variant relationships (Kaynak, 2003, Meyer and Collier 2001, Pannirselvam and Ferguson 2001, Wilson and Collier 2000).

6. Conclusions, limitations and future research

The research in this paper has investigated the complex relationship among manufacturing practice and manufacturing performance. It is based on a structural model that incorporates all the possible linkages among practices and performance. The research may have the following contribution to the literature on practice-performance linkage. First, the research is based on the conceptual framework of manufacturing strategy, therefore, the model prevents from ignoring any possible linkages. Second, the data analysis is conducted with all available methods so that differences and limitations of simple and multiple correlation analysis are identified. Finally, the research produces several different results which are worthwhile to be considered in research in operations management.

The main message from this research is that not all practices may directly contribute to performance. It is the structural programmes that directly contribute to performance. Whatever other programs or technologies or organizational practices to be implemented, the final goal is to improve the manufacturing process. If the process is not improved, the contribution of other practices may not be realized.

Since the research aims to be comprehensive and holistic, the scope of the paper is pretty wide. The ten hypotheses may not be fully discussed conceptually. The implications are not fully explored for each sub-relationship. Page and words limitation may also contribute to this weakness. However, in future research which looks at a sub-relationship, for example, between technology and structural programs, the conceptual part should be enhanced.

Some of the sub-relationships have been well studied. For example, the relationship between technology and HRS/organization has been studied insensitively in the past decades. However, future research may include the following topics, the relationship between technology and structural programmes, the relationship between technology and infrastructural programs, as well as the relationship between structural and infrastructural programs.

The research provides a conceptual model and data analysis approach for investigating practice-performance relationships. Triangulation research based on the model is welcomed and appreciated to cross-proof the validity of the research method. Based on this method, a series of comparative studies can be conducted, for example, between mass and job-shop process, between Small and Media Enterprises (SME) and larger companies, and between developed and developing countries.

7. Appendix: Questions

PT3. Please indicate to what extent your activity uses one of the following process types: (indicate percentage of total volume)

Process type	
one of a kind	*BPT3a* %
batches	*BPT3b* %
mass production	*BPT3c* %
	100 %

T1.   Please indicate to what extent the operational activity is performed using the following technologies:

		No use			High use	
Stand-alone/NC machines	*BT11*	1	2	3	4	5
Machining centres	*BT12*	1	2	3	4	5
CNC-DNC	*BT13*	1	2	3	4	5
Automated tool change - parts loading/unloading	*BT14*	1	2	3	4	5
Robots	*BT15*	1	2	3	4	5
Automated guided vehicles (AGVs)	*BT16*	1	2	3	4	5
Automated storage-retrieval systems (AS/RS)	*BT17*	1	2	3	4	5
Flexible manufacturing/assembly systems – cells (FMS/FAS/FMC)	*BT18*	1	2	3	4	5
Computer-aided inspection/ testing/ tracking	*BT19*	1	2	3	4	5
Computer aided design/engineering (CAD; CAE)	*BT110*	1	2	3	4	5

Integrated design-processing systems (CAD-CAE-CAM-CAPP)	BT111	1	2	3	4	5
Engineering databases, Product Data Management systems	BT112	1	2	3	4	5
LAN-WAN/ Intranet / Shared databases/Internet	BT113	1	2	3	4	5

T2.   To what extent are the following management areas software supported through the use of <u>Enterprise Resource Planning</u> systems?

		No use			High use	
Material management	BT21	1	2	3	4	5
Production planning and control	BT22	1	2	3	4	5
Purchasing and supply management	BT23	1	2	3	4	5
Sales and distribution management	BT24	1	2	3	4	5
Accounting and finance	BT25	1	2	3	4	5
Human Resources management	BT26	1	2	3	4	5
Project Management	BT27	1	2	3	4	5
Other (please specify____BT28a____)	BT28b	1	2	3	4	5

O1.   At the end of the last fiscal year, in your business unit you had:
a. __BO1a1__ employees in total, of which__ BO1a2 __were salaried employees,
b. __BO1b__ % of salaried employees belonging to a union or similar workers associations.
c. __BO1c__ % of employees in total who are temporary (i.e. not permanent) workers

O2.   How many organizational levels do you have (plant manager to first-line supervisors)?__ BO2__

O3.   How many employees are under the responsibility of one of your line supervisors (on average)?
__BO3a__ in <u>Fabrication</u>   __BO3b__ in <u>Assembly</u>

O4.   a. What proportion of your direct employees are payed on incentives? __BO4a__ % employees
b.   Among which   (please select all relevant alternatives)
Work Group incentive   ☐ **BO4b1,** Individual incentive   ☐   **BO4b2,**   Companywide incentive   ☐ **BO4b3**

O5.   To what extent are your employees giving <u>suggestions</u> for product and process improvement?

No suggestions			**High number of suggestions**	
1	2	3	4	5

O6.   a. What proportion of your total work force work in teams? (*):
in <u>Fabrication</u> **BO6a** % in <u>Assembly</u> __BO6b__ %

O7.   How many hours of training are given to new production workers? __BO7____ hours per new worker

O8.   How many hours of training per year is regularly given to regular work-force?
__BO8__ hours per employee

O9.   How many of your <u>production workers</u> do you consider as being <u>multi-skilled</u>?(*)
        ___<u>BO9</u>___ % of total number of production workers.
        (*) Note: A multi-skilled operator is skilled in several operational tasks.

O10.   How <u>frequently</u> do your production workers <u>rotate</u> between jobs or tasks?

Never				**Frequently**
1	2	3	4	5

C5   This question explores the <u>action programs</u> * to which your company is now
        devoting high resource and innovation effort and on which is concentrated the
        management focus and commitment. Please indicate whether the program has
        been undertaken within the last three years. (* *By <u>action program</u> is meant a major
        project aimed at producing considerable changes in the company's management practices
        and organization*)

	Action programmes	Degree of use last 3 years
C51a	Updating your <u>process equipment</u> to industry standard or better	1 2 3 4 5
C52a	Expanding <u>manufacturing capacity</u> (e.g. buying new machines; hiring new people; building new facilities; etc.)	1 2 3 4 5
C53a	Engaging in <u>process automation</u> programs	1 2 3 4 5
C54a	Implementing <u>Information and Communication Technologies</u> and/or Enterprise Resource Planning software	1 2 3 4 5
C55a	Reorganizing your company towards <u>e-commerce</u> and/or <u>e-business</u> configurations	1 2 3 4 5
C56a	Rethinking and restructuring your <u>supply strategy</u> and the organization and management of your suppliers portfolio	1 2 3 4 5
C57a	Concentrating on your core activities and <u>outsourcing</u> support processes and activities (e.g. IS management, maintenance, material handling, etc.)	1 2 3 4 5
C58a	Restructuring your manufacturing processes and layout to obtain <u>process focus</u> and streamlining (e.g. reorganize plant-within -a-plant; cellular layout, etc.)	1 2 3 4 5
C59a	Undertaking actions to implement <u>pull production</u> (e.g. reducing batches, setup time, using kanban systems, etc.),	1 2 3 4 5
C510a	Undertaking programs for <u>quality improvement</u> and control (e.g. TQM programs, 6σ projects, quality circles, etc.)	1 2 3 4 5
C511a	Undertaking programs for the improvement of your <u>equipment productivity</u> (e.g. Total Productive Maintenance programs)	1 2 3 4 5
C512a	Implementing actions to increase the level of <u>delegation and knowledge of your workforce</u> (e.g. empowerment, training, improvement or autonomous teams, etc.)	1 2 3 4 5
C513a	Implementing actions to improve or speed-up you process of <u>new product development</u> through e.g. platform design, products modularization, components standardization, concurrent engineering, Quality Function Deployment, etc.	1 2 3 4 5

C514a Putting efforts and commitment on the improvement of your company's <u>environmental compatibility</u> and workplace <u>safety and healthy</u>   1 2 3 4 5

D2.   Please indicate the amount of change of the following performance dimensions over <u>the last three years</u>

		Strongly deteriorated		No change	Strongly improved	
Manufacturing conformance	D21	1	2	3	4	5
Product quality and reliability	D22	1	2	3	4	5
Product customization ability	D23	1	2	3	4	5
Volume flexibility	D24	1	2	3	4	5
Mix flexibility	D25	1	2	3	4	5
Time to market	D26	1	2	3	4	5
Customer satisfaction	D27	1	2	3	4	5
Delivery speed	D28	1	2	3	4	5
Delivery reliability	D29	1	2	3	4	5
Manufacturing lead time	D210	1	2	3	4	5
Procurement lead time	D211	1	2	3	4	5
Procurement costs	D212	1	2	3	4	5
Labor productivity	D213	1	2	3	4	5
Inventory turnover	D214	1	2	3	4	5
Capacity utilization	D215	1	2	3	4	5
Overhead costs	D216	1	2	3	4	5
Environmental performance	D217	1	2	3	4	5

8. References

Ayres, Robert (1991) *Computer Integrated Manufacturing Volume I: Revolution in Progress,* CHAPMAN & HALL, London.

Beaumont, N., Schroder, R. and Sohal, A. (2002) Do foreign-owned firms manage advanced manufacturing technology better? *International Journal of Operations & Production Management;Vol.* 22, No. 7/8, pg. 759-772.

Beekun, R. I. (1990) "Assessing the effectiveness of socio-technical interventions: Anitdore or fad?", *Human Relations*, Vol. 42, pp.887-897.

Beekun, R. I. (1990) "Assessing the effectiveness of socio-technical interventions: Anitdore or fad?", *Human Relations*, Vol. 42, pp.887-897.

Bessant, J. and Haywood, B., (1988), "Islands, archipelagos and continents: progress on the road to computer-integrated manufacturing", *Research Policy*, Vol. 17, 349-362.

Byrne, B.M., 2001. Structural Equation modeling with AMOS: Basic Concepts, Applications and Programming. Lawrence Erlbaum Associates, Publishers, NJ.

Chase, R. B. and Aquilano, N. (1997) *Production and Operations Management,* IRWIN, Chicago.

Chase, R. B., Aquilano, N. and Jacobs, R. (2001) Operations management for competitive advantage (9th ed.) Boston : McGraw-Hill/Irwin

Cohen, J. (1988), *Statistical power analysis for the behavioral sciences*, 2nd ed. Hillsdale, N.J. : L. Erlbaum Associates, 1988.

Cua, K. O. McKone, K. E. and Schroeder, R. G. (2001) Relationships between implementation of TQM, JIT, and TPM and manufacturing performance. Journal of Operations Management. Vol.19, No. 6, pp. 675-694.

Dasa, A. and R., Narasimhan (2001) Process-technology fit and its implications for manufacturing performance, Journal of Operations Management 19 (2001) 521-540

de Vaus, D. A., (1993), *Survey in Social Research* (3rd ed.), UCL PRESS.

Fix-sterz, Jutta, Gunter Lay, Rainer S., Jurgen W. (1987), *Flexible Manufacturing Systems and Cells in the Scope of New Production Systems in Germany*, FAST Occasional paper No.135.

Flynn, B. B., Schroeder, R. G., and Sakakibala, S. (1994), "A framework for quality management research and an associated measurement instrument", *Journal of Operations Management*, Vol. 11, pp.1339-1366.

Gerwin, D. and H. Kolodny (1992) *Management of advanced manufacturing technology: strategy, organisation, and innovation,* Wiley & Sons, N.Y.

Goetsch, D. L., (1990) *Advanced Manufacturing Technology*, Delmar Publisher Inc., New York

Greenwood, N. R. (1988) *Implementing Flexible Manufacturing System*, Macmillan Education, London.

Groover, M. P. (1987), *Automation, Production Systems, and Computer Integrated Manufacturing*, Prentice-Hall, Inc., New Jersey.

Hanson, P. and Voss, C.A., (1993), Made in Britain, the true state of Britain's manufacturing industry, IBM Ltd/London Business School, Warwick, UK.

Hayes, R. H., Wheelwright, S. C. and Clark, K. B., (1988), Dynamic manufacturing, creating the learning organization. The Free Press, New York, NY.

Hayes, R.H. and Wheelwright, S.C. (1984) Restoring Our Competitive Edge. Wiley, New York.

Hayes, R.H. and Wheelwright, S.C., (1984), Restoring our competitive edge, Wiley, New York.

Haywood, B. (1990) "CIM: technologies, organisations, and people in transition", *Proceedings of the final IIASA conference on CIM*, Luxembourg, Austria.

Hill, T., (1995), Manufacturing strategy: Text and cases, Macmillan Press, London.

Hu, L., Bentler, P.M., 1998. Fit indices in covariance structure modeling: Sensitivity to underparameterized model misspecification. Psychological Methods 3, 424-453.

Hu, L., Bentler, P.M., 1999. Cutoff criteria for fit indexes in covariance structure analysis: conventional criteria versus new alternatives. Structural Equation Modeling 6(1), 1-55.

Jöreskog, K.G. and Sörbom, D., 1984. LISREL-VI user's guide (3rd ed.). Scientific Software, Mooresville, IN.

Kaynak, H. 2003. The relationship between total quality management practices and their effects on firm performance. Journal of Operations Management 21, 405-435.

Kim J.S., Arnold, P., 1996. Operationalizaing manufacturing strategy: an exploratory study of constructs and linkage. International Journal of Operations & Production Management 16(12), 45-73.

Kopelman, R. E. (1986) *Managing Productivity in Organizations, McGraw, NY.*

Kotha, S and P M Swamidass (2000) Strategy, advanced manufacturing technology and performance: Empirical evidence from U.S. manufacturing firms, Journal of Operations Management. Vol. 18, No. 3: 257

Kotha, S. and P. M., Swamidass, (1998), "Advanced Manufacturing Technology uses: exploring the effect of the nationality variable", *International Journal of Production Research*, Vol. 11, pp.3135-3146.

Lee, R. J. V. and Leonard, R. (1990), "Changing role of humans within an integrated automated guided vehicle system", *Computer-Integrated manufacturing Systems,* Vol. 3. No.2, pp.115-120.

Leong, G.K. and Ward, P.T., (1995), The six Ps of manufacturing strategy, International Journal of Operations & Production Management, Vol. 15, No. 12, p 32-45.

Lewis, M.A. and Roehrich, J. (2009), ``Contracts, relationships and integration: towards a model of the procurement of complex performance", *International Journal of Procurement Management,* Vol. 2 No. 2, pp. 125-142.

Lindberg, P. (1990) *Manufacturing Strategy and Implementation of Advanced Manufacturing technology,* Ph.D. Dissertation, Chalmers University of Technology, Gothenburg, Sweden.

Lindberg, P., Voss, C.A., Blackmon, K.L. (Eds), 1998. International Manufacturing Strategies: Context, Content and Change. Kluwer Academic Publisher, Boston.

MacCallum, R., 1986. Specification searches in covariance structure modeling. Psychological Bulletin 100(1), 107-120.

MacCallum, R.C., James, T. A., 2000. Applications of structural equation modeling in psychological research. Annual Review of Psychology 51, 201.

Marsh, H.W. and Hocevar, D., 1985. Application of confirmatory factor analysis to the study of self-concept: first- and higher-order factor models and their invariance across groups. Psychological Bulletin 97, 562-582.

McDermott, C.M., Stock, G.N. (1999) Organizational culture and advanced manufacturing technology implementation. Journal of Operations Management 17 (5), 521–533.

McDermott, C.M., Stock, G.N., 1999. Organizational culture and advanced manufacturing technology implementation. Journal of Operations Management 17 (5), 521–533.

Meyer, S. M. and A. Collier. 2001. An empirical test of the causal relationships in the Baldrige Health Care Pilot Criteria. Journal of Operations Management 19(4), 403:425.

Ng, K.C. and Hung, I. W., (2001), A model for global manufacturing excellence, Work Study, Vol. 50, No. 2, p 63-68.

Noori, H. (1990) *Managing the Dynamics of New Technology, issues in manufacturing management,* Prentice Hall, Englewood Cliffs, N.J.

Nunnally, J. C. (1978), *Psychometric Theory,* McGraw-Hill Publishing Company, New Yor.

Nunnally, J.C., Bernstein I.H., 1994. Psychometric theory. McGraw-Hill, New York, 510-512.

Pannirselvam, G.P., Ferguson, L.A., 2001. A study of the relationships between the Baldrige categories. International Journal of Quality and Reliability Management 18 (1), 14–34.

Parthasarthy, R. and S. P. Sethi (1992) "The Impact of flexible automation on business strategy and organizational structure", *Academy of Management Review,* Vol. 17, No:1. pp.86-112.

Peterson, R.A., 1994. A meta-analysis of Cronbach's coefficient alpha. Journal of Consumer Research 21(2), 381-391.

Riis, O.J. (1992) "Integration and Manufacturing Strategy". Computer in Industry, vol. 19, 37-50.

Samson, D., Sohal, A. and Ramsay, E., (1993) "Human resources issues in manufacturing improvement initiatives: case study experiences in Australia", *The International Journal of Human Factors in Manufacturing,* Vol.3, No. 2, pp.153-152.

Saraph, J. V., and Sebastian, R. J. (1992), "Human resources strategies for effective introduction of advanced manufacturing technologies (AMT)", *Production and Inventory Management Journal*, Vol.33, pp.764-770.

Saraph, J. V., Benson, P. G., and Schoeder, R. G. (1989), "An instrument for measuring the critical factors of quality management", *Decision Sciences*, Vol. 20, pp.810-829.

Schonberger, R. J., (1982), Japanese Manufacturing Techniques, Nine Hidden Lessons in Simplicity, The Free Press, New York NY.

Schonberger, R.J., (1986), World Class Manufacturing: The Lessons of Simplicity Applied, The Free Press, New York, NY.

Schroeder, R.G., Anderson, J.C., and Cleveland, G., (1986), The content of manufacturing strategy: An empirical study, Journal of Operations Management, Vol. 6, No. 4, p 405-415

Singh, N. (1996) *System Approaches to Computer Integrated Design and Manufacturing*, John Wiley & Sons, Inc., N.Y.

Skinner, W., 1969. Manufacturing-missing link in corporate strategy, Harvard Business Review 47(3), 136-145.

Sobek II, D. K., Liker, J. K. and Ward, A. C. 1998, Another look at how Toyota integrate product development, *Harvard Business Review*, July-August issue, pp.69-78.

Sun, H. (2000) "Current and Future Patterns of Using Advanced Manufacturing Technologies", *Technovation, The International Journal of Technological Innovation and Entrepreneurship*, Vol.20, No.11, pp.631-641.

Sun, H. (2001) "Human Resources Development in Integrated Manufacturing Systems", *Integrated Manufacturing System*, Vol. 12, No.3, pp.195-204

Sun, H. and Cui, H. (2002) The alignment between manufacturing and business strategies: its influence on business performance. Technovation 22, 699-705.

Sundstrom, E., DeMeuse, K.P., and Futell, D. (1990), "Work teams", *American Psychologist*, Vol. 45, pp. 120-133.

Swamidass, P.M. and Newell, W.T. (1987) "Manufacturing strategy, environmental uncertainty and performance": a path analytic model. Management Science, vol.33, 509-524.

Swamidass, P.M., Kotha, S. (1998) Explaining manufacturing technology use, firm size and performance using a multidimensional view of technology. Journal of Operations Management 17 (1), 23–37.

Tseng, H.C., Ip, W.H., and Ng, K.C., (1999), A model for an integrated manufacturing system implementation in China: a case study, Journal of Engineering and Technology Management, 16, p83-101.

Tunalv, C. (1991) Manufacturing strategy in Sweden engineering industry", Ph.D. thesis, Chalmes University of Technology, Sweden.

Wheaton, B., Muthen, B., Alwin, D.F. and Summers, G.F., 1977. Assessing reliability and stability in panel model. In: Heise, D.R. (Ed), Sociological Methodology. Jossey-Bsaa, San Francisco, 84-136.

Wheelwright, S. C., (1984), Manufacturing strategy: Defining the missing link, Strategic management Journal, 1, p 77-91.

Wilson, D.D., Collier, D.A., 2000. An empirical investigation of the Malcolm Baldrige National Quality Award causal model. Decision Sciences 31 (2), 361–390.

Woodward, Joan (1965) *Industrial Organization: Theory and practice*. London: Oxford University press.

14

Incorporating Technological Innovation and Environmental Strategy: An Integrated View of Cognition and Action

Xuanwei Cao

Department of Management, Xi'an Jiaotong-Liverpool University, Suzhou, China

1. Introduction

Research on (natural) environment issues in the filed of strategic management has a long history. Before the forming of RBV, scholar had put forwarded the natural resource-based firm theory, incorporated the challenges of natural environment issues into the scope of strategic management research (Hart, 1995). Considering the complex relationship between firms and natural environment, how to obtain sustainable competitive advantage under intensive competitive context with serious environmental and energy challenges and crisis is becoming an important issue with rising research concerns. Studies in existing literatures have explored widely on this issue, including strategic proactivity and approach to natural environment (Aragón-Correa, 1998; Sharma, 2000), the selection of corporate environment strategy (Sharma, 2000), proactive environment strategy and organizational competitiveness (Sharma & Vredenburg, 1998; Aragón-Correa, 2003), factors impacting environment strategy (González-Benito & González-Benito, 2006), organization and environment (Etzion, 2007), impact of institutions on corporate environment strategy (Wahba, 2010) as well as impact of environment strategy on organizational performance in varied contexts (Aragón-Correa *et al*, 2008). However, these studies followed still the routine of swings of a pendulum in traditional strategy management, i.e., the swing of research perspectives between "looking outside-in" and "looking inside-out" without touching the micro foundation of strategy formation (Hoskisson *et al*, 1998). As a positive response to the initiative of Gavetti (2005), this paper tries to explain the formation and development of corporate environment strategy through combining the perspectives of cognition and action.

A conceptual model would be put forwarded to explain different patterns of corporate environment strategy under varied configuration of managerial cognition and strategic action. Based on this conceptual model, this paper would shed light on the dynamic change and development of corporate environment strategy. We try to offer a new perspective and understanding to important questions on corporate environment strategy. To our minds, in coping with environment challenges in corporation, three questions should be understood and answered by managers. The first is 'what' question, what kinds of measures (such as management system, technological innovation, etc.) should be adopted to deal with environment issues in business? The second is 'how' question, how to take these actions? Take organizational learning as example, the 'how' question relates to the selection of corporation in taking actions, such as global search or local search. And the third is 'when' question, which involves the time and timing factor in taking actions. For example, managers should consider the selection and adoption of different strategies in making innovation, such as keeping in-sync with competitors or following competitors. Thus, corporate environment strategy would be studied as a time dependent process by holding a "looking inside-out" perspective focusing on the mechanism of the interaction between cognition and action. In turmoil and uncertain environment, organizations need to develop their capabilities on strategic learning and strategic innovation for winning sustainable competitive advantage. In this context, time must be considered as an important factor to help us understand how to converge managerial cognition into strategic action. This paper is expected to contribute to understand how variance of managerial cognition on time could cause different orientations of corporate environment strategy. It also offers an effort towards understanding of subjective aspect of time and the psychological foundation of the origin of strategy. The second section investigates different research perspectives and focus of previous research on environment strategy, finding out the research gap. The third section illustrates the micro foundation of the formation of environment strategy, exploring the dynamic relationship between managerial cognition and action, pointing out the importance of incorporating time dimension into consideration of environment strategy. The fourth section concludes and indicates more meaningful research in future as well as managerial implications.

2. Corporate environment strategy

Environment issue in corporations is increasingly a focus of analytical interest in the study of corporate strategy. In particular, the notion of 'environment strategy' has penetrated into the scope of strategic management and come to prominence as an important challenge in front of managers and students of strategy research. A number of scholars have identified ambiguities and unresolved questions associated with the concept. Until now, there is no universally accepted definition of environment strategy. According to Sharma (2000), it can be thought of as "the mode of managing the interface between business and natural environment, a series of action results from adopting measures voluntarily to reduce negative impacts on environment". This definition follows the traditional path of strategy research, understanding strategy as a series of actions, focusing on firm level. In response to the recent initial research advice on exploring the micro foundations of strategy, environment strategy is thought to comprise both cognition and action aspects of managers when dealing with environment issues in pursuing sustained competitive advantage in business operation. It is time dependent and context dependent (i.e., industries, scopes, ownerships, countries, governance, etc.).

To understand better the fuzzy concept of environment strategy in previous research, Table 1 below illustrates the different research questions, focus, analytical perspectives and conclusions summarized from previous research on this issue. Through this way, we are aiming to expose the 'dilemma of innovation' of managers in front of environment challenges in practice.

Source	Research question	Focus	Perspective	Conclusion
Russo & Fouts (1997)	Associations between environment performance and economic performance	Organizational performance	Looking inside	Environment performance and economic performance are positive related, influenced from industry growth
Aragón-Correa (1998)	Relationship between corporate strategy proactivity and the attitude of corporate on natural environment	Strategy proactivity	Looking inside	It is inter-related between strategy proactivity and the attitude of corporate on natural environment
Sharma & Vredenburg (1998)	The role of Proactive environment strategy on organizational capability development	Proactive environment strategy	Looking inside	It is intensively inter-related between corporations' positive response to environment issues and organizational special capabilities (such as sustained innovation, integrating stakeholders, and organizational learning)
Rhee & Lee (2003)	Dynamic change of corporate environment strategy	'Ritual' and 'real' of corporate environment strategy	Looking inside	Internal factors in organizational play significant role on influencing corporate environment strategy
Aragón-Correa et al (2008)	SMEs' environment strategy	SMEs' environment strategy and economic performance	Looking outside-in	SMEs show varied patterns of environment strategy. More proactive environment strategy can promote and enhance organizational performance
Fraj-Andrés et al (2009)	Factors influencing corporate environment strategy	corporate environment strategy in Spanish context	Looking outside-in	Competitive dynamics and management commitment are important factors influencing the incorporation of environment issues into corporate strategy
Wahba (2010)	Institutional shareholders' control and influence on corporate environment strategy in different institutional contexts	Institutional shareholders	Looking outside-in	Different institutional shareholders have different impacts on corporate environment strategy
Buysse & Verbeke (2003)	Correlativity between corporate environment strategy and the management of stakeholders	stakeholders	Looking inside-out	The correlation between corporate environment strategy and stakeholders management is influenced from other external factors
Banerjee (2001)	Influence from managerial perception on environment-orientation and environment strategy	managerial perception	Looking inside-out	Perceptions towards regulations, public environment awareness, top-level managerial commitment and achievement of competitive advantage could be transferred into corporate environment strategy

Source	Research question	Focus	Perspective	Conclusion
Worthington & Patton (2005)	To what extent is corporate green environment management influenced by manager's strategic intention	strategic intention	Looking inside-out	Managers in SMEs lack strategic intention to make environment performance a source of competitive advantage
Sharma (2000)	Influence of managerial interpretation in specific context on corporate environment strategy	managerial interpretation	Looking inside-out & Looking outside-in	Managerial interpretation to environmental issues is influenced by contextual factors
Sharma, Pablo & Vredenburg (1999)	Association between managerial interpretation to environment issues and subsequent responses adopted	Internal and external factors influencing corporate environment response strategy	Looking inside-out & Looking outside-in	Managerial interpretation to environment issues (such as opportunity or threat; controllable or uncontrollable; benefits or losts) decide corporations' responses

Table 1. Research on Environment Strategy

Through a systematic literature review, the previous research on environment strategy could be divided to three categories. The first group studies concern about the antecedents of environment strategy; the research question concerned is what factors can influence the formation of corporate environment strategy. The second group studies concern about the ex post outcomes of corporate environment strategy, i.e., exploring the impact of corporate environment strategy on economic and environmental performance of organizations, as well as the causal relationship between those antecedent factors and ex post outcomes. The third group studies try to find out the mechanism for the formation of environment strategy, i.e., exploring the micro foundation of environment strategy from cognition aspect. However, studies on exploring how cognition is transferred to strategic action and how environment strategy is constructed in organization are still lacking. Studies in the first group and the second group try to link directly the causal chain between the antecedent factors influencing environment strategy and the ex post outcomes from environment strategy, ignoring yet the micro foundation of the formation of environment strategy. Figure 1 below illustrates the antecedents and ex post outcomes as well as the inner black box of the formation of environment strategy.

As illustrated in Figure 1, antecedent factors influencing environment strategy include stakeholders, institution environment, firm, and competitors. Many studies before have made analysis at industry level, exploring how different ownership and firm scope, different industries, different institution environment, and different governance structure influence corporate environment strategy. This research tradition focuses actually the force of external factors on organization, reflecting a research perspective of 'looking outside-in' and an epistemology of environment determinism. It holds a relative static perspective without touching the 'black box' of strategy. Recently, some scholars have advocated considering the impact of dynamic competition on organization strategic actions. In dynamic competition, varied environment performance from different environment strategy is influenced from corporation's response to competitor's actions.

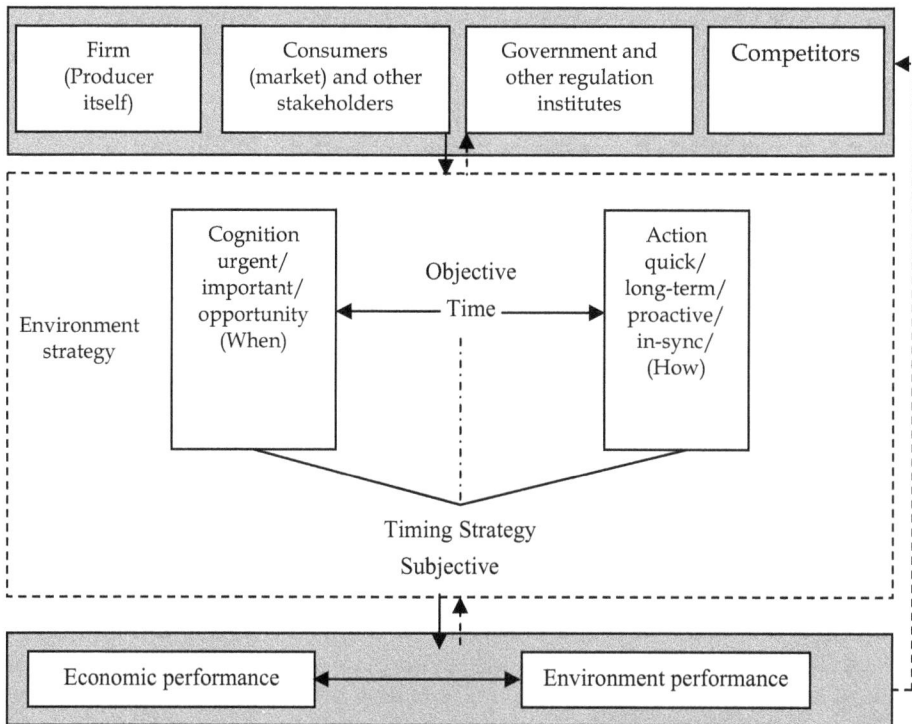

Fig. 1. The Antecedence and Outcome of Environmental Technological Strategy

With deepening research on the origin of strategy, scholars put forward to combine both 'looking outside-in' and 'looking inside-out' perspectives to explore how cognition and action at individual level impact strategy and dynamic capability at organization level.

In this paper, we make a try to explore how managerial cognition and action at individual level could impact the formation, implementation, renewal and innovation of corporate environment strategy at organization level.

3. A micro perspective on corporate environment strategy

3.1 Micro foundation of strategy

In recent years, studies on organizational performance, dynamic capabilities and strategy are presenting strong interest on exploring the micro foundation (Narayanan, Zane and Kemmerer, 2011). Attentions in studies on strategy has been experiencing the shift from concerning "strategic contents" to "strategic process" and "strategy-in-practice", focusing more intensively on answering the 'how' question, i.e., how does the formation, implementation and renewal (innovation) of strategy happen? At the same time, in turbulent times, the uncertainty of managers and their actions to ambiguous issues are impacting organizational strategy and performance prominently. The recent initiatives on calling for more investigations on the micro foundation for the formation of strategy open a new avenue to understand the strategy in practice (Lovallo et al, 2008; Jaworski, Balogun, and Seidl, 2007). Eisenhardt et al (2010) put forwarded that more research in strategy and organization should be taken to explore the micro foundations of organizational performance under dynamic environmental conditions. This is consistent with the suggestion from Meindl (1994) that the most important research in future must establish the links among cognition, behavior and organizational performance.

When studies penetrate into the inner mechanisms of the changing of strategy process, the dual characters of strategy was excavated. In their study of the origin of strategy, Gavetti & Rivkin (2007) pointed out that strategy is a unification of managerial cognition and action. On the one hand, strategy exists in the minds of managers, embodied as their perception towards the world and the position of their companies; on the other hand, strategy is represented as actions of companies, refined through specific activities, rules and routines. Since both cognition and action evolves over time, the strategic task for management is finally to keep an alignment between managerial cognition and action in dynamic changing situations. In this process, the managerial cognition framework and the interpretations to environment from managers compose the foundation for the actions of managers. Many studies have proven the relationship between managerial cognition and strategic action (Cho & Hambrick, 2006; Kaplan, 2008). Following this emerging mode on bridging the link between individual's micro cognition and organization's macro action, recent studies examined also the impact of CEO's attention on organization's adoption of new technology (Eggers & Kaplan, 2009). A noteworthy piece among those studies is the analysis conducted by Miller & Chen (1994). They pointed out that managers often tend to do what they have done the best, causing a 'downward spiral' of organizational development, and leading to a path of organization recession.

Although previous research have disclosed the inter-relationship between managerial cognition and strategic action and the impact of managerial on organizational performance, they presented few dynamic descriptions on cognition, without illustrating how cognition changes under influences from external factors over time in dynamic environment and the subsequent impacts on organization. Recently, however, some scholars have noticed the problem and made efforts towards understanding the dynamic process of strategy. Nadkarni & Barr(2008) tried to integrate studies on environment context, managerial cognition and strategic action to develop a more integrated and dynamic understanding of strategy process. McCarthy et al (2010) studied specifically the concept of environment velocity, considering in particular how environment velocity impacts strategic decision-making and new product development. These researches reflect the efforts of scholars to introduce dynamic perspective on understanding the change of cognition and the subsequent impact on actions at macro level. However, to a great extent the idea still follows a "looking outside-in" perspective, lacking stronger explanation power to the ultimate question of how cognition change over time.

In front of challenges from environment issues, a hard question for mangers in practice is to reckon *when and how* to conduct (environmental) technological innovation. Considering the dynamics of strategy, current research on organization and strategy is focusing on bridging the link between individual manager's cognition and organizational strategic actions as well as their interactions on organizational dynamic capabilities and organizational performance.

3.2 Environment strategy through the lens of cognition and action

Traditional study in the field of strategy tries to answer 'what' question (i.e., what is the right thing to do, what business to do) and 'how' question (i.e., how to take strategic actions to cope with external environment). The 'what' question pays attention to the contents of strategy, while the 'how' question concerns about the process of strategy. However, as environment is becoming more and more unstable and unpredictable, the traditional logic and perspective on strategy study could not offer effective solutions to give managers a clear guide on reckoning 'when' is suitable time (timing) to take strategic actions, such as a proactive environment strategy through technological innovation.

Because of the externality in making environment technological innovation, together with the imprinted perception of conflict and opposite between economic performance and environment performance, the problem of 'innovator's dilemma' is more prominent. For example, in coping with environment issues in business operation, managers in practice are always puzzled and confused by selecting a right way between radical innovation and incremental innovation, pursuing to be a technology leader or satisfying to be a technology follower, taking proactive actions or making response, creating a new path or sticking on path dependence. At the same time, these paradoxes are still hanging in doubt in academia. In front of these paradoxes, managers are far away from possessing the ambidexterity capability to solve such kind of ambiguous issue as environment strategy. What scholars observed of the variance of environment strategies in business could be tracked back actually to the variance of managerial cognition.

Sharma's study (2000) indicated that the selection of corporate environment strategy suffers from the impact of managerial cognition. Managers' interpretation of environment issues to be a threat or an opportunity influenced the selection of corporate environment strategy. Other studies considered also the influence of leadership styles on the environmental management (Bansal and Roth, 2000; Cordano and Frieze, 2000; Egri and Herman, 2000; Flannery and May, 2000; Sharma, 2000; Banerjee, 2001). Although this research tried to construct a link between individual behavior and organizational action, it didn't explore how contingent factors influenced managers' cognition and correspondent actions. Recently, studies on CEO's capability disclosed that the ambidexterity capability of decision-makers in dealing with ambiguous issues with both positive and negative meanings would decide the scope of action, risks and innovativeness (Plambeck & Weber, 2009, 2010). Corporate environment strategy could be regarded as an ambiguous issue since there are still lots of enterprises that have not incorporated environment strategy as an important foundation for their organizations' sustained competitive advantage. Among many other factors influencing the orientation of corporate environment strategy, it is managerial cognition that impacts organization's reply to the 'when' and 'how' questions. Research about the relationship between short-term performance and long-term survival of organizations has pointed out that this relationship is partly decided by actions the organization adopted in response to external environment; and organizational actions are also partly impacted by the purposive behaviors of individuals especially decision-makers at higher level in organization (Dutton & Jackson, 1987). Therefore, the study of organization strategy, in particular under ambiguity and uncertainty situations, must integrate individual actor's cognition to organization and environment with strategic actions at organizational level.

In their article, Buysse & Verbeke (2003) divided three types of environment strategy, namely reactive strategy, pollution prevention strategy, and environment leadership strategy. We would develop further a typology of environment strategy according to managers' cognition on environment issues and the actions they adopted. Figure 2 shows the four different types of actors and their correspondent environment strategies.

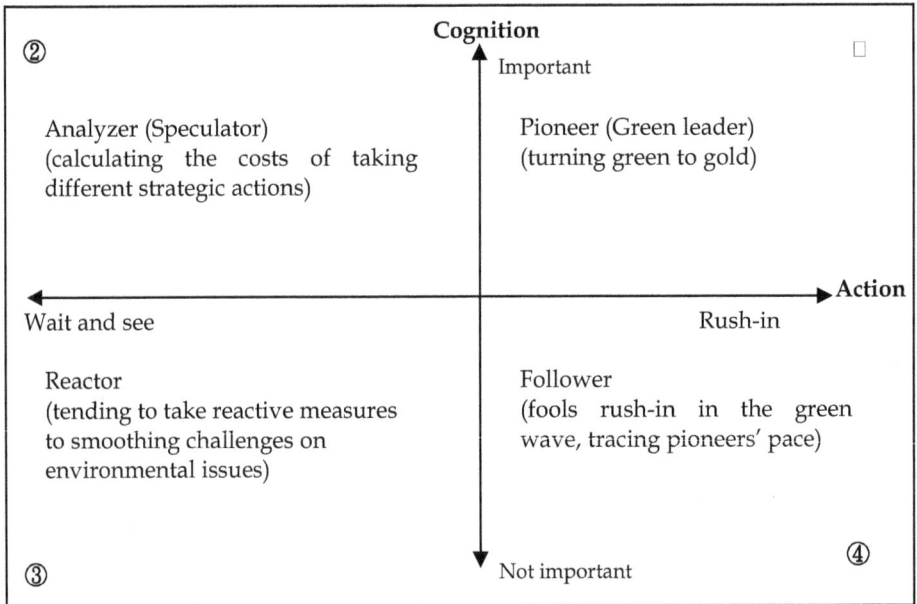

Fig. 2. A Typology of Environment Strategy Based on Cognition and Action

4. Time dependence of environment strategy

Strategy itself is time dependent (Eisenhardt, 2002). Considering the importance of environment issues on organization sustainable competitiveness, the formation, implementation and renewal of environment strategy could be varied on time-orientation. Due to different managerial perceptions towards organizational internal resources, organizational capabilities and development as well as pressures from external competitive environment and institution environment, corporate environment strategy in practice is actually embedded in a specific time frame which could be represented varied in terms of urgency vs. indifference, long-termism vs. short-termism, etc on cognitive aspect and proactivity vs. reactivity, path creation vs. path dependence, etc on action aspect. In uncertain and turmoil environment, it is harder and harder to pursue sustainable competitive advantage due to fast-paced competitive actions and counter responses among rivals. The requirement for more flexible strategy and strategic innovation raise an important issue for strategy study. That is how firms transition from one advantage to the next. When should they begin these transitions? How should managers manage time?

It is only in recent years that research on strategy management is becoming to make an echo to the initiative of Ancona *et al* (2001)'s research to make time a new research lens. Through the temporal lens, individual's time urgency and time perspective could be observed and integrated into a more holistic understanding of strategy, to achieve an ideal status of strategy management by doing right things right at right time. It is worthy to note that our understanding on the origin of strategy under the lens of an integrative perspective of cognition and action must be developed further with more consideration of the dynamic change of strategy. For example, the behavior of rivals' impacts on actor's cognition and action should be considered. In particular, the temporary component of competitive advantage, such as time pacing, sequence, frequency, time-orientation, etc., must be considered when making, implementing, renewing strategies in hypercompetition. The recent study of Katila & Chen (2008) pointed out that it is actually search timing compared to competitors instead of competition that cause difference of innovation among organizations. In terms of the search timing, Katila & Chen (2008) showed three strategies for organizations. The first is to keep ahead of competitors, exploring new field; the second is to keep in-sync with competitors, competing with rivals to provide new products to markets; the third is to keep following competitors, aiming to catch-up competitors in later time. This valuable research introduced a dynamic perspective towards understanding of strategic actions under varied contexts; however, we are still lacking knowledge on how cognition is transferred to strategic action in organization.

Nadkarni & Barr(2008) made a try to explain the dynamic interactions between managerial cognition and strategic action through the moderating variable of speed of industry change. This opens a new path to bridge the link between managerial cognition and strategic action through the factor of time. Time is a variable with diversified meanings. In previous research, too often time is portrayed and interpreted based on the measured, linear, forward-moving, and exact clock time. In fact, time could also be reflected as the subjective experience of individual actors. Therefore, future exploration on manager's time cognition and its impact on strategic actions and structure of organization could help open the black box of strategy process and strategy in practice.

Considering the ambiguity and complexity of corporate environment strategy in the minds of managers in practice, we suppose this issue has different time frame in managers. According to the behavior theory of the firm, organization is a problem-solving entity with limited attentional capability (Cyert & March, 1963). Compared to other issues with more directed influences on economic performance of organization, normally environment issues are allocated with less attention from managers. However, due to increasing pressures from other external factors, including regulators, stakeholders, and competitors, environment issues are also interpreted varied to some extent by managers due to their different managerial cognition. Among which, time is an important factor differentiating managerial cognition and consequent actions. Based on the research analysis on how different time cognition influences the formation, implementation and renewal of corporate environment strategy, we propose the following.

PROPOSITION 1: *If a manager posses a smooth cognition on time, he tends to adopt a responsive corporate environment strategy.*

PROPOSITION 2: *If a manager posses an urgent cognition on time, he tends to adopt a proactive corporate environment strategy.*

PROPOSITION 3: *If a manager predicts rival's innovation activities on environment issues, his time cognition of urgency would be reinforced.*

PROPOSITION 4: *If a manager pays more attention on emerging new technologies, the firm tends to be future-oriented, with earlier adoption of innovative technologies and earlier timing of entry into new market.*

The examination of managers' different cognition on time offers a good point to disentangle the interactions between managerial cognition and strategic actions as well as their influence on organizational strategy and organizational performance. This is consistent with the recent initiative from Lovallo *et al* (2008) to explore the psychological foundation of strategy management.

5. Discussion and implications

We try to offer a new perspective and understanding to important questions on corporate environment strategy. To our minds, in coping with environment challenges in corporation, three questions should be understood and answered by managers. The first is *'what'* question, what kinds of measures (such as management system, technological innovation, etc.) should be adopted to deal with environment issues in business? The second is *'how'* question, how to take these actions? Take organizational learning as example, the 'how' question relates to the selection mechanism in taking actions, such as global search or local search. And the third is *'when'* question, which involves the time and timing factor in taking actions. For example, managers should consider the selection and adoption of different strategies in making innovation, such as keeping in-sync with competitors or following competitors. Thus, corporate environment strategy would be studied as a time dependent process by holding a "looking inside-out" perspective focusing on the mechanism of the interaction between cognition and action.

From the attention based view of the firm (Ocasio, 1997), managers would focus more on economic performance when shareholders put more attention on the pressure of performance evaluations in organization. In this context, more attention on economic performance could create short-termism of managers instead of trying to maintain an ambidextrous trade-off between economic and environmental performance. Therefore, in practice, we need to combine both "looking outside-in" and "looking inside-out" perspectives to consider the impact of external factors on managers' cognition and action as well as the impact of individual manager's cognition on organizational strategic orientation. For example, with deepening localization of Foreign Invested Enterprises (FIEs) in China, indigenous managers in localized FIEs tend to put more attention on improving economic performance while reducing input on corporate environment issues. In this process, managers' cognition on the urgency and importance of environment issues could impact their temporal orientation on making, implementing and renewing corporate environment strategy. Therefore, it is meaningful and valuable to investigate how contextual factors could influence manager's cognition and its consequent strategic actions in future studies.

An integrated analysis of corporate environment strategy from both cognition and action aspects could help understand the micro mechanism for the formation, implementation and renewal of strategy, indicating possible solutions for managers to break the 'innovator's dilemma'. The introduction of time dimension into the analysis of dynamic change of cognition could increase our understanding on the interactive relationship between managerial cognition and strategic actions. For example, we can investigate the causal relationship between managerial cognition and actions on environment strategy through examining the degree of urgency of manager's cognition on environment issues, the temporal orientation of manager towards environment issues, the timing of taking actions with consideration of rival's behavior.

In view of the emerging shift of strategy research to strategy as practice, a time-based perspective on analyzing strategic activities ignites actually further thinking and research on solving the dilemmas in managerial practices, such as the 'when' and 'how' questions for managers in front of the serious challenges of triple E crisis (i.e., economy, environment, energy). Until now, environment technological innovation is still regarded widely as a paradox both in theory and practice. The tensions between radical innovation vs. incremental innovation, exploration vs. exploitation, proactive vs. reactive, paradigm shift vs. technological trajectory, first-mover vs. catch-up, hidden actually managers' different cognition on time. With increasing uncertainty in environment as well as fast-pacing strategic change and business model innovation of competitors, a 'right' time strategy to guide innovation is without doubt more and more important in terms of its prominent value on improving organization's competitiveness.

In terms of the methodology to measure cognition, a popular method is to apply the content analysis method based on the materials of Letter to Shareholders (LTS). Compared with the interview method, it could avoid the subjective reconstruction with an ex post analysis, benefiting a longitudinal analysis. Considering the emerging development and interest on measuring cognition in other related disciplines, it is really helpful to strengthen and integrate other methods into the study, such as the latest collaborative research from organizational strategy and organizational cognitive neuroscience (Senior, Lee, and Butler, 2011). To our mind, a reasonable and effective measurement of managers' cognition on time is an interesting but arduous challenge for future research. To understand well the micro psychological foundation of strategy formation, we encourage scholars from varied fields including but not least to those from psychology, politics and behavior research to improve our understanding on the dynamic link between managers' cognition and actions at micro level and the evolution and renewal of corporations' environment strategy at macro level.

6. Acknowledgement

The author wishes to thank Dr. Hua Zhang for his valuable suggestions for the earlier versions of the manuscript. Financial support by the National Natural Science Foundation of China (NSFC) to the Youth Scholar Research Project (70902021) is gratefully acknowledged. This paper is part of the research outcomes of the research project (70902021).

7. References

[1] Hart, S.L. A Natural-Resource-Based View of the Firm. The Academy of Management Review, 1995, 20(4): 986-1014.

[2] Aragón-Correa, J.A. Strategic Proactivity and Firm Approach to the Natural Environment. The Academy of Management Journal, 1998, 41(5): 556-567.

[3] Sharma, S. Managerial Interpretations and Organizational Context as Predictors of Corporate Choice of Environmental Strategy. The Academy of Management Journal, 2000, 43(4): 681-697.

[4] Sharma, S., Vredenburg, H. Proactive corporate environmental strategy and the development of competitively valuable organizational capabilities. Strategic Management Journal, 1998, 19(8):729-753.

[5] Aragón-Correa, J.A., Sharma, S. A Contingent Resource-Based View of Proactive Corporate Environmental Strategy. The Academy of Management Review, 2003, 28(1): 71-88.

[6] González-Benito, J., González-Benito, Ó. A review of determinant factors of environmental proactivity. Business Strategy and the Environment, 2006, 15(2):87-102.

[7] Etzion, D. Research on Organizations and the Natural Environment, 1992-Present: A Review. Journal of Management, 2007, 33(4): 637-664.

[8] Wahba, H. How Do Institutional Shareholders Manipulate Corporate Environmental Strategy to Protect Their Values. Business Strategy and the Environment, Business Strategy & Environment, 2010, 19(8): 495-511.

[9] Aragón-Correa, J.A, Hurtado-Torres, N., Sharma, S., Garcia-Morales, V.J. Environmental strategy and performance in small firms: A resource-based perspective. Journal of Environment Management, 2008, 86(1): 88-103.

[10] Hoskisson, R.E., Hitt, M.A., Wan, W.P., Yiu, D. Theory and research in strategic management: Swings of a pendulum[J]. Journal of Management, 1999, 25(3): 417-456.

[11] Gavetti, G. Cognition and hierarchy: Rethinking the microfoundations of capabilities' development. Organization Science, 2005, 16(6): 599-617.

[12] Russo, M.V., Fouts, P.A. A Resource-Based Perspective on Corporate Environmental Performance and Profitability. Academy of Management Journal 1997, 40(3): 534-559.

[13] Rhee, S.K., Lee, S.Y. Dynamic change of corporate environmental strategy: rhetoric and reality. Business Strategy and the Environment. 2003, 12(3), 175-190.

[14] Fraj-Andrés, E., Martínez-Salinas, E., Matute-Vallejo, J. Factors affecting corporate environmental strategy in Spanish industrial firms. 2009, 18, 500-514.

[15] Buysse, K., Verbeke, A. Proactive environmental strategies: a stakeholder management perspective. Strategic Management Journal, 2003, 24(5): 453-470.

[16] Banerjee, S.B. Managerial perceptions of corporate environmentalism: interpretations from industry and strategic implications for organizations. Journal of Management Studies, 2001, 38(4): 489-513.

[17] Worthington, I., Patton, D. Strategic intent in the management of the green environment within SMEs: An analysis of the UK screen-printing sector. Long Range Planning, 2005, 38(2): 197-212.

[18] Sharma, S., Pablo, A.L., Vredenburg, H. Corporate Environmental Responsiveness Strategies: The Importance of Issue Interpretation and Organizational Context. Journal of Applied Behavioral Science, 1999, 35(1): 87-108.

[19] Narayanan, V.K., Zane L.J., Kemmerer, B. The Cognitive Perspective in Strategy: An Integrative Review. Journal of Management, 2011, 37(1): 307-351.

[20] Gavetti, G., Rivkin, J.W. On the Origin of Strategy: Action and Cognition over Time. Organization Science, 2007, 18(3): 420-439.

[21] Eisenhardt, K.M., Furr, N.R., Bingham, C.B. Microfoundations of Performance: Balancing Efficiency and Flexibility in Dynamic Environments. Organization Science, 2010, 21(6): 1263-1273.

[22] Laamanen, T., Wallin, J. Cognitive Dynamics of Capability Development Paths. Journal of Management Studies, 2009, 46(6): 950-981.

[23] Cockburn, I.M., Henderson, R.M. and Stern, S. 2000. Untangling the origins of competitive advantage. Strategic Management Journal, 21(10-11), 1123-1145.

[24] Meindl, J. R., Stubbart, C., & Porac, J. F. 1994. Cognition within and between organizations: five key questions. Organization Science, 5(3): 289-293.

[25] Ginsberg, A., Venkatraman, N. 1995. Institutional Initiatives for Technological-Change-from Issue Interpretation to Strategic Choice. Organization Studies, 16(3): 425-448.

[26] Jenkins, M., Johnson, G. Linking Managerial Cognition and Organizational Performance: A Preliminary Investigation Using Causal Maps. British Journal of Management, 1997, 8(s1): 77-90.

[27] Kiesler, S., Sproull, L. Managerial Response to Changing Environments: Perspectives on Problem Sensing from Social Cognition. Administrative Science Quarterly, 1982, 27(4): 548-570.

[28] Barr, P.S., Stimpert, J.L, Huff, A.S. Cognitive change, strategic action, and organizational renewal. Strategic Management Journal, 1992, 13(s1):15-36.

[29] Miller D, Chen M J. Sources and Consequences of Competitive Inertia: A Study of the US Airline Industry. Administrative Science Quarterly, 1994, 39(1):1-23.

[30] Cho, T. S. & Hambrick, D. C. 2006. Attention as the Mediator Between Top Management Team Characteristics and Strategic Change: The Case of Airline Deregulation. Organization Science, 17(4): 453-469.

[31] Kaplan, S. Cognition, Capabilities, and Incentives: Assessing Firm Response to the Fiber-Optic Revolution. Academy of Management Journal, 2008, 51(4): 672-695.

[32] Eggers, J.P., Kaplan, S. Cognition and Renewal: Comparing CEO and Organizational Effects on Incumbent Adaptation to Technical Change. Organization Science, 2009, 20(2): 461-477.

[33] Nadkarni, S., Barr, P. S. Environmental context, managerial cognition, and strategic action: an integrated view. Strategic Management Journal, 2008, 29(13): 1395-1427.

[34] McCarthy, I.P., Lawrence, T.B., Wixted, B. and Gordon, B.R. A Multidimensional Conceptualization of Environmental Velocity. The Academy of Management Review, 2010, 35(4): 604-626.

[35] Plambeck, N., Weber, K. CEO Ambivalence and Responses to Strategic Issues. Organization Science, 2009, 20(6):993-1010.

[36] Dutton, J.E., Jackson, S.E. Categorizing Strategic Issues: Links to Organizational Action. The Academy of Management Review, 1987, 12(1): 76-90.

[37] Katila, R. Chen, E.L. Effects of Search Timing on Innovation: The Value of Not Being in Sync with Rivals. Administrative Science Quarterly, 2008, 53: 593-625.

[38] Gavetti, G., Levinthal, D. Looking forward and looking backward: Cognitive and experiential search. Administrative Science Quarterly, 2000, 45: 113-137.

[39] Mosakowski, E., Earley, P.C. A selective review of time assumptions in strategy research. Academy of Management Review, 2000, 25(4): 796-812.

[40] D. Lovallo, T.C. Powell, C.R. Fox, and D.J.Teece. Strategic Management Journal Special Issue Call For Papers. Psychological Foundations of Strategic Management. Journal of Strategic Management, 2008, 1-3.

[41] Duriau, V.J., Reger, R.K., Pfarrer, M.D. A Content Analysis of the Content Analysis Literature in Organization Studies: Research Themes, Data Sources, and Methodological Refinements. Organizational Research Methods, 2007, 10(1): 5-34.

[42] Senior, C., Lee, N., and Butler, M. Perspective—Organizational Cognitive Neuroscience. Organization Science, 2011, 22(3): 804-815.

[43] Deborah G. Ancona, Paul S. Goodman, Barbara S. Lawrence, Michael L. Tushman. Time: A New Research Lens. Academy of Management Review, 2001, 26(4): 645-663.

[44] Eisenhardt, K.M. Has strategy changed. MIT Sloan Management Review, 2002, 88-91.

[45] Ocasio, W. Towards an attention-based theory of the firm. Strategic Management Journal, 1997, 18 (S1): 187-206.

Permissions

The contributors of this book come from diverse backgrounds, making this book a truly international effort. This book will bring forth new frontiers with its revolutionizing research information and detailed analysis of the nascent developments around the world.

We would like to thank Hongyi Sun, PhD, for lending his expertise to make the book truly unique. He has played a crucial role in the development of this book. Without his invaluable contribution this book wouldn't have been possible. He has made vital efforts to compile up to date information on the varied aspects of this subject to make this book a valuable addition to the collection of many professionals and students.

This book was conceptualized with the vision of imparting up-to-date information and advanced data in this field. To ensure the same, a matchless editorial board was set up. Every individual on the board went through rigorous rounds of assessment to prove their worth. After which they invested a large part of their time researching and compiling the most relevant data for our readers. Conferences and sessions were held from time to time between the editorial board and the contributing authors to present the data in the most comprehensible form. The editorial team has worked tirelessly to provide valuable and valid information to help people across the globe.

Every chapter published in this book has been scrutinized by our experts. Their significance has been extensively debated. The topics covered herein carry significant findings which will fuel the growth of the discipline. They may even be implemented as practical applications or may be referred to as a beginning point for another development. Chapters in this book were first published by InTech; hereby published with permission under the Creative Commons Attribution License or equivalent.

The editorial board has been involved in producing this book since its inception. They have spent rigorous hours researching and exploring the diverse topics which have resulted in the successful publishing of this book. They have passed on their knowledge of decades through this book. To expedite this challenging task, the publisher supported the team at every step. A small team of assistant editors was also appointed to further simplify the editing procedure and attain best results for the readers.

Our editorial team has been hand-picked from every corner of the world. Their multi-ethnicity adds dynamic inputs to the discussions which result in innovative outcomes. These outcomes are then further discussed with the researchers and contributors who give their valuable feedback and opinion regarding the same. The feedback is then collaborated with the researches and they are edited in a comprehensive manner to aid the understanding of the subject.

Apart from the editorial board, the designing team has also invested a significant amount of their time in understanding the subject and creating the most relevant covers. They scrutinized every image to scout for the most suitable representation of the subject and create an appropriate cover for the book.

The publishing team has been involved in this book since its early stages. They were actively engaged in every process, be it collecting the data, connecting with the contributors or procuring relevant information. The team has been an ardent support to the editorial, designing and production team. Their endless efforts to recruit the best for this project, has resulted in the accomplishment of this book. They are a veteran in the field of academics and their pool of knowledge is as vast as their experience in printing. Their expertise and guidance has proved useful at every step. Their uncompromising quality standards have made this book an exceptional effort. Their encouragement from time to time has been an inspiration for everyone.

The publisher and the editorial board hope that this book will prove to be a valuable piece of knowledge for researchers, students, practitioners and scholars across the globe.

List of Contributors

Gheorghe Grigoraş and Gheorghe Cârţină
"Gheorghe Asachi" Technical University of Iasi

Elena-Crenguţa Bobric
"Stefan cel Mare" University of Suceava, Romania

Claudia Rinaldi, Fabio Graziosi, Luigi Pomante and Francesco Tarquini
Center of Excellence DEWS, University of L'Aquila, Italy

Mazanai Musara
University of Fort Hare, South Africa

Meine Pieter van Dijk
UNESCO-IHE Institute for Water Education, DA Delft, The Netherlands

Dale-Marie Wilson
University of North Carolina at Charlotte, USA

Aqueasha M. Martin and Juan E. Gilbert
Clemson University, USA

David Domonkos
TMTT Doctoral School of BME, Gedeon Richter Plc., Hungary

Imre Hronszky
Faculty of Economic and Social Sciences, Budapest University of Technology and Economics (BME), Hungary

Anna Comacchio and Sara Bonesso
Ca' Foscari University of Venice, Italy

Urvashi Sharma, Julie Barnett and Malcolm Clarke
Brunel University, Department of Information Systems and Computing, United Kingdom

Isaac O. Abereijo and Matthew O. Ilori
Obafemi Awolowo University, Ile-Ife, Nigeria

Alfredo De Massis
Università di Bergamo, Center for Young & Family Enterprise (CYFE), Italy

Valentina Lazzarotti and Emanuele Pizzurno
Università Carlo Cattaneo – LIUC, Italy

Enrico Salzillo
Business Integration Partners (BIP), Italy

Ana Gargallo-Castel and Carmen Galve-Górriz
University of Zaragoza, Spain

Fredy Becerra-Rodríguez
Industrial Engineering Department, National University of Colombia-Manizales
Department of Business Administration, Manizales University, Colombia

Hongyi Sun
Department of Systems Engineering and Engineering Management, City University of Hong Kong, China

Xuanwei Cao
Department of Management, Xi'an Jiaotong-Liverpool University, Suzhou, China